Intergenerational Space

Intergenerational Space offers insight into the transforming relationships between younger and older members of contemporary societies. The chapter selection brings together scholars from around the world in order to address pressing questions about both the nature of contemporary generational divisions as well as the complex ways in which members of different generations are (and can be) involved in each other's lives. These questions include: how do particular kinds of spaces and spatial arrangements (e.g. cities, neighbourhoods, institutions, leisure sites) facilitate and limit intergenerational contact and encounters? What processes and spaces influence the intergenerational negotiation and contestation of values, beliefs and social memory, producing patterns of both continuity and change? And if generational separation and segregation are in fact significant social problems across a range of contexts – as a significant body of research and commentary attests – how can this be ameliorated? The chapters in this collection make original contributions to these debates drawing on original research from Belgium, China, Finland, Poland, Senegal, Singapore, Tanzania, Uganda, the United States and the United Kingdom.

Robert M. Vanderbeck is Senior Lecturer in Human Geography at the University of Leeds, UK. His research focuses on contemporary processes of social exclusion and inclusion, with particular emphases on childhood, youth and intergenerational relations; sexualities; religion; and urban public spaces. His work has been published in numerous journals including *Annals of the AAG*, *Transactions of the IBG*, *Urban Geography*, *Urban Studies*, *Religion*, *Sexualities* and *Children's Geographies*.

Nancy Worth is a Banting Fellow in the School of Geography and Earth Sciences at McMaster University, Canada. Her current research examines work and social life with millennial women. She has published on temporality, sociality, mobilities, lifecourse research praxis and young people's transition to adulthood in journals such as *Area*, *Geoforum*, the *Journal of Geography in Higher Education*, *Social & Cultural Geography* and *Urban Studies*.

Routledge studies in human geography

This series provides a forum for innovative, vibrant and critical debate within Human Geography. Titles will reflect the wealth of research which is taking place in this diverse and ever-expanding field. Contributions will be drawn from the main sub-disciplines and from innovative areas of work which have no particular sub-disciplinary allegiances.

Published:

Intergenerational Space

**Edited by Robert M. Vanderbeck
and Nancy Worth**

Routledge
Taylor & Francis Group

LONDON AND NEW YORK

First published 2015
by Routledge
2 Park Square, Milton Park, Abingdon, Oxon OX14 4RN

and by Routledge
52 Vanderbilt Avenue, New York, NY 10017

First issued in paperback 2020

Routledge is an imprint of the Taylor & Francis Group, an informa business

British Library Cataloguing in Publication Data
A catalogue record for this book is available from the British Library

Library of Congress Cataloging in Publication Data
Intergenerational space / edited by Robert Vanderbeck, Nancy Worth.
   pages cm. – (Routledge studies in human geography ; 50)
   Includes bibliographical references and index.
   1. Intergenerational relations. 2. Intergenerational communication.
   3. Conflict of generations. 4. Social policy. I. Vanderbeck, Robert M.
   II. Worth, Nancy.
   HM726.I475 2014
   305.2–dc23   2014022442

ISBN 13: 978-0-367-66932-4 (pbk)
ISBN 13: 978-0-415-85531-0 (hbk)

Typeset in Times New Roman
by Wearset Ltd, Boldon, Tyne and Wear

For Paul Johnson and Mimi Terry (Robert)
For Jane and Brian Worth (Nancy)

Contents

Plates and figure

Plates

Figure

Tables

Contributors

Adefemi Adekunle is a detached youth worker and social/policy researcher at University College London. He was affiliated to the Runnymede Trust but now works freelance pursuing his research interests in the numerous intersections of youth, place and identity. His particular focus at this stage is finding innovative participatory methods of how young people react both positively and negatively to the part of the city that they call home using mobile phones and participatory GIS.

Flavian Bifandimu is a Programme Manager at HelpAge International in Tanzania. He is involved in advocating for older people's rights at all levels, contributing to the *Convention for the Elimination of All Forms of Discrimination Against Women* shadow report (2008) and the *Universal Periodic Review* report (2011) for the United Nations. He has also assisted with proposal development for the European Commission, Comic Relief, Big Lottery Fund (UK) and UN Women.

Tine Buffel is an associate member of Sociology at the University of Manchester where she has a Marie Curie Fellowship. Her research in Manchester is examining older people's experiences of social exclusion and inclusion in urban neighbourhoods and it builds upon previous work at the Free University of Brussels (VUB). She is also a visiting professor at the VUB where she is involved in a number of research projects in the context of the Belgian Ageing Studies. Before coming to the University of Manchester, she was a research associate at the VUB's Department of Adult Educational Sciences, where she obtained her PhD focusing on experiences of place and neighbourhood in later life.

Chen Fengbo is an Associate Professor of Agricultural Economics at South China Agricultural University in Guangzhou, China. His publications are in the areas of transition of the Chinese peasant and rice economy. He has authored *Changes in Peasant Household Behaviors and the Rural Economics Development: Experience from the Jianghan Plain* (in Chinese, China Agriculture Press, 2007). Recent publications include 'Changing rice cropping patterns: Evidence from the Yangtze River Valley, China' in the journal *Outlook on Agriculture*.

Luke Dickens is a Research Associate in the Department of Geography at the Open University. His current research is concerned with the intersections between young people's cultural practices, politics and senses of belonging and identity; and the use of digital and audio-visual technologies to both capture and create experiences of place. He has a number of recent publications exploring these themes, including in *Emotion, Space, Society* and *ACME: An International e-Journal in Critical Geographies*.

Ruth Evans is Associate Professor in Human Geography, University of Reading. She teaches and researches in the area of social and cultural and development geography. She is the author of *Children Caring for Parents with HIV and AIDS: Global Issues and Policy Responses* (with Saul Becker, Policy Press, 2009) and journal articles including those focused on sibling caringscapes, emotional interactions in families affected by HIV and young people's responses to parental death.

Joyce Gilbert works as Education Officer for the Royal Scottish Geographical Society (RSGS) based in Perth, Scotland. She has worked in the field of environmental education for over 25 years and has a particular interest in how people can reconnect with their cultural and natural heritage through first-hand, outdoor experiences. In her current role with RSGS, she has been exploring the use of story and journey to design projects where learners of all ages can encounter geography in new ways.

Mark Gorman is Director of Strategic Development at HelpAge International. He joined HelpAge in 1988 and was Deputy Chief Executive between 1991 and 2007. His work focuses on the development of organizational strategy in ageing and health, and the impact of environmental change on ageing populations. He has a degree in history from the University of Cambridge, and holds Master's Degrees from Cambridge and Bristol Universities. He has published a number of articles on aspects of ageing in the developing world. He was awarded an MBE in the Queen's New Year's Honours List in 2008.

Jawaid Haider is Professor of Architecture at The Pennsylvania State University and has practiced and taught in Asia, Australia, Europe and the United States. His interdisciplinary research underscores how the designed environment can be child-friendly, elder-friendly and, potentially, provide improved intergenerational interaction and relationship formation in civic spaces. He has lectured and offered workshops throughout the world and has been interviewed on television and radio in the United States and elsewhere.

Irene Hardill is Professor of Public Policy and Director of the Centre for Civil Society and Citizenship, Northumbria University, UK. Her publications include *Gender, Migration and the Dual Career Household* (Routledge, 2002), *Enterprising Care: Unpaid Voluntary Action in the 21st Century* (with Susan Baines (MMU, UK), Policy Press, 2011) and a special issue of the journal *Contemporary Social Science* on knowledge exchange and impact with Professor Jon Bannister (MMU, UK).

Alan Hatton-Yeo worked for 16 years as CEO of the Beth Johnson Foundation and retired in January 2014 to focus on his interest in intergenerational work and age-friendly communities. He continues to work with the Beth Johnson Foundation as Principal Advisor with a particular focus on work in Wales and Europe. He has an international reputation for his work in the fields of Intergenerational Practice and Ageing and has written extensively. In 2012 he was awarded an MBE for his contribution to the development of intergenerational practice.

Amanda Heslop is a trainer and researcher specializing in ageing and intergenerational issues. She worked with HelpAge International as Research & Training Manager (1992–2006). Recent research studies include 'Salt, soap and shoes for school: the impact of pensions on older people and grandchildren in the KwaWazee project in Tanzania's Kagera region', 2008, and 'Towards universal pensions in Tanzania: opportunities and challenges from a remote area, Ngenge Ward, Kagera', 2014 (HelpAge International and Kwa Wazee Switzerland).

Kirsi Pauliina Kallio is Research Fellow with the Academy of Finland. She is affiliated with the Academy of Finland Centre of Excellence in Research on the Relational and Territorial Politics of Bordering, Identities and Transnationalization (RELATE), working in the Space and Political Agency Research Group (SPARG) in the University of Tampere, Finland. Kallio's research concentrates on youthful political agency, political subjectivity, spatial socialization, children's rights, and child and youth policy.

Matthew Kaplan is Professor of Intergenerational Programs and Aging at Penn State University. In this position, he conducts research, develops curricular resources and provides leadership in the development and evaluation of intergenerational programmes. His overall research programme aims to identify the key principles and basic values that apply to intergenerational work conducted across disciplines, sectors and settings. In recent years, he has been increasingly involved in work that explores the international dimension of intergenerational engagement.

Michael Leyshon is a Senior Lecturer in Social Geography at the University of Exeter. He has published widely in the area of young people's geographies and has a particular interest in theories of embodiment and identity formation. His research explores, and seeks solutions to, the ways in which young people become socially, politically and economically excluded from society.

Helen Lomax is Senior Lecturer in the Faculty of Health and Social Care at the Open University. Her research interests include the role of policy and popular culture in shaping parenting and childhoods. She has a long-standing interest in working inclusively with communities to understand health and social inequalities.

Richard MacDonald is a lecturer in the department of Media and Communications, Goldsmiths, University of London. He researches amateur film, vernacular photography and civic archives in relation to collective memory practices. His monograph on civic cinephilia *The Appreciation of Film: The Film Society Movement and Film Study in Britain* is published by University of Exeter Press.

Dawn Mannay is a Lecturer in the School of Social Sciences at Cardiff University, Wales, UK. She works in the areas of visual methods, ethnography, inequality, social class, nationality, gender, intergenerational studies and domestic violence. She has published in a range of international interdisciplinary and social sciences journals including *Qualitative Research*, *Qualitative Inquiry*, *Visual Studies* and *Gender and Education*, and is currently writing the sole authored book *Visual, Narrative and Creative Research Methods* for Routledge.

Greg Mannion works as a Senior Lecturer in the School of Education, University of Stirling, Scotland. His research brings together understandings of the links between nature and culture, to consider the way places, particularly places other than classrooms, can be sites of learning and participation. Currently, his research considers the way schools might harness natural places and community into education for sustainability, place-based education and intergenerational learning.

Julie Melville is an associate researcher and the European Intergenerational Projects Manager at the Beth Johnson Foundation. She recently finished her doctorate degree at Keele University studying the UK's first purpose-built Intergenerational Centre in London. She currently has a number of publications in press, including a review of the use theory in the field of intergenerational research and reflections on Intergenerational Learning in Europe

Cheryl Morse is Assistant Professor of Geography at the University of Vermont, and former director of the Center for Research on Vermont. She works in the fields of rural studies, nature-culture theory and youth geographies. Her research focuses on the co-production of space and identity in rural communities and her work has been published in journals including *Health & Place*, *Gender, Place & Culture* and *Children's Geographies*.

Dorothy Moss is a Principal Lecturer in Childhood Studies at Leeds Metropolitan University. She teaches and researches in the sociology of childhood and youth. Publications include *Children and Social Change: Memories of Diverse Childhoods* (Continuum, 2011), *Gender, Space and Time: Women and Higher Education* (Rowman and Littlefield, 2006) and articles in the international journals *Children & Society*, *Childhood* and, with Ingrid Richter, *Young*.

Elisha Sibale Mwamkinga is the founder and Executive Director of the Good Samaritan Social Services Trust, which he has been heading since 1998. Prior to starting GSSST, he worked as a Social Worker at Mirembe Hospital, Regional Personnel Officer with the Tanzania Electrical Power Company

(TANESCO), as a Hospital Administrator and Project Coordinator at Marie Stopes International. He has a MA in Social Work from the Open University of Tanzania.

Chris Phillipson is Professor of Sociology and Social Gerontology at the University of Manchester where he is Director of the Manchester Institute for Collaborative Research on Ageing. His research has focused on a number of themes connected with ageing and urbanization as well as issues concerned with the impact of globalization on older people and social theory and ageing. He is the co-editor of the *Sage Handbook of Social Gerontology* (Sage, 2010) and co-editor of *Ageing, Meaning and Social Structure* (Polity Press, 2013).

Gina Porter works in the Department of Anthropology, Durham University and also advises on transport services for the DFID-funded Africa Community Access Programme (AFCAP). She has researched and written widely on the mobility issues facing commonly excluded populations (women, children, older people) in sub-Saharan Africa. She is currently leading an interdisciplinary research study on the impact of mobile phones on young people's lives in Ghana, Malawi and South Africa.

Samantha Punch is Professor of Sociology at the University of Stirling. Her research interests include siblings and birth order, childhoods and youth transitions in Latin America, rural livelihoods in China, Vietnam and India, and children's food practices in Scotland. She is co-editor of *Global Perspectives on Rural Childhood and Youth* (Routledge, 2007), *Children's Food Practices in Families and Institutions* (Routledge, 2011) and *Children and Young People's Relationships: Learning Across Majority and Minority Worlds* (Routledge, 2013).

Paul Simpson is a Lecturer in Sociology at the University of Manchester and specializes in ageing sexuality, masculinities and ethnographic methods. He is currently writing a book (due to be published 2015) entitled *Middle-Aged Gay Men and Ageing: Over the Rainbow?* He is currently leading an interdisciplinary team researching into old people and sexual/intimate citizenship and, individually, is investigating masculinities in service sector workplaces.

Anna Tarrant is a Research Associate at Open University. She is a feminist geographer interested in how the intersections of gender and age structure the lives of older men, including grandfathers. She has published a number of journal articles and book chapters about the contemporary experiences of grandfathers and is now developing her research interests in relation to young masculinities and men's relationships with staff in different welfare settings.

Amleset Tewodros has been working as the Country Director of HelpAge International in Tanzania since January 2012 where she has been steering the work of HelpAge in partnership with a range of local organizations of older people, managing human resources, representing the organization in a number of forums and forging new partnerships and mobilizing resources. She has a BA in

Management and Public Administration from the University of Addis Ababa in Ethiopia and an MA in Medical Sociology from Nairobi University of Kenya.

Leng Leng Thang is Associate Professor at the Department of Japanese Studies, National University of Singapore. She is a socio-cultural anthropologist with research interests on ageing, intergenerational approaches and relationships, gender and family. She has numerous publications relating to Asia, especially Japan and Singapore. Her latest book publication is titled *Experiencing Grandparenthood: An Asian Perspective* (co-edited with Kalyani Mehta, Springer Publishing, 2012). She is also co-editor of the *Journal of Intergenerational Relationships* (Taylor and Francis, USA).

Tea Tverin is a doctoral candidate in the College of Life and Environmental Science at the University of Exeter. She is currently completing her thesis on the way in which adolescents and children organize their memories and thoughts about holiday experiences in the countryside.

Gill Valentine is Professor and Pro-Vice Chancellor for the Faculty of Social Sciences at the University of Sheffield. Her research interests are focused in three interconnected areas: social identities and belonging; childhood, parenting and family life; and urban cultures and consumption. Although trained as a geographer, much of her work has been interdisciplinary involving collaboration with colleagues from Departments of Education, Sociology and Social Policy, Information Studies and Theology and Religious Studies. Her research has been supported by the award of 14 grants (value of >£4 million) from the European Research Council, Economic and Social Research Council, Arts and Humanities Research Council, Joseph Rowntree Foundation and The Leverhulme Trust, as well as research contracts from government departments and NGOs.

Robert M. Vanderbeck is Senior Lecturer in Human Geography at the University of Leeds. His research focuses on contemporary processes of social exclusion and inclusion, with particular emphases on childhood, youth and intergenerational relations; sexualities; religion; racialization; urban public spaces; and qualitative methodologies. He is the recent author of *Law, Religion and Homosexuality* (with Paul Johnson, Routledge, 2014). He is currently involved in a large, collaborative programme of research funded by the AHRC involving international comparative research on the theme of intergenerational justice.

Nancy Worth is a Banting Fellow in the School of Geography and Earth Sciences at McMaster University, Canada. Her current research examines work and social life with millennial women. She has published on temporality, sociality, mobilities and young people's transition to adulthood in journals such as *Area*, *Urban Studies* and *Geoforum*. She is also interested in lifecourse research praxis and epistemology, co-editing the forthcoming *Researching the Lifecourse: Critical Reflections from the Social Sciences* for Policy Press.

Qiong Xu is a research assistant in University of Worcester. She received her PhD in sociology of family from the Institute of Education in 2012. Her thesis explored Chinese adolescent girls' relationships with their fathers. Her published research includes *Journal of Family Issues* and *Educate~*. She has presented papers in many international conferences including the British Sociological Association Annual Conference and the American Association for Asian Studies.

Acknowledgements

We are grateful to the support of the Citizenship and Belonging Research Cluster in the School of Geography at the University of Leeds. The cluster supported the conference 'Intergenerational Geographies: Spaces, Identities, Relationships, Encounters', 10–11 May 2012, from which the idea for this collection emerged. Financial support from the cluster allowed us to keep the event affordable and accessible to academics, students and members of the public. Cluster members who made essential contributions to organizing and executing the conference include Tom Collins, Ayona Datta, Kristina Diprose, Anna Gawlewicz, Kate Kiping, Hannah Lewis, Rosa Mas Giralt, Ying Nan, Deborah Phillips, Louise Waite and Nichola Wood. Special thanks to Calum Carson for his excellent administrative support of the event.

Acknowledgements

1 Introduction

Robert M. Vanderbeck and Nancy Worth

There is growing recognition of the need for research that offers insights into the transforming relationships between younger and older members of contemporary societies. In contexts across both the global North and the global South one can find commentators describing different generations as 'clashing' (Zemke *et al.* 2000) or 'colliding' (Lancaster and Stillman 2009) – or having the potential to do so – in a variety of sites and in relation to a range of issues. A large number of current social, political, economic and environmental concerns are characterized as having significant 'generational' or 'intergenerational' dimensions. For example, academics and policy makers frequently express worries about the future of the implicit 'intergenerational contract' that underpins the provision of social welfare (Marcum and Treas 2013); movements to combat climate change and reduce rates of natural resource depletion are often underpinned by calls for 'intergenerational justice' (Gardiner 2006); and many efforts to promote forms of 'intergenerational practice' (Moore and Statham 2006) have emerged as mechanisms to improve relations between the young and the old. Discussions of the sources of social and political unrest often invoke 'generational inequality' as an explanatory factor, such as in many commentaries on the 2011 London riots (e.g. Berry 2011). Although there is good cause to critique the language of conflict, division and separation that seems to permeate much recent public discourse about intergenerationality, the pervasiveness of this rhetoric is nevertheless indicative of the importance of research that offers new insights into contemporary intergenerational relationships.

The term *intergenerational space* – which forms the title of this collection – is most commonly understood to denote a site that has been designed for the purpose of facilitating and promoting interaction between members of different generational groups (most commonly the young and the old). These kinds of intentionally produced sites and spaces feature prominently in a number of the original contributions to this collection. However, the vision of intergenerational space we adopt here is also much more expansive in recognition of the inherently geographical nature of social life, a point frequently stressed by – amongst others – human geographers and other scholars influenced by the so-called 'spatial turn' in social theory (Massey 2005). Spaces and places are not merely static arenas in which relationships between people transpire; rather, they are both constituted by and constitutive of social relations, including relations of age and generation. As such, the chapters in this collection approach questions of intergenerationality in relation to diverse contemporary geographical concerns. These include the relationship between places and identities; processes of segregation and integration; the socio-spatial organization of practices of care; encounters with social difference; the influence of cyberspace and new technologies on people's social networks; and the relationship between space and embodiment.

In this introduction, we begin by first placing the study of intergenerational space within a wider academic context. We then turn to a discussion of key themes that cross-cut many of the chapters in this collection, followed by an overview of the thematic organization of the book and the contributions of each chapter.

Studying the spaces of intergenerationality

The term 'generation' is employed in a variety of different ways both colloquially and in academic discourse. As such, the concept of 'intergenerationality' is a complex and sometimes confusing one, given that it potentially invokes a wide range of different kinds of relationships, interactions and encounters both within and beyond families. Theorists of age and ageing have advanced a three-fold understanding of the concept of generation in terms of life stages (e.g. 'child', 'teenager', 'middle-aged adult', or 'pensioner'/'retiree'), membership of a birth cohort (which is often ascribed particular characteristics and dispositions based on shared historical position and experience) and positions within a family structure (see Hagestad and Uhlenberg 2005; Vanderbeck 2007). Although in many respects analytically distinct, these three notions of generation intersect within the context of individual biographies and, as argued by Biggs and Lowenstein (2011), contribute to the production of generational statuses that are not necessarily experienced or understood by individuals in terms of separate dimensions. Rather, Biggs and Lowenstein suggest, generational status is experienced phenomenologically 'as an undifferentiated whole, all in one go, as part of who one is' (2011: 6), and the social effects of life stage, cohort and family position are often not easily differentiable (if at all). While the question of whether or not generational status is necessarily experienced as 'an undifferentiated whole' is an empirical one that might be answered differently in different contexts, nevertheless in practice the significance of various dimensions of generational status can be difficult to disentangle. And, as we elaborate below, many of our contributors are not interested solely in a single dimension of intergenerationality but rather recognize how in particular situations multiple notions of intergenerationality come into play.

The diversity of usages of the term 'generation' (and, hence, 'intergenerational') in the social sciences has proven a source of critique, with some scholars (e.g. Närvänen and Näsman 2004) arguing that use of the term should be more narrowly restricted to provide greater analytical precision and avoid potential confusion. However, while we would agree that it is important that scholars delineate how they are using the terms 'generation' and 'intergenerational', it is *not* our goal in this collection to attempt to impose a particular fixed understanding of these concepts. Rather, we seek to provide a multidisciplinary (and often interdisciplinary) showcase for recent scholarship that attempts to address important conceptual, theoretical, methodological and empirical challenges related to the sites and spaces of intergenerational conflict, cohesion and change (broadly conceived). In doing so, we hope to contribute to bringing different traditions and strands of work into conversation with one another.

The idea for this collection originated in a conference hosted by the Citizenship and Belonging research cluster of the School of Geography at the University of Leeds entitled 'Intergenerational Geographies: Spaces, Identities, Relationships, Encounters' in May 2012. In organizing the event, we certainly hoped to attract some interest from beyond human geography in the discipline's expanding conversation on intergenerationality, but we were pleasantly surprised that roughly half of the submitted abstracts originated from outside of geography departments. This suggests not only that issues of space (and other geographical ideas) are of potentially wide interest for research on intergenerationality, but also that there is a need for venues for these multi- and interdisciplinary conversations about the geographies of intergenerationality to transpire.

It has been argued that research related to age had developed in a highly compartmentalized fashion within the discipline of geography (Vanderbeck 2007), with quite separate literatures having evolved in relation to childhood/youth (an area of research that had seen substantial growth over the previous decades) and old age (an area of research that has experienced relative contraction). Despite an evident shared concern in these literatures with issues of generational separation, segregation and conflict, these literatures rarely spoke directly to one another. Further, while the tendency in geographical research related to age has been to focus on the 'bookend' (Hagestad and Uhlenberg 2005: 350) generations (i.e. the young and the old), the category 'adult' holds an ambivalent position. In one sense, 'adults' are arguably (and implicitly) the primary actors of concern in the majority of human geographical research (as is true in most other social science areas); however, their generational status as adults is infrequently problematized or conceptualized. Certain 'adult' identities and experiences are interrogated in relation to specific roles, such as in often feminist-influenced research on parenting practices and cultures (e.g. Katz and Monk 1993; Holloway 1998; Aitken 2009) or on the care responsibilities of adult children towards ageing parents (e.g. Bailey *et al.* 2004). Nevertheless, adulthood as a position and life stage remains somewhat under conceptualized in much human geographical research (Vanderbeck 2010).

There have been recent efforts by geographers to view age through a more relational, less compartmentalized lens (Hopkins and Pain 2007), and there is an evident growing interest in ideas regarding intergenerationality in the discipline (e.g. Tarrant 2010; Hopkins *et al.* 2011; Waite and Cook 2011; Valentine *et al.* 2012; Mitchell and Elwood 2013; but see Horton and Kraftl 2008). While other disciplines and research areas have their own specific histories and trajectories, a tendency towards compartmentalization in age-related research has not been solely evident in geography. Hagestad and Uhlenberg (2005: 346), for example, in reviewing scholarship on the theme of age segregation (which features prominently in this collection, as we elaborate below) note their puzzlement that 'the literature ... never combined an interest in both young and old'. There certainly remains scope for greater engagement and dialogue between social gerontology and childhood/youth studies (but see Hockey and James 1995; Settersten 2005, 2007). There are also established traditions of research on issues of intergenerational space that resonate with the concerns of geographers and with which there is greater scope for cross-disciplinary engagement. For example, the *Journal of Intergenerational Relationships* published its first issue in 2003 and routinely features work on themes related to intergenerational space, although direct engagement with the corpus of geographical thinking is not all that common (but see, for example, Kinoshita 2009; Mannion 2012). While one certainly can find many examples of interchange and collaboration, there is some truth in the claim that 'the different disciplines which are engaged with the concept of generation, such as sociology, psychology, medicine, [and] geography, rarely cross-communicate' (Biggs and Lowenstein 2011: 3).

Cross-cutting issues

Although drawn from diverse disciplines, contexts and perspectives, the contributions to this volume show evidence of a number of shared concerns. These cross-cutting concerns include the following.

Age segregation and the promotion of age integration

Many of the chapters in this collection share an implicit or explicit concern with patterns and processes of age segregation and integration. Age segregation has myriad social and geographical manifestations (Hagestad and Uhlenberg 2005; Vanderbeck 2007; Rogoff *et al.* 2010; Winkler 2013), including at the level of everyday interactions, neighbourhoods and institutions (e.g. the sorting of children and young people into age-segregated educational institutions, the placement of many older people who require high degrees of care into nursing homes, and the segregation of both younger and older people from workplace environments). Patterns of age segregation have been both produced and reinforced by approaches to urban and regional planning that have contributed to the production of spaces – such as city centres – that can prove relatively inaccessible or unwelcoming to people at particular life stages.

Age segregation has been considered a problem from a variety of angles. Commentators have suggested that although a degree of age segregation may be necessary or appropriate in particular contexts, the extent of contemporary age segregation has harmful social consequences. For example, age segregation – it is often argued – produces environments in which ageism and age-based stereotypes can proliferate, given the circumscribed opportunities that members of different generational groups have to interact with and know one another on an interpersonal level (Hagestad and Uhlenberg 2005). Patterns of age segregation can impede the development of what has been called 'generational intelligence' (Biggs and Lowenstein 2011), or the ability to take into account the vantage point of people from different generations when acting in the world.

In response to these types of concerns, a growing variety of efforts have emerged designed to promote age integration, including the growing field of intergenerational practice (Hatton Yeo 2005; Pain 2005; Seedsman 2013), intergenerational education (Mannion 2012), efforts to promote 'intergenerational cities' (van Vliet 2011) and calls by the United Nations to produce 'a society for all ages' (United Nations Economic Commission for Europe 2012). However, it is also important to stress that narratives of generational segregation are not equally evident in all societies, and the emphasis on age segregation in much writing on age relations can 'reflect the interests of research and policy in the global North', as Ruth Evans (this volume) suggests in her work on intergenerational relationships in diverse African contexts. Additionally, while there is a considerable need for research on age segregation and related issues, there has also been a marked tendency, as Kjørholt (2003: 264) argues, for social scientists to focus disproportionately on 'generational discreteness and difference such that disrupture and discontinuity ... [are] stressed at the expense of continuity'.

Geography and the construction of generational statuses and identities

A concern with how space and place are implicated in the construction of generational statuses and identities is also evident in many of the contributions to this volume. As other scholars have stressed, generational statuses and identities are inherently geographical phenomena (Christensen and Prout 2003; Schwanen *et al.* 2012). This point has been central to research both on the geographies of childhood/youth and the geographies of old age. For example, geographical research on childhood and youth has recognized the pervasive influence of spatial discourses that construct particular spaces as appropriate for children and young people, while other spaces are seen as inappropriate, harmful or conflicting with the requirements of a 'modern' childhood (Holloway and Valentine 2000). Contemporary children's rights discourses, for instance, often construct schools as essential sites of child development, while involving children to any great extent in the workplace is constructed as a form of exploitation (Vanderbeck 2005).

Similarly, geographical research on old age has helped shape a view that later life is fundamentally emplaced (Laws 1995). As Kevin McHugh (2000, 106) suggests, 'The adage "act your age" carries an implicit, unspoken spatial extension: "And do so in the proper and suitable surroundings"'. It has been argued that '(s) paces have their own age identities ... [T]he home, the neighbourhood, and particular sites such as parks, shopping malls and clubs acquire meanings of age' (Pain 2001: 156–157). The construction of individual generational subjectivities (by which we mean individual knowledges and understandings of one's generational status) is intimately linked to discursive practices that construct particular individuals and groups as 'in place' and 'out of place' in particular contexts.

Spaces of intergenerational transmission, contestation and negotiation

A number of contributions to this volume are also explicitly interested in questions related to how values, beliefs and ideas circulate amongst and between generations. While in many Western countries discussions of a cultural 'generation gap' (Mead 1970) are more muted now than in the 1960s and 1970s – when this form of rhetoric was at its most prominent – there are nevertheless indications that there can be important differences in values and attitudes between younger and older generations on some important issues, such as those related to ethnic, racial, sexual and other forms of diversity. Within the context of sometimes rapid economic and environmental transformations in both the global North and South, different generational views can emerge on issues such as intergenerational justice and fairness due to the differential ways in which generational groups benefit or suffer as a result of changing conditions and landscapes.

A concern with questions of intergenerational transmission has been a long-standing one for researchers and commentators. For a number of scholars, the key problem with contemporary patterns of generational differentiation and segregation has been one of cultural transmission and reproduction. Bronfenbrenner (quoted in Elder 1975: 174), for instance, has argued that 'If children have contact only with their own age-mates, there is no possibility for learning culturally established patterns of cooperation and mutual concern'. As others have stressed, however, this view needs to be tempered with the recognition that transmission of all aspects of a culture may not necessarily be desirable, and also that younger generations have much to teach older generations (Mead 1970; Thang 2001). Additionally, the notion of intergenerational transmission – which implies a relatively unidirectional flow from older to younger generations – has been justly critiqued for not recognizing the more fluid and reciprocal nature of intergenerational negotiations and contestations both within and beyond families (e.g. Hopkins *et al.* 2011).

In the next section, we introduce the thematic organization of the collection and explore in more depth the specific contributions of the 21 chapters of the volume that follow, reflecting on how they serve – both individually and collectively – to advance current discussions about intergenerational space.

Thematic organization and chapter contributions

This collection brings together a diverse group of scholars who are seeking to address pressing questions about both the nature of contemporary generational divisions as well as the complex ways in which members of different generations are (and can be) involved in each other's lives. The contributions focus on diverse geographical contexts, spanning countries including Belgium, China, Finland, Poland, Senegal, Singapore, Tanzania, Uganda, the United States and the United Kingdom. The collection is divided into five parts that constitute particular nexuses of interest in contemporary intergenerational research: (I) spaces of intergenerational encounter; (II) memory and intergenerational (dis)continuities; (III) the negotiation of values, beliefs and politics; (IV) education, work and care; and (V) intergenerationality and ageing.

Spaces of intergenerational encounter

The first part starts off the collection with a set of chapters that make new contributions to our understanding of intentional efforts to create spaces that promote age integration and positive forms of intergenerational encounter. In Chapter 2, Leng Leng Thang discusses the opportunities and challenges of intergenerational design and the complexities of producing age-inclusive cities. She focuses on a case study of the co-location of facilities for both the young and old in high-rise public housing estates in Singapore. These spaces provide venues for extrafamilial encounters between children and older people in neighbourhoods, but interestingly also afford opportunities for new kinds of grandparent/grandchild interactions. Identifying both strengths and potential limitations of this model, she argues that shared sites need to be integrated into wider urban planning initiatives to encourage success. This would allow shared spaces to harmonize with the surrounding community and potentially afford synergies from corresponding public spaces. In Chapter 3, Matthew Kaplan and Jawaid Haider continue with this theme, articulating ideas of intergenerational design and its potential to promote active living, where healthy lifestyles are integrated into daily life through design principles. In particular, they consider how physical spaces (parks and pavements (sidewalks) as well as purpose-built age-shared centres) can promote informal, unstructured intergenerational encounters.

The focus on shared spaces continues in Chapter 4 with a contribution from Julie Melville and Alan Hatton-Yeo of the Beth Johnson Foundation's Centre for Intergenerational Practice. They examine recent thinking on intergenerational shared spaces, which seek to counter age segregation by involving older and young generations in scheduled programmes as well as more informal socializing. While many shared spaces have been created, few have been evaluated and this chapter offers useful policy suggestions for making the most out of these initiatives. Helen Lomax's work concludes the part, extending the idea of shared space to community research and the 'ethnographic encounter', detailing a project where both younger and older people worked together to study their local neighbourhood (Chapter 5). In positioning the research process as an opportunity for intergenerational dialogue, Lomax makes a nuanced argument for the recognition of how different voices are heard/silenced in research.

Memory and intergenerational (dis)continuities

The second part of the book considers how individual and collective memories are constructed, transmitted and contested intergenerationally. In developing the notion of the 'intergenerational self', Fivush *et al.* (2005: 131) argue that memory is critical to self-identity: 'We are who we are because of what we have experienced and what we have been told; our sense of self is constructed from both personal history and the social cultural history in which our personal history is embedded'. In Chapter 6, Luke Dickens and Richard MacDonald examine institutional memories and extrafamilial intergenerational relationships at the annual camp of the Salford Lads Club in the UK. Using a participatory story-telling method, they highlight stories and memories of the club's long history to show how the camp's traditions and routines, as well as its physical dislocation from everyday life, allowed generational barriers to shift. The space of possibility that is offered by camp is further examined in Chapter 7, where Michael Leyshon and Tea Tverin make a case for memory as a practice and a tool to promote well-being. Working with a camp for disadvantaged young people in the UK, they discuss how building happy memories at camp and the temporary forgetting of hardship happens through a 'flat ontology' where both campers and adult counsellors mutually create opportunities for care. The memories produced by these camp experiences, Leyshon and Tverin argue, can act as a future resource on which young people can draw in attempting to navigate challenging life situations and circumstances.

Life story research demonstrates how much practices of reminiscence about the past can influence the present (Pillemer 2000; Bertaux and Thompson 2005). In Chapter 8, Dawn Mannay uses a case study of a mother and daughter to reflect on the intergenerational legacies that define our connections to home and neighbourhood. Mannay uses a 'possible selves' framework, eliciting positive and negative versions of both the future and the past, to develop understanding of intergenerational spaces of working-class femininity. Finally, drawing on Halbwachs' (1992 [1925]) groundbreaking work *On Collective Memory*, Dorothy Moss examines children's memories of war (Chapter 9). Using retrospective narrative accounts, Moss considers the effect of memories transmitted across generations, arguing that indirect experiences of war have a profound effect on children.

The negotiation of values, beliefs and politics

The third part considers the intergenerational negotiation of values, beliefs and politics. Aspects of this area of intergenerational research are well established in fields including social psychology and education, with past research having focused on diverse topics including environmental values (Grønhøj and Thøgersen 2009), alcohol consumption (Valentine *et al.* 2012), gender roles (Farré and Vella 2013) and religious beliefs (Hopkins *et al.* 2011; Min *et al.* 2012). Much of this previous research focuses on the context of the family; the chapters in this part include the family but also make an important contribution to the literature by considering how values, beliefs and politics are intergenerationally negotiated and contested within diverse contexts. In Chapter 10, Kirsi Pauliina Kallio develops the concept of 'intergenerational recognition' for understanding Finnish young people's political agency in everyday life. Building on Hannah Arendt's work in political philosophy, Kallio uses ethnographic methods to explore different contexts of intergenerational recognition, including home and school, demonstrating that children are both subjects and objects of recognition. Next, in Chapter 11, Gill Valentine draws on findings from a large-scale comparative study of the UK and Poland on the theme 'Living with Difference'. Her chapter examines why older generations are reported as having more prejudices than young generations (Zick *et al.* 2011). Rather than ascribe to the theory that older people are 'living in a different time', as is implied by some previous research and a number of her younger respondents, the study discusses how older people have fewer opportunities to encounter difference, as they live more segregated lives. The chapter also examines generational prejudice, suggesting that stereotypes held by both young and old need to be challenged by meaningful encounters between generations.

Building on the previous chapter's focus on prejudice, Adefemi Adekunle explores generational experiences of racism in Birmingham, UK (Chapter 12). Adekunle discusses the community work of the Runnymede Trust, focusing on changing generational understandings of 'race' between older and young people. The chapter also highlights the valuable intergenerational dialogue generated by the research as an example of community work that connects generations in an attempt to 'end racism in a generation'. The part concludes with Qiong Xu's study of fathers and daughters in Shanghai (Chapter 13). She investigates the negotiation and contestation of parental control and young people's spatial autonomy, both within and beyond the home. She find that while fathers often endeavour to protect their daughters by controlling their physical mobility, new technology has increased the power struggle been fathers and daughters by bringing independence inside the home.

Education, work and care

The chapters in Part IV are interested in how different generations engage with each other in the realms of economic production and social reproduction. Education is often primarily considered a domain of children and young people, yet it also provides opportunities for interaction with adult teachers and community members. Work is often considered the territory of the 'middle generation' of adults (particularly in the global North), yet times of economic uncertainty can result in transformations in the nature of the contributions children and older people are expected to make to formal and domestic economies. Finally care is often centred on the old and young, with adult carers often framed as part of the 'sandwich generation' (i.e. sandwiched between young and old, providing care to both). However, children and young people are increasingly taking up caring roles in diverse contexts (Bowlby *et al.* 2011; Rose and Cohen 2010), and older people often play a key role in providing child care and other support in families. Therefore, the theme 'education, work and care' highlights shifting patterns of responsibility and obligation between generations across the lifecourse.

In the first chapter of this theme, Ruth Evans draws attention to increasing pressures on the 'generational bargain' (where the transfer of wealth to children from economically active adults is expected to be returned in their old age) for families affected by HIV from Tanzania, Uganda and Senegal (Chapter 14). While practices of care still centre on the family, generational categories of childhood, youth and adulthood are dynamic and Evans contends that understanding *intra*-generational (sibling) relationships is increasingly critical as the 'middle generation' is affected by the HIV epidemic. In Chapter 15, Cheryl Morse explores extrafamilial practices of care at a wilderness therapy camp in the US state of Vermont, where the campers (who were primarily adjudicated youth) and staff developed their own form of 'family' relations. Critical here is the concept of a 'splintered generation' where young people are of a similar age, but lack shared experience (campers were underprivileged and often displayed anti-social behaviour while 'chiefs' were only a few years older, but more educated, middle class and courteous). Camp was not just a place, but a site of meaningful connection and care, that 'upend[ed] normalized and static notions of generation, family, and home' (Morse, this volume).

Turning to education, in Chapter 16 Greg Mannion and Joyce Gilbert outline a theoretical framework for place-responsive intergenerational education. Their approach outlines two premises (1) that people and places are reciprocally enmeshed and (2) that people learn from responding to difference. The aim of their chapter is to highlight the importance of place in learning, as well as the possibilities of intergenerational engagement. Finally, Chen Fengbo and Samantha Punch's research with fisher households in southern China examines the effect of the end of the fishing industry and the economically driven relocation of family members (Chapter 17). They argue that a change in livelihood has different generational effects, with the older generation feeling left behind while the younger generation leave for the city to pursue work. They also highlight a range of intergenerational interdependencies that contribute to economic survival, including grandparental care, that demonstrate the importance of a three-generational approach to understanding employment change.

Intergenerationality and ageing

The final part of the collection focuses specifically on intergenerationality and the process of ageing. Here, while the five chapters all focus on (dis)connections between generations, the experiences of older people are highlighted to draw attention to the generational bias that constructs older people as out of place in particular environments (e.g. the digital realm, the gay village, the inner city). In Chapter 18, Gina Porter, Amanda Heslop, Flavian Bifandimu, Elisha Sibale Mwamkinga, Amleset Tewodros and Mark Gorman report on a participatory project investigating ageing in rural Kibaha, Tanzania, where co-investigation with older people was a critical part of the research design. The chapter considers the challenges and opportunities of an intergenerational research praxis where older people led aspects of the data collection.

In Chapter 19, Irene Hardill discusses intergenerational connections facilitated by information and communication technologies (ICT). She contends that sustaining technology use is not related to age, but is connected to intergenerational support and affective motivations, including bonding with grandchildren. Anna Tarrant's research with grandfathers in Chapter 20 also picks up on the importance of technology to promote intergenerational contact, especially amongst families. Her research works against some of the age and gender expectations of family caringscapes, as grandfathers are keen to use technology to be part of their grandchildren's lives. In Chapter 21, Paul Simpson proposes an analytical framework for understanding the narratives of middle-aged gay men who use the often multigenerational spaces of the gay village in Manchester, UK, where (like many spaces of leisure) youthfulness and physical attractiveness are often privileged characteristics. The approach uses a dialectic between structural constraints (e.g. gay ageism) and men's everyday agential choices to more expressively capture the complexity of lived experience of ageing. Finally, Tine Buffel and Chris Phillipson (Chapter 22) trace ethnic and generational boundaries in deprived urban communities in England and Belgium. Phillipson and Buffel focus on deprived older people, who are often displaced in processes of gentrification and cannot participate fully in the 'new consumerism' of cities. In poor neighbourhoods, where urban change is often swift, they argue that it is vital to adopt an integrated approach to demographic and urban change.

References

Aitken, S.C. (2009) *The Awkward Spaces of Fathering*, Burlington, VT: Ashgate.

Bailey, A.J., Blake, M.K. and Cooke, T.J. (2004) 'Migration, care and the linked lives of dual-earner households', *Environment and Planning A*, 36(9): 1617–1632.

Berry, C. (2011) 'How generational inequality helped set England's cities alight', *Inequalities: Research and Reflections from Both Sides of the Atlantic*, online, available at: http://inequalitiesblog.wordpress.com/2011/08/23/how-generational-inequality-helped-set-england's-cities-alight/ (accessed 10 January 2014).

Bertaux, D. and Thompson, P. (eds) (2005) *Between Generations: Family Models, Myths and Memories*, Piscataway, NJ: Transaction Publishers.

Biggs, S. and Lowenstein, A. (2011) *Generational Intelligence: A Critical Approach to Age Relations*, Abingdon: Routledge.

Bowlby, S., McKie, L., Gregory, D. and MacPherson, I. (2011) *Interdependency and Care Across the Lifecourse*, Abingdon: Routledge.

Christensen, P. and Prout, A. (2003) 'Children, place, space and generation', in B. Mayall and H. Zeiher (eds) *Childhood in Generational Perspective*, London: Institute of Education, University of London, pp. 133–156.

Elder, G. (1975) 'Age differentiation and the life course', *Annual Review of Sociology*, 1: 165–910.

Farré, L. and Vella, F. (2013) 'The intergenerational transmission of gender role attitudes and its implications for female labour force participation', *Economica*, 80(318): 219–247.

Fivush, R., Bohanek, J.G. and Duke, M. (2005) 'The intergenerational self: Subjective perspective and family history', in F. Sani (ed.) *Self Continuity: Individual and Collective Perspectives*, Mahwah, NJ: Erlbaum, pp. 131–144.

Gardiner, S.M. (2006) 'A perfect moral storm: Climate change, intergenerational ethics and the problem of moral corruption', *Environmental Values*, 15(3): 397–413.

Grønhøj, A. and Thøgersen, J. (2009) 'Like father, like son? Intergenerational transmission of values, attitudes, and behaviours in the environmental domain', *Journal of Environmental Psychology*, 29(4): 414–421.

Hagestad, G. and Uhlenberg, P. (2005) 'The social separation of old and young: A root of ageism', *Journal of Social Issues*, 61(2): 343–360.

Halbwachs, M. (1992 [1925]) *On Collective Memory*, Chicago, IL: University of Chicago Press.

Hatton Yeo, A. (2006) 'Intergenerational practice: Active participation across the generations', Stoke-on-Trent: Beth Johnson Foundation, online, available at: www.centreforip.org.uk/res/documents/publication/intergenerational_practice_report.pdf (accessed 15 January 2014).

Hockey, J. and James, A. (1995) 'Imaging second childhood', in M. Fetherstone and A. Wernick (eds) *Images of Aging: Cultural Representations of Later Life*, London: Routledge, pp. 133–148.

Holloway, S.L. (1998) 'Local childcare culture: Moral geographies of mothering and the social organisation of pre-school education', *Gender, Place & Culture*, 5(1): 29–53.

Holloway, S.L. and Valentine, G. (2000) 'Children's geographies and the new social studies of childhood', in S.L. Holloway and G. Valentine (eds) *Children's Geographies: Playing, Living, and Learning*, London: Routledge, pp. 1–28.

Hopkins, P. and Pain, R. (2007) 'Geographies of age: Thinking relationally', *Area*, 39(3): 287–294.

Hopkins, P., Olson, E., Pain, R. and Vincett, G. (2011) 'Mapping intergenerationalities: The formation of youthful religiosities', *Transactions of the Institute of British Geographers*, 36(2): 314–327.

Horton, J. and Kraftl, P. (2008) 'Reflections on geographies of age: A response to Hopkins and Pain', *Area*, 40(2): 284–288.

Katz, C. and Monk, J. (eds) (1993) *Full Circles: Geographies of Women over the Life Course*, London: Routledge.

Kinoshita, I. (2009) 'Charting generational differences in conceptions and opportunities for play in a Japanese neighborhood', *Journal of Intergenerational Relationships*, 7(1): 53–77.

Kjørholt, A.T. (2003) '"Creating a place to belong": Girls' and boys' hut-building as a site for understanding discourses on childhood and generational relations in a Norwegian community', *Children's Geographies*, 1(1): 261–279.

Lancaster, L.C. and Stillman, D. (2009) *When Generations Collide*, New York, NY: HarperCollins.

Laws, G. (1995) 'Embodiment and emplacement: Identities, representation and landscape in Sun City retirement communities', *International Journal of Aging and Human Development*, 40(4): 253–280.

Mannion, G. (2012) 'Intergenerational education: The significance of reciprocity and place', *Journal of Intergenerational Relationships*, 10(4): 386–399.

Marcum, C.S. and Treas, J. (2013) 'The intergenerational social contract revisited', in M. Silverstein and R. Giarrusso (eds) *Kinship and Cohort in an Aging Society: From Generation to Generation*, Baltimore, MD: The Johns Hopkins University Press, pp. 293–313.

Massey, D. (2005) *For Space*, London: Sage.

McHugh, K.E. (2000) 'The "ageless self": Emplacement and identities in Sun Belt retirement communities', *Journal of Aging Studies*, 14(1): 103–115.

Mead, M. (1970) *Culture and Commitment: A Study of the Generation Gap*, New York, NY: Natural History Press.

Min, J., Silverstein, M. and Lendon, J.P. (2012) 'Intergenerational transmission of values over the family life course', *Advances in Life Course Research*, 17(3): 112–120.

Mitchell, K. and Elwood, S. (2013) 'Intergenerational mapping and the cultural politics of memory', *Space and Polity*, 17(1): 33–52.

Moore, S. and Statham, E. (2006) 'Can intergenerational practice offer a way of limiting anti-social behaviour and fear of crime?', *The Howard Journal of Criminal Justice*, 45(5): 468–484.

Närvänen, A.L. and Näsman, E. (2004) 'Childhood as generation or life phase', *Young: Nordic Journal of Youth Research*, 12(1): 71–91.

Pain, R. (2001) 'Age, generation and life course', in R. Pain, M. Barke, D. Fuller, J. Gough, R. MacFarlane and G. Mowl (eds) *Introducing Social Geographies*, London: Arnold, pp. 141–163.

Pain, R. (2005) *Intergenerational Relations and Practice in the Development of Sustainable Communities*, background paper for the Office of the Deputy Prime Minister, Durham: International Centre for the Regional Regeneration and Development Studies, Durham University.

Pillemer, D.B. (2000) *Momentous Events, Vivid Memories*, Cambridge, MA: Harvard University Press.

Rogoff, B., Morelli, G.A. and Chavajay, P. (2010) 'Children's integration in communities and segregation from people of differing ages', *Perspectives on Psychological Science*, 5(4): 431–440.

Rose, H.D. and Cohen, K. (2010) 'The experiences of young carers: A meta-synthesis of qualitative findings', *Journal of Youth Studies*, 13(4): 473–487.

Schwanen, T., Hardill, I. and Lucas, S. (2012) 'Spatialities of ageing: The co-construction and co-evolution of old age and space', *Geoforum*, 43(6): 1291–1295.

Seedsman, T. (2013) 'A mosaic of contemporary issues, insights, and perspectives surrounding intergenerational practice', *Journal of Intergenerational Relationships*, 11(4): 347–355.

Settersten, R.A. Jr. (2005) 'Linking the two ends of life: What gerontology can learn from childhood studies', *The Journals of Gerontology Series B: Psychological Sciences and Social Sciences*, 60(4): S173–S180.

Settersten Jr, R.A. Jr. (2007) 'The new landscape of adult life: Road maps, signposts, and speed lines', *Research in Human Development*, 4(3/4): 239–252.

Tarrant, A. (2010) 'Constructing a social geography of grandparenthood: A new focus for intergenerationality', *Area*, 42(2): 190–197.

Thang, L.L. (2001) *Generations in Touch: Linking the Old and Young in a Tokyo Neighborhood*, Cornell, NJ: Cornell University Press.

United Nations Economic Commission for Europe (2012) 'Ensuring a society for all ages: Promoting quality of life and active ageing: synthesis report on the implementation of the Madrid International Plan of Action on Ageing in the UNECE region', online, available at: www.un.org/esa/socdev/ageing/documents/Review_and_Appraisal/ECERe-port.pdf (accessed 1 February 2014).

Valentine, G., Jayne, M. and Gould, M. (2012) 'Do as I say, not as I do: The affective space of family life and the generational transmission of drinking cultures', *Environment and Planning A*, 44(4): 776–792.

van Vliet, W. (2011) 'Intergenerational cities: A framework for policies and programs', *Journal of Intergenerational Relationships*, 9(4): 348–365.

Vanderbeck, R.M. (2005) 'Anti-nomadism, institutions, and the geographies of childhood', *Environment and Planning D: Society and Space*, 23(1): 71–94.

Vanderbeck, R.M. (2007) 'Intergenerational geographies: Age relations, segregation and re-engagements', *Geography Compass*, 1(2): 200–221.

Vanderbeck, R.M. (2010) 'Kompetentteja sosiaalisia toimijoita? Pohdintoja lapsuuden tutkimuksen "aikuisista"' ['Competent social agents? Reflections on the "adult" of childhood studies'], in K.P. Kallio, A. Ritala-Koskinen and N. Rutanen (eds) *Missä Lapsuutta Tehdää?* [Where is Childhood Made?] Helsinki: Finnish Youth Research Network, the Finnish Society of Childhood Studies, and the Childhood and Family Research Unit of the University of Tampere, pp. 33–50.

Waite, L. and Cook, J. (2011) 'Belonging among diasporic African communities in the UK: Plurilocal homes and simultaneity of place attachments', *Emotion, Space and Society*, 4(4): 238–248.

Winkler, R. (2013) 'Research note: Segregated by age: are we becoming more divided?' *Population Research and Policy Review*, 32(5): 717–727.

Zemke, R., Raines, C. and Filipczak, B. (2000) *Generations at Work: Managing the Clash of Veterans, Boomers, Xers, and Nexters in Your Workplace*, New York, NY: AMACOM Div American Mgmt Assn.

Zick, A., Beate, K. and Hövermann, A. (2011) *Intolerance, Prejudice and Discrimination: A European Report*, Berlin: Forum.

Part I

Spaces of intergenerational encounter

2 Creating an intergenerational contact zone

Encounters in public spaces within Singapore's public housing neighbourhoods

Leng Leng Thang

Introduction

In recent decades there has been growing attention from multiple perspectives to the need to promote age-inclusive environments, especially in cities with rapidly ageing populations. Proposed frameworks such as creating liveable cities for all ages (van Vliet 2009), the World Health Organization's (WHO) guide to global age-friendly cities (2007), its subsequent WHO Global Network of Age-Friendly Cities and Communities (GNAFCC) and the numerous age-friendly city and regional initiatives provide notable examples of efforts which seek to promote age-friendly cities, communities and societies.

Although environments that are characterized as 'age friendly' or 'for all ages' should in theory be inclusive for everyone regardless of age, these concepts as policy ideas have in practice overwhelmingly emphasized 'elder friendliness'. Buffel *et al.* (2012), in a critical discussion on 'age-friendly cities', have traced the WHO Global Age-Friendly Cities project to the idea of 'active ageing' first developed in 1999 United Nations' Year of Older People. They also suggest that ideas about age-friendly cities are linked to concepts such as 'sustainable' and 'harmonious cities' in models of urban development produced during the 1990s and early 2000s, giving evidence of such influences in the UK in the concepts of 'lifetime homes' and 'lifetime neighbourhoods' which are aimed at supporting population ageing in the community (pp. 599–600).

Nevertheless, as van Vliet (2011) reminds us in his call for convergence in efforts to promote environments that are friendly for children/youth and older people, the characteristics of elder-friendly cities overlap considerably with those involved in creating child/youth-friendly cities: 'All benefit from neighborhoods that are safe and walkable and housing that is affordable and near shops, neighbors, and services, and easy access to public spaces for social interactions' (p. 354). His proposal that intergenerational integration necessitates synergistic efforts to achieve a harmonious living of all ages brings forth the vision of creating an all-inclusive environment that is genuinely *age-friendly* and liveable for all ages.

The drive towards an intergenerational perspective is not new. As early as 1949, Lewis Mumford explored the need for an 'organic conception of city planning' that urges one to establish balance within the urban community 'in time through inter-relationship between the phases of life' (p. 16). He further proposes more age integration both in the context of family settings as well as neighbourhood development and relationship building (Mumford 1956). While such aspirations may be inadequately pursued due to the dominant mono-generational focus in environment planning and design, more recently calls such as van Vliet's (2009, 2011) are indications of emerging concerns for more inter-generational meeting places in public spaces, suggesting how spontaneous inter-generational gatherings and informal interactions could be stimulated with the creation of conducive spaces (Vanderbeck 2007; Kaplan *et al.* 2007). This chapter explores the potential of such public spaces in urban neighbourhoods to contribute towards intergenerational integration. Examples from Singapore, an Asian city-state, will be the focus of the chapter. As one of the world's most urbanized states with a total population of 5.3 million on 715.7 sq km of land (as of 2012), an eye-catching and ubiquitous feature of Singapore's landscape is its high-rise public housing which caters to more than 80 per cent of its population. What are the potential intergenerational meeting places in such public housing estates? This chapter has chosen to emphasize the co-location of public ameni-ties for different generations, using the case example of a playground built alongside a fitness station for older adults situated in the open among the housing blocks. Data for the study is taken from a pilot investigation where participant observations were conducted at intervals from 2011 to 2013 in the co-located space. They were mostly carried out during the early evenings when this public space becomes the most vibrant spot in the neighbourhood, drawing together residents of different ages. How effective are these co-located public amenities in promoting intergenerational encounters and integration? Will they promote parallel co-existence more than the mixing of old and young? What is the poten-tial of such purposeful inter/multigenerational design to promote community bonding and well-being of the generations?

In examining these questions, I draw upon Pratt's (1991) concept of 'contact zone' to understand co-located spaces. Pratt has situated the idea in the area of language, communication and culture, intending it to in part 'contrast with ideas of community' (p. 37), for 'contact zone' is a space for cultures to meet, clash and grapple with each other (p. 34). In an example she gives of a college class on multiculturalism as a contact zone, chaos was aroused when students were confronted with cultures, ideas and values vastly different from their own. She concludes that 'along with rage, incomprehension, and pain, there were exhila-rating moments of wonder and revelation, mutual understanding, and new wisdom – the joys of the contact zone' (p. 39). I contend that the concept of the contact zone can be usefully extended to co-located public spaces where dif-ferent people gather, although of course public spaces lack formal curricula which readily facilitate and necessitate interaction.

I will begin by providing an overview of the intergenerational field which provides the framework for the subsequent discussion. I will then provide back-ground on intergenerational and community building efforts in Singapore's public housing, followed by an examination of a co-located intergenerational leisure space to exploring the potential of such a space to serve as a form of contact zone.

The intergenerational field and intergenerational integration

The idea that intergenerational approaches can contribute to social and community cohesiveness has garnered increasing attention in the context of rapid population ageing, the consequences of which are aggravated by the age segregation which is especially prevalent in industrialized societies (Hagestad and Uhlenberg 2005; Vanderbeck 2007). Age segregation is known to have undesirable consequences at the level of individuals, neighbourhoods and societies, contributing to age discrimination, age-based conflicts, increased vulnerabilities, and social isolation, among other problems (Kaplan 1993; Foner 2000). Van Vliet (2011)'s article 'Intergenerational cities' opens with the example of a tragedy in Tokyo where a man had been dead for three years without his neighbours noticing. Although this is a case of social isolation in the extreme, it hardly is inconceivable when socially segregated neighbourhoods deprived of mutual support and interaction become the norm in a generational/age-segregated environment. To counter social isolation and other problems facing age segregation, van Vliet (2011) has cited various successful intergenerational programmes from the US and other countries as evidence of the benefits of intergenerational integration.

Intergenerational programmes and activities, said to have been first conceptualized in the US in the 1960s, were defined by the US National Council on Aging in 1985 as 'activities or programs that increase cooperation, interaction or exchange between any two generations' (Thorp 1985: 3, cited in Kaplan *et al.* 2007). Newman *et al.* (1997: 56) further specify the extrafamilial links that characterize such interactions, defining the initiatives as 'designed to engage non-biologically linked older and younger persons in interactions that encourage cross-generational bonding, promote cultural exchange, and provide positive support systems that help to maintain the well-being and security of the older and younger generations'. A subsequent definition of intergenerational programmes as 'social vehicles that create purposeful and ongoing exchange of resource and learning among older and younger generations' by the International Consortium of Intergenerational Programmes (based in the UK) has since shifted and expanded the conceptual idea beyond structured programming to also include 'cultural and communal practices for bringing the generations together' (Kaplan *et al.* 2007: 84). More recently, as alluded to in discussions in this and other chapters in the volume, the increasing attention on building conducive intergenerational spaces has also extended the intergenerational field to concerns with how to create intergenerational environment in public spaces and environmental design practices that will promote intergenerational connections (Kaplan *et al.* 2007; Thang and Kaplan 2013).

Here, a distinction between multigenerational and intergenerational settings should be made whereby a multigenerational design would accommodate the physical and psychological needs of people of different ages and abilities with inclusion of 'universal design' and 'inclusive design' specifications. In contrast, an intergenerational design concept, besides creating environments that are accessible for multiple generations, would also be concerned with creating conducive environments where opportunities for meaningful engagement between members of different generations can take place, i.e. emphasizing the 'relational' aspect where communication and relationships develop between individuals of different generations (Thang and Kaplan 2013).

Among the array of intergenerational spaces possible, the co-located, shared sites or age-integrated centres are among the most researched intergenerational shared spaces (see for example Jarrot *et al.*'s edited special issue on shared sites, 2011; Hatton-Yeo and Melville, this volume). Compared with these purpose-built shared sites often complemented with programmes and activities, intergenerational environments in public spaces that have the potential to encourage spontaneous interaction still remain a little-explored aspect in the field. The attempt in this chapter to focus on public spaces – such as the co-located leisure facilities of a children's playground and fitness corner built with exercise stations for older people in an urban neighbourhood – thus seeks to offer new insights into the potential of environment design to promote spontaneous intergenerational interactions.

A focus on the community and spaces of encounters in public housing: the role of the Housing and Development Board in Singapore

Created in the early 1960s with the objective of clearing the slums and building mass low-cost housing to meet the urgent housing shortage facing the state, the Housing and Development Board (HDB) – the agency responsible for Singapore's public housing programme – has achieved remarkable success in providing near universal housing, enabling home ownership for both low-income and middle-income groups.[1] As a provider of over one million flats in Singapore, over the years, HDB has evolved from meeting the needs of basic shelter to today's emphasis on providing a total living environment and community development (Wong *et al.* 1985; Yuen 1995). HDB's priorities and roles today are thus three-fold. The first of these involves providing affordable, quality homes, ensuring vibrant towns with the provision of commercial, recreational and social facilities and amenities. The second involves transforming the living environment with upgrades and, renewal and rejuvenation programmes. The final role is 'focusing on the community', including the creation of community spaces for residents to interact so as to promote cohesive communities.[2]

To achieve the third key priority, in 2009 a division called the Community Relations Group (CRG) was set up in HDB with a focus on strengthening community ties, nurturing a more caring and close-knit heartland community with a stronger sense of community identity. Since then, various initiatives have been implemented, such as the Good Neighbour Awards, welcome parties for residents at newly completed housing blocks, and a community week inaugurated in May 2012 with a series of community-bonding activities (HDB 2012).

Appropriately titled *Our Homes: 50 Years of Housing a Nation*, the book commemorating the jubilee anniversary of HDB (Fernandez 2011) has succinctly captured the process by which HDB has moved beyond merely building the 'hardware' (houses) to include emphasis on the 'software' of fostering communities and promoting bonds between people. This included accounts of the integral role that HDB plays in national policy efforts by utilizing its unique position as the sole provider of public housing to shape community attitudes and further social goals. Housing policies such as allowing parents and their married children to apply together for adjoining flats, priority allocation for multigenerational families and housing grants for children applying for a flat near their parents (and vice versa) not only promoted intergenerational support among extended family members by living together or close to each other, but have also resulted in a greater mix of different age groups in the living environment.[3] HDB surveys have shown that more married children are living with or in the same estates as their parents between 2003 and 2008, an increase from 29.3 per cent to 35.5 per cent during the two survey period (Fernandez 2011: 207).

Mixing is also prevalent in terms of racial mixing through the ethnic integration policy (1989) which imposed an ethnic quota in both the allocation of new flats and the sale and purchase of resale flats for each neighbourhood or block. Such a measure, requiring the ethnic mix be roughly equivalent to the national ethnic proportion of 75 per cent Chinese, 13 per cent Malay, 9 per cent Indian and 3 per cent other races, has effectively prevented the development of ethnic enclaves, and promoted the common sights of different groups living next to each other in the housing estates. In 2010, a Singapore Permanent Resident (SPR) quota system was further introduced to ensure social cohesion and the integration of permanent resident families into local communities.[4]

In addition to the mixing of age groups and ethnicities is the mixing of different socio-economic groups to foster integration, common community and national identity. This is achieved by building flats of different sizes in close proximity to each other, as well as sometimes including a range of different flat types and sizes within a single block. The public housing estates are not confined only to public housing as a small percentage of private housing developments are also often developed side-by-side with the public flats, contributing further to the mixing of different socio-economic groups within the same vicinity.

How can interaction be fostered in the face of diversity amongst residents? In focusing on encounters in public spaces, the remainder of the section will delineate two public spaces in the housing blocks which are known to facilitate encounters and informal gatherings of the residents, namely (1) the void decks and (2) leisure/recreational spaces such as small landscaped gardens, playgrounds, fitness areas and hard courts for ball games.

Void decks for interaction

Void decks (Plate 2.1) are unique to the HDB flats built from 1970 onwards as a space 'deemed necessary to promote social cohesion and bonds among neighbors, just as there used to be in the kampongs (villages) that many lived in before moving into their HDB flats' (Fernandez 2011: 109). As empty/void spaces on the ground floor of a block of flats, they serve various purposes. They are practical for promoting good airflow, natural light and short cuts for people walking through the estates. They also serve social functions such as a place for residents to hold weddings and funeral wakes. Void decks may be entirely empty, but more often there will be at least portions of the spaces that are equipped with benches and/or tables and chairs (some tables come with chess boards). While children may use the empty spaces to play soccer and other ball games, the areas with seats serve as rest spots, places for games and casual meeting places where neighbours gather to chat. Sometimes, a small playground for toddlers and children under five is built on void deck spaces. Kiosks selling sundries, commonly called *ma-ma* stalls (from Tamil language 'mamak', which means uncle or elder) traditionally tended by Indians, provide another occasional presence at the void decks (Bachtiar 1992).

The 2008 HDB survey indicates that void decks are used by one-third of the households in the sample, and older people tend to use them more often (HDB 2010: 102). Wong *et al.*'s (1985) lifestyle study of HDB residents reveals that although young people also frequent the void decks to gather and play games,

Plate 2.1 A typical void deck with seating (photograph © author).

they usually do not mingle with the older residents who tend to occupy the void deck seats. One elderly respondent's quotation suggests the norm of age segregation: 'if the old people are using the void deck seats, the young people won't come here' (quoted in Wong *et al.* 1985: 481). However, on the other hand, Wong *et al.* also provide examples of intergenerational encounters, as in the case of a 15-year-old boy's leisure pattern. He usually spends his evening leisure time with his three neighbours, including an elderly man, either at the void deck or nearby provision shop and kiosk, 'chatting and listening to bird calls'. Songbird rearing is a traditional form of hobby in which some Chinese and Malay men take an interest (Lai 2010). The boy was taught the art of distinguishing bird calls from his elderly neighbour (Wong *et al.* 1985: 468). Void decks thus have the potential to create opportunities for various types of exchanges, including intergenerational ones.

Since 1973, after the first void deck space was rented out for conversion into an education centre, the presence of services such as child-care centres, kindergartens, senior citizens' clubs, neighbourhood police posts, Residents' Committees' offices, and family service centres have become a mainstay in some void decks.[5] This has sparked disapproval from some residents who suggest that these services should be allocated on the second level and above so that the void deck could remain genuinely 'void' to provide space for social interaction (Wee 2012). However, the director of Community Relations Group of HDB referred to the changes in void space availability as an evolution in public space design, with the availability of more interactive spaces in the new and renewed/upgraded estates, such as at the bigger precinct pavilions for residents' events and covered link ways and at drop-off porches which provide incidental space for interaction (Woo 2013).

Recreational spaces in the precinct

A precinct is conceptualized as the smallest unit of the hierarchy of the new town-neighbourhood-precinct model (parallel with the central place theories) adopted in the late 1970s. Whereas a new town is made up of about five to six neighbourhoods of 4,000 to 6,000 flats provided by a cluster of services for each neighbourhood, a precinct within a neighbourhood refers to a cluster of HDB blocks of about 400 to 800 flats. While the new town provides recreational facilities of a larger scale such as town parks, stadiums and swimming pool complexes, the precinct provides a 'door-step' approach to recreational opportunities, including landscaping gardens (which can be at the rooftop level as well) and providing local play facilities such as a children's playground or a small hard court for sports such as basketball, badminton and volleyball (Yuen 1995). Where spaces and interest are available, HDB has also worked with Residents' Committees in the precincts and neighbourhoods to build community garden projects tended by local residents as an effort to recapture 'kampong (village) spirit' (Fernandez 2011; Yuen 1995). The availability and sharing of common facilities at the precinct level is important in the promotion of social integration

(Fernandez 2011: 96; Yuen 1995: 246). Opportunities afforded by available precinct public spaces encourage interactions with those who live close to each other, and hopefully lead to a higher likelihood of fostering ties and stimulating a common community identity as residents of that vicinity.

It is inevitable for exchanges to be fostered more commonly among certain groups of residents who tend to routinely frequent the facilities more than the others. Wong *et al.* (1985) distinguish between the 'localites' (residents like housewives, seniors and primary school children who mostly move within the estate) and 'urbanites' (who tend to locate their activities outside the estate, such as working adults and youths) among the residents, noting a much higher usage of estate facilities and daily activities within the estate among the 'localites'. Similarly, the 2008 HDB survey of the usage levels of the various facilities in the living environment reveals differing usage depending on the life-cycle stage of families. This is especially obvious among some age-specific facilities. For example, playgrounds are used most frequently by families with young children (44.3 per cent), and least frequently by families without children (5.3 per cent) and non-family households (3.5 per cent) (HDB 2010: 102). Noting the specific family lifecourse moments where playgrounds may matter to adults, McKendrick (1999) has noted that a playground is often perceived as one of the least important sites in the locale for majority of a neighbourhood's adult populace (p. 5).

However, the newer co-located facilities found in HDB estates may alter the perception of playgrounds from a marginalized site to the centre of a space important for multiple generations. In considering new ways to bond the generations in the community setting, HDB has introduced co-located facilities conceptualized as 3-Generation (3G) facilities, comprising a playground, adult fitness corner and fitness corner for older people with exercise stations catering to their wellness needs. 3G facilities are now commonly seen in housing estates and built usually at the heart of the precincts to facilitate the congregation of residents. The acting director of the design development of HDB has highlighted the catering of such facilities to multigenerational families (The Straits Times 2013). This is evidenced in stories of how the availability of such multigenerational facilities have promoted opportunities to spend time together in play and exercise, as well as for the making of new friends through frequenting the same 3G facilities (Desai 2013; Goh *et al.* 2012).

In focusing particularly on encounters between old people with the young, the discussion below will focus on a co-located site consisting of a playground placed next to a fitness corner for older people, perhaps more appropriately termed 2G (two generations) in the context of the 3G facilities.

Contact zone: co-located playground and exercise stations for older people

The co-located playground and fitness corner for older people chosen for pilot investigation is situated in one of the two precincts belonging to a small neighbourhood of less than 2 sq km known commonly as St George's estate, bounded on the south-east side by the Whompoa River and the north by Central Express Highway. This neighbourhood is situated less than 5 km from the city centre. It consists of 24 blocks of flats of various sizes and heights, ranging from four-level walk-up blocks (usually two or three bedrooms with one living room types, to high-rise blocks of 25 levels which are larger units of three bedrooms, living room and dining room). The neighbourhood is situated within the Whampo/Kallang estate, which is classified as a mature estate as many of the flats were originally built in the 1970s. At the St George's neighbourhood, the flats were developed over three periods, with the blocks in the St George's West precinct mostly completed earlier in 1975 and 1978, and blocks in the St George's East precinct completed at a later phase around 1983 and 1984. As with other mature estates, the St George's neighbourhood frequently undergoes upgrading and renewal. Over the years, many blocks have taken turns to be provided with lift (elevator) upgrading to allow all levels of a flat to have a lift landing, covered link ways to shield people from sun and rain when moving between the blocks, and playgrounds upgraded with new padded flooring, new colourful plastic play fixtures and additions to become co-located facilities and so on. Residents have easy access to facilities within the neighbourhood, such as eating places, provision stores, clinics, places of religious worship such as churches and Buddhist temples, kindergartens, child care centres, and schools. The neighbourhood also has a sports stadium with tennis courts, squash courts and football fields, in addition to the availability of the school hall and field for sports rental. A supermarket and mass transit station are situated in the next neighbourhood a short walking distance away. A private housing estate with a mix of landed houses, private apartments and condominiums is situated right next to the St George's neighbourhood.

Various recreational facilities dot the public spaces of the neighbourhood. There are four playgrounds, of which two are 2G – situated near fitness corners for older people, and one is 3G, which also has an adult fitness corner next to the fitness corner for older people (Plate 2.2), and only one is a stand-alone playground, but with a basketball court close by. There are three other adult fitness corners in the neighbourhood, two are located as stand-alone, and one adjoining a basketball court. Besides one more basketball court next to a playground as mentioned earlier, there is another stand-alone hard court for games such as badminton.

Generally, the important and common gathering nodes in the neighbourhoods consists of the markets, food centres and coffee shops – more affectionately called *kopitiams*[6] in the local context. In addition, playgrounds and fitness corners are also increasingly recognized to be spaces of community encounters; the roles of playgrounds, for example, have been noted for being a space not only for the children, but also for adults (McKendrick 1999).

The site selected for this study has been observed to be one of the most popular gathering spots for residents in the neighbourhood, probably due to its central location in the precinct. The site is a typical open space in the precinct planning, being clustered by four blocks of flats. It is also situated next to a

Plate 2.2 A 3G recreational facility: the equipment for adults is on the left in the fore-
front, the fitness corner for older people is on the right, and behind them is the
playground for children (photograph © author).

child-care centre, and at the nexus where there are link ways, two pavilions, a
ma-ma stall at a void deck block, and seats at another void deck all near to the
playground. These other amenities and facilities have all contributed towards the
site's vibrancy, which is unparalleled by the other co-located playgrounds and
fitness corners in the neighbourhood.

On a typical evening, children and adults (mostly caregivers of the children)
start to congregate at the site from 5:30 PM onwards, and could swell to more
than 30 people packed in a small area, with children playing, usually small children using the slides and play equipment, while older children of ten years and
above scream and run around the playground. Sometimes, some big children
and teenagers arrive on bicycles and leave their bicycles at the side to gather and
play with others. Children can also be seen playing soccer at a pavilion without
chairs next to the playground and link way. Around 6 PM, some children who are
from the child-care centre just next to the playground can be seen joining in the
play, as their fathers, mothers or grandparents look on, sitting at the side.
Fathers, for example, are seen coming with newspapers to read while their children play, and mothers may gather in twos or threes to chat. Other adults around
include the common caregivers of children: grandparents and live-in foreign

female domestic workers.[7] Although some domestic workers may tend to the toddlers and young children under their charge playing in the playground, it is more common for them to be sitting on a bench chatting away with each other or on mobile phones. Some of them may also be at the playground together with the grandparents.

This open-air space nestled within the high-rise blocks on three sides and a car park on the remaining side is also a passing path for many who are walking to their homes after work or school. Domestic workers and mothers who bring their children to the provision stores nearby may pass by and stop awhile to allow their children to play en route home, too. The busy sights and sounds slowly taper off when the sky turns dark close to 7 PM, whereupon children and adults start to head home.

What can be observed about intergenerational contacts from the everyday activities at the co-located site? It is indeed a site where intergenerational contact is evident, albeit within certain parameters. In terms of the use of facilities by different generations, cross-use of space appears common. It should be noted that although this is a co-located site, distinct notices are set up by the town council at each facility to specify the users suitable for each. On the playground where a big combo colourful Little Tikes play system is erected, the notice states that 'This playground is for children aged two to 12 years old only'. At the edge of the playground just behind the long wooden bench stood a notice for the fitness corner with six exercise stations, states that 'The fitness corner is meant for senior citizens. Please exercise with care'. However, despite the age distinction, they seem to be perceived as one integrated area by the users. Children casually flow from the playground to the exercise stations and vice versa. Although we may argue that grandparents who bring their grandchildren to the playground will do the same whether it is a co-located site or otherwise, by staying with the children and going inside the playground area to support their young grandchildren at play, nonetheless, having the exercise stations for older people co-located does encourage older people who may not otherwise try exercising to try it out casually at the fitness corner while their grandchildren may either stay with them or play in the playground.

At first glance, age integration is obvious at the site with cross-use and the presence of older people with children. However, activity here is characterized by *familial* interaction between grandparents with their young grandchildren, suggesting the significance of such a space in facilitating play and contact between grandchildren and their grandparents, who are likely to be caregivers to their grandchildren when their parents are at work. Much less communication has been observed between young and old unless the older people know the children, such as instances where two neighbours chat at the playground, and the older woman would be seen patting the children when chatting with the mother. Overall, parallel co-existence is most common. Older people are an expected presence in the co-located site's vicinity: they are seen sitting at the pavilion next to the playground, or in the chairs at the void deck in the nearby block, or sitting on the benches lining the two sides of the playground, as well as using seats at the side of the fitness corner for older people. Generally, they stay at a distance from the children and chat to each other while looking onto the lively playground. At times, there are also the more frail older people in wheelchairs at the pavilion accompanied by their domestic worker caregiver, who sometimes feed them, or at other times chat with other domestic workers. During the earlier stage of observation, a pair of older couples used to frequent the playground, where the husband using a wheelchair would spend time with his wife sitting on the bench immersed in the sights and sounds of the playground.

Besides the parallel co-existence of older people and children, similar parallel patterns are also found between different ethnicities and cultures, and between citizens and foreign residents. The daily gathering of people at the playground in some ways resembles a microcosm of Singapore society, including Chinese, Malays and Indians, foreigners from mainland China, South Asia and other countries who have settled into local society with young children. The foreign domestic workers are a prominent group at the playground area, often outnumbering local mothers, especially during weekdays. However, the adults usually gather and interact within their own groups, such as the Chinese mothers and grandmothers chatting among themselves, or Indian older people sitting in the pavilion while Chinese older people sit at the void deck. Occasionally, cross-cultural intergenerational interaction is evident when older local residents are seen chatting with the domestic workers sitting at the bench, with exchanges that can vary from local cooking to caring for children. Children, on the other hand, tend to integrate more naturally. For example, Indian boys were seen inviting a Chinese boy to join them in playing a 'catching' game that involved running around the playground, and small children sharing the playground equipment easily play with others next to them regardless of their race and ethnicity. The children also serve to help stimulate conversations between adults.

As a public space which comes alive every day for about one to two hours in the evening, the co-located site has provided a natural space which facilitates the gathering of different people in the community. On the one hand, this site plays the role of a contact zone – not only one for different generations, but also in terms of the larger multi-ethnic and multicultural make-up of the society. However, the intensity of contacts differs. Contact generally consists of parallel co-existence between different age groups and ethnic groups, while there is seemingly potential to develop more interactive, intermingling contact within the same group, such as among the children, older people, domestic workers, and young mothers with children of similar ages. Nonetheless, the parallel co-existence is dynamic, although the contact zone in this context may not function beyond knowing the existence of each other. The encounters have arguably at least turned total strangers into somewhat familiar faces in the neighbourhood, and paved the way for more opportunities for spontaneous befriending and new social networks during other chance encounters, such as at the void decks, inside lifts, at lift landings, markets, schools and *kopitiam*.

Conclusion

This chapter has explored a little-studied form of intergenerational meeting place, namely the co-located playground for children and fitness corners for older people, designed to promote intergenerational encounters and integration. The site was selected for preliminary investigation because of its vibrancy and accommodation of different generations in the neighbourhood. While the analysis shows the vibrant multi-ethnic, multicultural and multigenerational character of the site, which increases the potential for intergenerational encounters between different cultures and ethnicities, it also reveals limits to such co-located public spaces as contact zones for the old and the young, although benefits for grandparents interacting with their grandchildren were observed.

However, despite the limitations, the existence of such a co-located space remains important for incidental meetings, suggesting its potential as an intergenerational contact zone to promote age integration in the community. While the preliminary observations suggest the need for further research, including surveys of the neighbourhood and detailed ethnographic fieldwork to enrich our understanding in the topic, I would like to conclude with some suggestions based on the lessons of this research.

First, more effort should be invested in conceptualizing an intergenerational design. For example, creative play equipment and exercise stations with the objective of encouraging meaningful engagements between the old and the young users should be designed to produce a more conducive intergenerational environment.

Second, the co-located sites should be strategically located and be complemented by other intergenerational efforts. Compared with the other 2G and 3G co-located facilities in the neighbourhood, the vibrant case study site is strategically positioned near a child-care centre and various rest spots such as the pavilion with benches, the void deck with seats where residents – especially older residents – naturally congregate. If the child-care centres could be more aware of strategies to promote intergenerational integration within the community, there could be efforts such as frequent mini-outings in the precinct for children at child care which would facilitate children meeting with older people in the community. The resulting enhanced familiarity would lead to more chances for exchanges when the old and young meet at nearby co-located facilities for their own play and exercise.

Third, following on from the above, the creation of intergenerational spaces should be more comprehensively incorporated into town planning and design from the initial stages so that co-located facilities in public spaces could seamlessly complement, and be complemented by, the surrounding activities and other spaces. An example of this would be the co-location of playgrounds and play spaces with other non-recreational spaces that older generations frequent, such as near to *kopitiams*, day-care services for older people and so forth. There is vast potential for innovative co-location of community and even commercial facilities intergenerationally that could serve as important intergenerational contact zones (Thang and Kaplan 2013).

As Buffel and Phillipson (this volume) remind us, cities have the potential to enhance the quality of life for all age groups. In an era of population ageing where competition over urban spaces will become intensified between the generations, there is an urgent need to incorporate purposeful inter/multigenerational design when conceptualizing age-inclusive future environments that provide a better quality of life for all.

Notes

1 About 80 per cent of Singaporean residents live in HDB flats, and about 90 per cent of the HDB residents own their HDB flats. Although a public housing scheme, the government has introduced the home ownership scheme in 1964, where the purchase of flats commonly comes with 99-year lease tenure (HDB 2013).
2 See HDB (2014a) for details of the role of HDB.
3 Although older/mature estates have larger ageing demographics than other, younger, estates, there have been numerous efforts to try to improve the demographic balance in the mature estates such as building new flats in these areas so that more younger families would move in to these neighbourhoods, rejuvenating the towns under HDB's Remaking Our Heartland plans to make these attractive residential areas for both the young and old families (Lee 2013).
4 The new regulation imposes a cap of 8 per cent set for permanent resident households (except Malaysians) in a block and 5 per cent in each neighbourhood (HDB 2014b). However, there have been concerns that enclaves may still be formed since there are no restrictions in rental to foreigners (Chang 2010).
5 HDB allows for up to two-thirds of the free space at the void decks of residential blocks for conversion into services for the residents (Fernandez 2011: 101).
6 *Kopitiams* are widely regarded as a quintessential part of Singapore's everyday life and public culture, a manifestation of the nation's multiculturalism and migration process (Lai 2010).
7 To encourage women in Singapore to contribute to the work force, the Singapore government started the foreign domestic worker scheme in 1978, where women (usually young women) mostly from neighbouring countries such as the Philippines and Indonesia, and a lesser number from Myanmar, India, Nepal and Sri Lanka came to work as live-in maids with work contract on a two-year renewal basis. They provide labour for domestic realm such as house cleaning, babysitting, child care and elder care. In 2012, slightly more than 200,000 domestic workers live in Singapore, averaging to about one in every five households (MOM 2014).

References

Bachtiar, I. (1992) 'The oldest ma-ma shop', *The Straits Times*, September 17.
Buffel, T., Phillipson, C. and Scharf, T. (2012) 'Ageing in urban environments: Developing "age-friendly" cities', *Critical Social Policy*, 32(4): 597–617.
Chang, R. (2010) 'HDB quota for PRs may not avoid enclaves', *The Straits Times*, 31 January.
Desai, A. (2013) 'All-inclusive play!', *MyNiceHome*, 23 May 2013, online, available at: http://mynicehome.sg/2013/05/28/all-inclusive-play/#sthash.6JqMg50L.dpuf (accessed 1 July 2013).
Fernandez, W. (2011) *Our Homes: 50 Years of Housing a Nation*, Singapore: Straits Times Press.

Foner, A. (2000) 'Age integration or age conflict as society ages?' *The Gerontologist*, 40(3): 202–275.

Goh, S.T., Giam, S. and Teo, W.G. (2012) 'The changing face of playgrounds: Play areas today feature sophisticated attractions and cater to 3 generations', *The Straits Times*, 15 June.

Hagestad, G.O. and Uhlenberg, P. (2005) 'The social separation of old and young: A root of ageism', *Journal of Social Issues*, 61(2): 343–360.

HDB (Housing and Development Board) (2010) *Public Housing in Singapore Residents' Profile, Housing Satisfaction and Preferences, HDB sample household survey 2008*, Singapore: HDB.

HDB (Housing and Development Board) (2012) 'Let's "shine": Share in neigbourliness', 20 May 2012, online, available at: www.hdb.gov.sg/fi10/ fi10296p.nsf/PressReleases/E CA23F76E79E41C948257A060030ED9E?OpenDocument (accessed 10 July 2013).

HDB (Housing and Development Board) (2013) 'Public housing in Singapore', *HDB InfoWEB*, updated 23 September 2013, online, available at: www.hdb.gov.sg/ fi10/ fi10320p.nsf/w/AboutUsPublicHousing?OpenDocument#ownership (accessed 12 March 2014).

HDB (Housing and Development Board) (2014a) 'Our role', HDB InfoWEB, updated 28 February 2014, online, available at: www.hdb.gov.sg/fi10/fi10320p.nsf/w/ AboutUsOur Role?OpenDocument (accessed 12 March 2014).

HDB (Housing and Development Board) (2014b) 'Ethnic integration policy and SPR quota', *HDB InfoWEB*, updated 7 January 2014, online, available at: www.hdb.gov.sg/fi10/ fi10321p.nsf/w/BuyResaleFlatEthnicIntegrationPolicy_EIP (accessed 12 March 2014).

Jarrott, S.E., Kaplan, M.S. and Steinig, S.Y. (eds) (2011) 'Special issue: Shared site intergenerational programs: Common space, common ground', *Journal of Intergenerational Relationships*, 9(4): 343–734.

Kaplan, M.S. (1993) 'Recruiting senior adult volunteers for intergenerational programs: Working into create a "jump on the bandwagon" effect', *Journal of Applied Gerontology*, 12(1): 71–82.

Kaplan, M.S., Haider, J., Cohen, U. and Turner, D. (2007) 'Environmental design perspectives on intergenerational programs and practices: An emergent conceptual framework', *Journal of Intergenerational Relationships*, 5(2): 81–110.

Lai, A.E. (2010) *The Kopitiam in Singapore: An Evolving Story About Migration and Cultural Diversity*, Asia Research Institute Working Paper Series No 132, Singapore: National University of Singapore.

Lee, Y.S. (2013) 'Better towns for all ages', Ministry of National Development, online, available at: www.mnd.gov.sg/budgetdebate2013/speech_lys.htm (accessed on 1 September 2013).

McKendrick, J.H. (1999) 'Playground in the built environment', *Built Environment*, 25(1): 5–10.

MOM (Ministry of Manpower) (2014) *Foreign Workforce Number*, updated 28 January 2014, online, available at: www.mom.gov.sg/statistics-publications/others/statistics/ Pages/ForeignWorkforceNumbers.aspx (accessed 12 March 2014).

Mumford, L. (1949) 'Planning for the phases of life', *The Town Planning Review*, 20(1): 5–16.

Mumford, L. (1956) 'For older people – not segregation but integration', *Architectural Record*, 119: 191–194.

Newman, S., Ward, C.R., Smith, T.B., Wilson, J.O. and McCrea, J.M. (1997) *Intergenerational Programs: Past, Present and Future*, Washington, DC: Taylor and Francis.

Pratt, M.L. (1991) 'Arts of the contact zone', *Profession*, 33–40.

Thang, L.L. and Kaplan, M.S. (2013) 'Intergenerational pathways for building relational spaces and places', in G.D. Rowles and M. Bernard (ed.) *Environmental Gerontology: Making Meaningful Places in Old Age*, New York, NY: Springer.

The Straits Times (2013) 'Multi-generation facilities to cater to changing needs: HDB', 18 June.

Thorp, K. (ed.) (1985) *Intergenerational Programs: A Resource for Community Renewal*, Madison, WI: Wisconsin Positive Youth Development Initiative, Inc.

van Vliet, W. (2009) 'Creating livable cities for all ages: Intergenerational strategies and initiatives', Working Paper CYE-WP1-2009, Children, Youth and Environments Center, University of Colorado, online, available at: www.colorado.edu/cye/sites/default/files/attached-files/CYE-WP1-2009%20website%20verson.pdf (accessed 2 June 2014).

van Vliet, W. (2011) 'Intergenerational cities: A framework for policies and programs', *Journal of Intergenerational Relationships*, 9(4): 348–365.

Vanderbeck, R.M. (2007) 'Intergenerational geographies: Age relations, segregation and re-engagements', *Geography Compass*, 1(2): 200–221.

Wee, A. (2012) 'Free up void decks by building amenities on levels above', *The Straits Times*, 16 February.

Wong, A., Ooi, G.L. and Ponniah, R.S. (1985) 'Dimensions of HDB community', in A. Wong and S.H.K. Yeh (eds), *Housing a Nation: 25 Years of Public Housing in Singapore*, Singapore: Maruzen Asia for Housing and Development Board of Singapore.

Woo, J. (2013) 'Q&A with HDB community relationships group director of policy and planning', *My Paper*, 30 May.

WHO (World Health Organization) (2007) *Global Age-Friendly Cities: A Guide*, World Health Organization.

Yuen, B. (1995) 'Public housing-led recreation development in Singapore', *Habitat International*, 19(3): 239–252.

3 Creating intergenerational spaces that promote health and well-being

Matthew Kaplan and Jawaid Haider

Introduction

This chapter aims to describe some design approaches for creating intergenerational indoor and outdoor environments that are conducive to healthy, active lifestyles for people of all ages. We zero in on the intersection of three areas of inquiry and intervention, each with its own multidisciplinary roots and connections: intergenerational engagement, environmental design and health promotion.

To define intergenerational engagement, we draw on Beth Johnson Foundation's broad definition for the term 'intergenerational practice' (2001):

> Intergenerational practice aims to bring people together in purposeful, mutually beneficial activities which promote greater understanding and respect between generations and contribute to building more cohesive communities. Intergenerational practice is inclusive, building on the positive resources that the young and old have to offer each other and those around them.

While this definition is commonly used in the context of intergenerational *programmes* implemented in schools, community centres, retirement communities and other community settings, it also encompasses public policies and environmental design.

Environmental design deals with the effect of the built environment on people and the natural world and establishes viable strategies for physical interventions. More specifically, it refers to a comprehensive approach to the design of the built environment that embraces architecture, planning, landscape architecture, conservation, engineering and the social and behavioural sciences. Understanding of environmental design is based on a holistic vision of the environment and the quality of life it affords to its inhabitants. It includes the following critical elements:

- a holistic understanding of the built environment
- an interdisciplinary methodology
- sustainability and ecological concerns
- social and behavioural factors
- implementation strategies.

Our focus on environmental design interventions is in the context of facilitating intergenerational engagement around the general topic of health promotion. Environmental design interventions across indoor and outdoor environments, public and private spaces and environments involve a wide range of factors at multiple scales: micro, meso and macro (Bronfenbrenner 1979). The micro-scale is analogous to the individual at the neighbourhood level and is the smallest scale of study. The meso-scale is a middle ground, concerned with communities of individuals. The largest scale is the macro-scale, which recognizes the role of institutions and policies. The multiple scales suggested by the socio-ecological perspective are particularly suited for active living strategies and their ability to provide additional insights into intergenerational design because of the place-based nature of physical activity. For the purposes of this chapter, 'intergenerational design' alludes to the challenge of designing physical environments that are conducive to intergenerational engagement and cooperation while also reflecting programmematic, organizational, socio-cultural, political and economic goals and realities (Kaplan *et al.* 2007).

We also make a distinction between the goals of developing 'multigenerational' versus 'intergenerational' settings. The former term refers to environmental design driven by the goal of accommodating the needs and interests of multigenerational groups of users; calls for 'universal design' and 'inclusive design' fit into this framework (Preiser and Smith 2011; Steinfeld and Maisel 2012). Intergenerational designs do more than enable people of different generations to occupy the same spaces; they also generate opportunities for meaningful engagement between the generations.

The third area of emphasis for this chapter is health promotion. We take an ecological approach to understanding health insofar as we view an individual's health as being nested within a family, community, ecosystem and society. We also draw from developments in the social sciences and public health professions that provide support for a wellness-oriented model of health promotion as an alternative to an illness-oriented model of disease treatment.[1]

The main question we pose in this chapter is: 'How can intergenerational design be framed in ways that promote healthy lifestyles, primarily in terms of healthier diets and increased physical activity for people across the lifespan?' In this context, we define and draw heavily on the multifaceted notion of 'active living'.

This chapter identifies critical components of the built environment to encourage active living and promote healthy lifestyles. We explore issues related to creating parks and playgrounds, streets and sidewalks (pavements), purpose-built age-integrated centres ('intergenerational shared site' facilities) and other community settings where the generations can readily meet, interact and work together to adopt healthier lifestyle behaviours. Beyond a strict focus on the physical environment, we also consider issues related to establishing intergenerational programmes, policies and partnerships that have a bearing on individual and community health practices and outcomes.

Health and wellness across the lifespan

The need to promote health and wellness for all ages has acquired a sense of urgency in contemporary society. As we experience the 'longevity revolution', with people living longer and more productive lives, it is important that they remain physically, emotionally and socially healthy across the lifespan. Understanding the complexity of factors that impinge on healthy living is necessary to meet this formidable challenge.

Why an intergenerational approach to health promotion?

Concerns about health transcend generational position: healthy communities for young people are also healthy communities for older adults. The safe and supportive environments needed for 'healthy ageing' intersect with the safe and supportive environments that we need for 'healthy youth development'. Both age groups need places to play and be active and have access to affordable healthy foods. This orientation for considering health from a lifespan development perspective is gaining attention in the public health community as is evident by the title chosen for the American Public Health Association's 2012 annual meeting – 'Prevention and Wellness Across the Life Span'.

Intergenerational relationships can serve as a powerful form of social influence, informing people about lifestyle behaviours that can improve their health and motivating them to make health-related lifestyle changes. We also know that intergenerational programmes can be quite effective in influencing people's health-related behaviours. For example, OASIS Institute's intergenerational health programmes for convincing young people to adopt healthier diets have shown promising results. Their *CATCH Healthy Habits* programme (2011–), which builds upon the Institute's earlier *Active Generations* programme (2006–2010), has been found to enhance children's knowledge about nutrition and fitness, increase their fruit and vegetable consumption and increase physical activity (for the senior volunteers as well as the children) (Teufel *et al.* 2012).

Another line of promising work in this area is taking root at the Intergenerational School, which operates as a public charter school in Cleveland, Ohio and offers a range of on-site services and mentoring opportunities for older adults (some with dementia). The Intergenerational School is launching several new programmes focused on health and wellness, including an 'Edible Forest Garden Project', which offers hands-on ways for youth and older adults to learn together about healthy foods and lifestyle behaviours (George *et al.* 2011).

Conceptualizing active living

Consistent with our ecological approach to understanding health as being tied to the family, community and society, the role of physical activity in addition to healthy food choices needs to be highlighted. In this context, it is useful to introduce the concept of active living in order to create a basic understanding of how physical activity, health and the built environment are interrelated. It can potentially provide a conceptual basis for intergenerational planning and design strategies. Active living integrates physical activity into daily routines (Sallis *et al.* 2006). It is broader as a concept than physical activity, in the sense that it takes into account physical activity in all domains of life (work, play, transportation and household activities, rather than focusing only on leisure or recreational physical activity). When viewed through the lens of active living, health benefits associated with physical activity can be gained in a variety of ways, such as walking or bicycling for transportation, exercise or pleasure, playing outdoors, working in the garden, climbing stairs and using recreational facilities.

According to the Ecological Model of Four Domains of Active Living developed by Sallis *et al.* (2006), active living occurs in four domains of life: (1) active recreation, (2) active transport, (3) occupational activities and (4) household activities. Active recreation can take place in neighbourhood public spaces and parks in a variety of settings that include athletic fields, building structures for recreational activities, community gardens, sports fields, children's play areas and so on. Active transport refers to any method of travel that utilizes human energy, but most commonly refers to walking and biking. Household activities are located in the home environment and include a wide range of activities. Some household tasks involve physical activity, like household chores and gardening. The use of labour-saving devices and electronics (computers, gaming systems and television) decrease physical activity and promote sedentary behaviour (Lanningham-Foster *et al.* 2003). Occupational activities usually refer to the work environment for adults and school environments for children, youth and young adults.

Based on this four-domain model of active living, the framework (as highlighted in Figure 3.1) delineates the connection of the four domains of active living with individual/group behavioural characteristics and the environment at various scales. Behaviour that occurs at the intersection of people and place is of particular interest in this context. Drawing from a socio-ecological perspective, place is conceptualized in terms of behaviour settings (Barker 1968; Rapoport 1990). Behaviour settings are bounded places where particular activities occur and should include physical qualities that relate to active living. These settings provide cues or signals for appropriate rules and behaviour within the setting. Both objective and perceived features of the built environment are called into play in transmitting cues/signals about appropriate behaviour. Characteristics, such as the size, scale and the nature of the physical elements in the setting, are referred to as objective features, while the perceived safety and/or comfort of a particular setting is referred to as subjective features of the built environment. The subjective features are more often linked to the larger socio-cultural context within which settings are embedded and vary between people or groups. In this context, understanding individual and group behavioural characteristics, such as personal preferences, self-efficacy, attitude, motivation and social networks are crucial. The physical and perceived qualities have implications for active living and responsive intergenerational design. Thus, in examining behaviour in the context of settings, the larger socio-cultural milieu of the people/groups involved plays a significant role and should be taken into account at the micro, meso and macro scales while planning and designing active living environments for different age groups.

The crucial relationship between active living and the built environment is often understated. As noted in Kent *et al.* (2011: 45):

> The form of the built environment, such as residential and commercial density, land use mix, connectivity and accessibility, also influences the way we move and what we do within that environment. In particular, the built environment can shape travel behaviour, including the quantity of walking, cycling, public transport and car travel, as well as the amount of leisure time that is available for other healthy pursuits. The built environment can also facilitate opportunities for recreational physical activity, by providing well maintained and useful open spaces, in addition to safe and amenable streets for non-utilitarian walking and cycling.

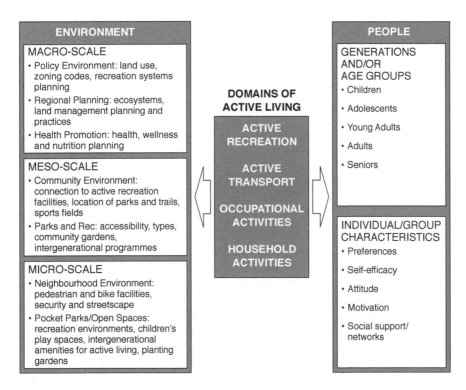

Figure 3.1 A framework for intergenerational design based on domains of active living.

Intergenerational design for healthy living: a multidimensional approach to environmental design

Designing environments with particular behavioural outcomes in mind – such as increased walking or gardening – is a complex, multidimensional undertaking. It is not as simple as designing, building and providing access to environments that afford opportunities for physical activity. If it were as simple as this, the US would not have 40 per cent of its population living sedentary lifestyles (Caballero 2007). Haider *et al.* (2010) have shown in their study of the Pottstown Area parks and recreation systems that those who are physically inactive or marginally active (about 50 per cent of the population of the target area) do not necessarily require more parks for physical activity and there are many other significant factors that impact active living.

Human motivation is complex, as is the challenge of designing environments that are conducive to intergenerational engagement and physical activity. There are many factors to consider, including: the relationship between the natural and the built environment; development of design criteria; alignment of the environmental design process and products with the goals, objectives and operational features of local programmes and policies; the degree to which intergenerational designs for healthy living reflect local history and cultural traditions of the user groups; and the level of involvement of user groups in the design process.

The role of the natural environment

In recent years, there has been increased attention to the health benefits associated with spending time in nature and of nature-based play and recreation (McCurdy *et al.* 2010). Such findings have implications for medical practice, with some medical doctors even writing 'park prescriptions' for their patients (Institute at the Golden Gate 2010).

Contact with nature need not be limited to the traditional interpretation of nature as a 'wild' encounter, but should also consider other forms of natural contact: gardening, native plants and unique natural features. Some natural features can reasonably be expected to be included in any park site. Haider *et al.* (2010) have identified specific strategies to increase contact with nature at parks, which have implications for intergenerational design:

- Improve local ecology to make parks more attractive and part of a functioning larger ecosystem. Parks are often created on surplus remnants or land that cannot be used for other development because of floodplains or difficult terrain. These features are indicators of larger ecosystems and landscapes, which include not only vegetation but also animal and avian biodiversity.
- Provide opportunities for diverse natural landscapes, rather than simply providing open green spaces. Diversity in features, uses and landscapes creates successful park spaces that attract and sustain use throughout the seasons. This includes a richness in texture, smells, colour and forms.

- Include planting gardens as focal elements. Planting gardens – for either ornamentals or vegetables – serve as a focus not only for food production, but also a sense of community and engagement. Gardens often are managed and cared for by the people who use them. They are also valued as places for hobby work and for socializing with others.

Clarifying design criteria

An early step in design is to determine the values that will guide the design process. For example, should the site be oriented to any specific type of recreational activity? Should there be a mixture of active and passive recreational activities? Should those who will use the site be involved in its design and maintenance?

If, for example, the design goal is to create an environment that encourages walking, design criteria should be oriented towards creating 'walkable' sidewalks, pathways, trails and other communal spaces. It would also be important to include design features that accommodate variation in people's abilities, perceptions and preferences with regard to how they walk. For example, benches can be strategically placed along walking paths to provide respite for those who need periodic rests.

If there is interest in providing residents with community garden plot allotments as a strategy for improving their access to fresh produce, a process needs to be put in place for building gardens on sites that are deemed 'garden-able' (e.g. with safe, toxic-free soil), creating garden plots that accommodate variation in people's ability to garden (e.g. include raised beds for those participants who experience difficulty or inability to bend to the ground) and assigning lots to community residents in an equitable manner.

In some cases the behaviour-change goal might be compelling but amorphous, such as 'increasing opportunities for intergenerational play'. In the midst of growing recognition of how shared play experiences can serve as a bridge between generations (Davis *et al.* 2002; Generations United 2008; Haider and Haenen 2004), this goal can be achieved in many ways, including with a focus on 'opening up' play at playgrounds to help older adults be more active and to facilitate cross-generational meetings (Sillito 2006). To accommodate individual variation in terms of ability level (e.g. in terms of balance, speed and coordination), it is helpful to provide different levels of challenge to match different levels of abilities and to provide a mix of active and passive play opportunities, from passive observation to intensive immersion (Kaplan *et al.* 2007).

It is also useful to consider ways to avoid or modulate certain design features that seem inimical to the desired goal of promoting intergenerational participation in active living. Expanding security zones and shrinking public space are examples of design features that limit opportunities for residents in the setting to see, find, approach, get to know and be physically active with one another.

Aligning design with programmes and policies

For designers pursuing the goal of encouraging people to walk more, there are advantages to aligning physical environmental interventions (such as building paths) with programmes and policies that are conducive to walking. For example, a local school might establish a 'walking buddies' programme in which older adult volunteers accompany students on their walks to and from school. Such programmes tend to offer increased safety, as well as encouragement to use and care for walking pathways.

An example of intervention at the policy level is the 'Complete Streets' ordinances that have been adopted in hundreds of communities. Complete Streets refer to streets that are designed to work for all users, including walkers, bikers, drivers and transit users (Smart Growth America 2014). There is considerable evidence that street networks are critical in promoting active transport especially for children and the elderly (Kent *et al.* 2011).

Tuning in to the socio-historical context

Alongside the physical parameters of the environment are social factors that are embedded in the local context, such as social norms, cultural values and traditions and existing interpersonal relationships. Connections with local history, culture and traditions can give character to public spaces such as parks and make them desirable destinations. These features also help to distinguish unique sites and tie them into a larger social, cultural and historical context (Haider *et al.* 2010).

As an illustration of the need to consider variation across cultural groups, consider the deep cultural significance that indigenous Australians attribute to walking. Walking is the primary activity in the Yiriman project, an intergenerational programme developed by Aboriginal elders in Western Australia to help young people who are harming themselves with drugs and alcohol (National Youth Affairs Research Scheme 2006: 115):

> For those involved in the Yiriman project, walking is not simply a recreational activity or something that just involves physical exercise. Here, walking is also a means through which the young get exposed to education, hunt and collect food, meet other groups, travel to and carry out ceremonies, burn areas of land and carry out other land management practices, send messages and communicate, 'freshen up' paintings, collect and produce material culture such as tools and other implements, 'map' boundaries and collect intelligence and build knowledge.

The NYARS (2006: 117) report also emphasizes the intergenerational relationship-building characteristics of walking in the Aboriginal culture:

> As one of the Yiriman workers point out, walking or travelling together for a substantial period of time often encourages a deep and complex level of

interaction. He says of the depth of intimacy, 'often the most intense and powerful relationship building goes on while you are walking ... there is something about walking together for a week that brings you into a real closeness ... one of the truly great things about the Yiriman walks is that young people get to spend quality and intense time with other members of the community ... we can't underscore how powerful and long lasting this is ... communities talk about it for months afterwards.' In this way walking together symbolizes the bringing of people into communion with those who are different. It stands as a powerful metaphor for intergenerational exchange as a means to build social contact between groups of people who otherwise might have limited dealings with each other.

Gardening is another healthy lifestyle activity rooted in socio-cultural context and meaning: selection of plants and fruits and vegetables should be consistent with residents' traditions and cultural heritage with regard to horticultural practices, food preferences and practices that are organized around food.

The importance of participatory design

Some environmental designers embrace participatory methods, based on the notion that when people have a hand in planning, managing and evaluating their local environment (facilities, outdoor sites, etc.), they are more likely to use and care for it (Hou *et al.* 2005)

An inclusive intergenerational process for developing a walking trail, for example, might involve bringing an age-integrated group of community residents together to plan, build and later assess the walking trail. An example of a participatory approach for planning an intergenerational walking tour is the 'Walk-About-Talk-About' activity that is conducted as part of 'Neighborhoods-2000' projects.[2] This entails having multi-age groups walk around the space/community in question and stop at every place or niche that holds meaning for one or more of the participants. Participants can then discuss how they perceive and experience that place and how it could be designed to better meet their needs.

There are other techniques for involving residents and consulting with stakeholders in the design process, such as inviting mixed-age groups of residents to join an environmental/facility 'design team' which organizes community collaborative meetings or 'charettes' in which design professionals help participants articulate, operationalize and visualize what they hope to accomplish in the environment. This includes values and goals related to intergenerational communication and cooperation. The collective vision may be solidified by looking at photographs of other sites and inviting participants to respond to what they notice, what they foresee as problematic and what they appreciate in the design. As participants see their ideas taking shape, they can further discuss, debate and further refine details of the environment's design elements.

Weaving an intergenerational component into such participatory design processes is likely to show participants and professional designers how the priorities and issues that affect different age groups intersect at many points, as should be the case with the programmes, policies and places developed to address those issues.

Components of multifaceted intergenerational health promotion initiatives

Intergenerational professionals (practitioners, specialists, researchers) tend to focus on programme-related variables such as training for the participants and the selection of programme activities that facilitate intergenerational communication and cooperation, while ignoring the environmental context in which behaviour occurs. Approaches to intergenerational health initiatives can create a stronger sense of community and behaviour change by integrating the physical context.

Establishing planning teams with intergenerational programming and environmental design expertise

Insofar as the physical environment clearly plays a major role in influencing patterns of intergenerational engagement, intergenerational practitioners need to have at least a moderate level of *design literacy*. At the same time, environmental designers need more of a sense of *intergenerational literacy*, if they are to be successful in designing effective *intergenerational settings*. We therefore suggest establishing programme planning teams which include individuals with expertise in both areas, intergenerational programming and environmental design.

Training for programme participants: A key component of most intergenerational programmes involves providing participants with some information on health and human development across the lifespan. For programmes with health promotion themes in particular, this includes providing some training on the well-established relationship between regular physical activity and healthy ageing (Centers for Disease Control and Prevention 2010). Participants learn that health does not just happen at discrete periods of life; it is something that needs to be nurtured across the lifespan. Lifestyle behaviours at a young age can have lifelong consequences; however, there is also some plasticity and health can potentially be improved via lifestyle changes at an older age.

To increase attention to the role of the built environment in terms of its impact on the food choices and activity options available for community residents, there should be an *environmental literacy* component in the training that is offered for programme participants, staff and volunteers. Important issues that should be addressed through the environmental literacy programme include the manner in which culture shapes people's overall food choices and behaviour, as well as the fundamental relationship between food choices and health.

Transforming knowledge into action

Intergenerational programmes focused on health promotion are action-oriented. Participants engage in community health education projects and other efforts aimed at improving their own health and at ameliorating local health-related problems. Such actions can be focused on: changing the attitudes and behaviours of programme participants, educating the public, improving local environments through community service projects, advocating for policies and other actions to address local health-related concerns and creating new settings that encourage and support residents in their efforts to adopt healthier lifestyles.

Programme participants tend not to be satisfied just learning and talking about lifestyle changes that can improve health. They embrace activities that involve developing and enacting plans to incorporate exercise regimes into their daily activities (e.g. aerobics, resistance training, stretching, walking, etc.) and more nutritious foods into their diets.

Here are some strategies that have proven useful in creating healthy living social marketing and multimedia campaigns aimed at promoting public participation in targeted physical activity and nutrition programmes:

- *Provide an empowerment message:* Overweight and obesity among children, youth and adults can be *prevented* by increasing physical activity and ensuring a healthier diet.
- *Convey a positive view about ageing:* Much of the media attention to issues related to older adults' health is negative, emphasizing the growing social and fiscal costs associated with providing additional health care. More attention needs to be drawn to a competing storyline, one in which older adults are portrayed as seeking ways to remain productive and live lives that matter. One way to do this is to highlight projects in which older adult volunteers work side-by-side with local youth to address community health challenges affecting all generations. One such example is the *Harvest 4-Health* programme (part of the Pennsylvania 4-H youth programme), in which gardens are created to increase community access to healthy and affordable foods.
- *Recommend clear behaviour-change goals:* An example provided by Glickman *et al.* (2002) is to encourage children and adolescents to reduce calories by eating less of foods that are high in sugar and fat and replace them with foods they generally under-consume such as fruits, vegetables and whole grains.
- *Use multimedia strategies for educating the public,* via art, theatrical performances, exhibits, social media communications and other forms of expression.

There are various other ways to frame intergenerational health promotion programmes. A good example is the 'Generations for a Healthy Community' programme developed and conducted by OASIS, local Girl Scout clubs and two schools in Eastern Missouri. This programme brings adults over the age of 50 together with primary school students (9–11-year-olds) to learn about the importance of healthy lifestyles, connect with community members and take action by calling for policies that promote healthy eating and physical activity. In this advocacy-oriented programme, participants learn about community health issues through photography, map-making and songs/raps about health. They also conduct meetings with local leaders to share their ideas and recommendations for improving the health of their community (National Center for Creative Aging, n.d.).

Creating shared sites (age-integrated facilities) and shared spaces

Intergenerational *shared-site* programmes are usually defined as programmes in which children and/or youth and older adults participate in ongoing services and programmes concurrently at the same site (or on the same campus within close proximity) and where participants interact during planned intergenerational activities as well as through informal encounters (Goyer 2001). A more inclusive definition incorporates *shared spaces*, whether in outdoor settings, work environments, or individuals' homes (Jarrott *et al.* 2011).

There is a growing body of evidence which indicates that intergenerational shared-site facilities generate enriched learning experiences for children and adults, enhanced health for older adults, increased activity in the community and reduced financial expenditure (Jarrott 2007; Sullivan 2002; van Vliet 2011). Such purpose-built age-integrated centres have been found to be rich in psychosocial meaning; they provide inhabitants with opportunities for gaining a cognitive understanding and an emotional appreciation of age diversity and intergenerational communal experience (Thang and Kaplan 2013) and for promoting personal as well as local community health (George *et al.* 2011).

Some intergenerational shared-site facilities focus explicitly on health-oriented activities. One such example is JEWEL (Joining Elders With Early Learners), a large (22,000 square feet) shared-site facility consisting of the Mount Kisco Child Care Center and My Second Home, an adult day programme of Family Services of Westchester in New York State. This facility fields a wide variety of activities with healthy living themes, such as Breakfast Buddies, dancing, crafts and exercise sessions (Bellamy and Meyerski 2011).

Another such example is the Lucille W. Gorham Intergenerational Community Center (Greenville, North Carolina) which houses the FRESH (Food and Relationships for Equitable and Sustainable Health) programme. Through Project FRESH, youth package and organize the delivery of fresh produce and healthy recipe cards to homebound seniors, which are then delivered on foot by a cadre of youth and adults interested in increasing their exercise and engaging in community service.

The Meadows School Project is an initiative in which a sixth-grade class in British Columbia, Canada was relocated to a nearby assisted living facility for a five-week period.[3] In addition to the students working on mandated curriculum activities each day, there were many spin-off activities in which residents took part on a voluntary basis. During the relocation period, the students and residents sought one another out for walks and for participation in scheduled and unscheduled fitness activities. These activities included: armchair fitness classes, seasonal craft projects, science fairs, spelling bees, sing-alongs and celebrations. Interactions also included some residents helping to plan and lead activities and students taking on service tasks which included caring for the rabbits and chickens on the assisted living facility's spacious grounds (Carson *et al.* 2011).

One practice that emerged in the Meadows School Project which served as a prelude and motivator for intergenerational involvement in healthy living activities was 'door knocking':

> 'Door knocking' refers to a behaviour that researchers saw repeatedly whereby students initiated contact with residents by knocking on their doors to personally invite them to participate in activities. It represents a spontaneous desire to reach out to residents. It emerged very powerfully in interviews as a key factor in residents' participation in the project.
>
> (Carson *et al.* 2011: 411)

Germany's Ministry for Family Affairs, Senior Citizens, Women and Youth, with funding from the federal government and the European Social Fund, launched the Multigenerational Centres programme in 2006. As of 2012, there were a total of 500 multigenerational centres throughout Germany, with each centre receiving €40,000 per year for a five-year period. Although the Centres are diverse in terms of local partners, activities and physical environment, they basically function as public drop-in centres, with the general goal of promoting interaction, cooperation and mutual support between the generations. Activities span the areas of early childhood education and child care, family support services, mentoring for youth, social engagement activities targeting isolated older adults, gardening, theatre, music and other creative arts activities (Burchard *et al.* 2012).

Multigenerational Centres are designed to accommodate mono-generational activities and, with 90 per cent of the centres including 'open meeting points', intergenerational activities. Burchard *et al.* (2012: 302) note several features for facilitating both realms of activity:

> Different zones should be provided: a zone for children where they can crawl on the ground and spread out their toys and where the furniture has no dangerous edges and a zone for adults where they can have cups of coffee, sit together, talk to one another or just relax. Some multigenerational centers also try structures with a variety of opportunities to meet and arrange places where visitors are able to observe activities without participating themselves. The aim is that visitors should feel as if they were part of the activities without necessarily being at the center of the room. Some older adults

do not wish to communicate constantly but just want to be in contact with others, see something, observe, reflect a little and exchange a word every now and then. Structuring the open meeting point in such a way helps to facilitate a wide range of encounters between the generations without having to force such exchanges.

Conclusions

In framing our orientation for exploring and creating intergenerational settings for healthy living, we have alluded to several intergenerational design principles based on traditional models of person–environment relationships, as articulated in the conceptual framework laid out in Kaplan *et al.* (2007). In particular, we emphasized the importance of tuning in to the role of structured spaces in fostering informal, unstructured intergenerational encounters. This includes developing indoor and outdoor places with the intention of providing opportunities for spontaneous meetings, whether it is by inserting nooks and alcoves for impromptu gatherings outside and on the periphery of rooms used for structured activities, or the inclusion of small communal sitting areas placed on the outskirts of park areas used for active recreational purposes. Such spaces help invite intergenerational encounters and provide the spark for igniting meaningful relationships. At the same time, however, we emphasize the need to allow users to enter age-integrated facilities and other sites and participate in activities at their own pace and in a manner that respects their privacy and preferences for intergenerational engagement.

We have also emphasized the need to align environmental design with programmes and activities that are developed to provide participants with opportunities to find out about their intergenerational similarities as well as their differences (real and imagined). Such discussion and reflection helps pave the way for creating additional layers of social connectedness and community involvement in people's lives, two powerful determinates of health and well-being.

Environmental design, planning and policy interventions are generally recognized as the most promising avenues for creating substantial improvements in physical activity, healthy eating, weight status and social well-being. We propose an interdisciplinary and multi-level approach to promoting active living lifestyles consistent with social ecological principles. This holistic framework is consistent with emergent ideas on intergenerational design or planning. The process of designing an intergenerational environment is not antithetical to the creative process. It involves a flexible, iterative series of steps where multiple perspectives are brought together for listening carefully, proposing non-routine solutions to problems and reflecting on possibilities and preferences. The process is also inclusive, with a focus on exploring diverse perspectives in planning and generating questions to address in the design. Participants in the process should include not only designers, but also people of different age groups who will be using the space.

Notes

1 In support of a wellness-oriented model of health promotion, there is strong scientific evidence indicating that dietary changes and engaging in regular physical activity can help prevent obesity, which is correlated with major chronic diseases, such as type-2 diabetes and heart disease. For example, the Diabetes Prevention Program (DPP) demonstrated that lifestyle modifications that involve changes to diet, exercise and losing a little weight (about 10 per cent) can reduce the risk of type-2 diabetes by 58 per cent among those adults at high risk (Diabetes Prevention Program Research Group 2002). Regular physical activity has also been found to reduce rates of other chronic diseases like coronary heart disease, stroke, high blood pressure and some cancers (HHS 2008).
2 Neighborhoods-2000 was developed at the Center for Human Environments (City University of New York) in 1988. Children (usually fifth and sixth graders) and older adult volunteers work together on community exploration, visioning and advocacy-type activities (Kaplan 1994).
3 Technically, this is more of an 'immersion' model than a 'shared site' model insofar as the students were relocated rather than co-located with the older adults as is the case with the typical shared-site programme.

References

Barker, R.G. (1968) *Ecological Psychology: Concepts and Methods for Studying the Environment of Human Behaviour*, Stanford, CA: Stanford University Press.

Bellamy, P. and Meyerski, D. (2011) 'JEWEL: Joining elders with early learners', *Journal of Intergenerational Relationships*, 9(4): 462–465.

Beth Johnson Foundation (2001) 'Definition of intergenerational practice', Stoke-on-Trent: Centre for Intergenerational Practice, online, available at: www.centreforip.org.uk (accessed 4 December 2013).

Bronfenbrenner, U. (1979) *The Ecology of Human Development: Experiments by Nature and Design*, Cambridge, MA: Harvard University Press.

Burchard, R., Doubravova, D. and Oldenburg, A. (2012) 'Intergenerational encounters in multigenerational centers in Germany', *Journal of Intergenerational Relationships*, 10(3): 299–303.

Caballero, B. (2007) 'The global epidemic of obesity: An overview', *Epidemiologic Review*, 29(1): 1–5.

Carson, A.J., Kobayashi, K.M. and Kuehne, V.S. (2011) 'The Meadows School Project: Case study of a unique shared site intergenerational program', *Journal of Intergenerational Relationships*, 9(4): 405–417.

Centers for Disease Control and Prevention (2010) 'Physical activity for everyone: How much physical activity do older adults need?' online, available at: www.cdc.gov/physicalactivity/everyone/guidelines/olderadults.html (accessed 4 December 2013).

Davis, L., Larkin, E. and Graves, S.B. (2002) 'Intergenerational learning through play', *International Journal of Early Childhood*, 3(2): 42–49.

Diabetes Prevention Program Research Group (2002) 'Reduction in the incidence of type 2 diabetes with lifestyle intervention or metformin', *New England Journal of Medicine*, 346(6): 393–403.

Generations United (2008) *Play is Forever: The Benefits of Intergenerational Play* (factsheet), Washington, DC: GU.

George, D., Whitehouse, C. and Whitehouse, P. (2011) 'A model of intergenerativity: How the intergenerational school is bringing the generations together to foster collective wisdom and community health', *Journal of Intergenerational Relationships*, 9(4): 389–404.

Glickman, D., Parker, L., Sim, L.J., Del Valle Cook, H. and Miller, E.A. (eds) (2002) *Accelerating Progress in Obesity Prevention: Solving the Weight of the Nation*, Institute of Medicine of the National Academies, Washington, DC: The National Academies Press.

Goyer, A. (2001) *Intergenerational Shared Site and Shared Resource Programs: Current Models*, Generations United Project SHARE background paper, Washington, DC: Generations United.

Haider, J. and Haenen, J.P. (2004) 'Childhood experiences in the age of consumerism: Intergenerational play in designing public space', Congress proceedings, Second Child in the City Congress, London, October, pp. 147–152.

Haider, J., Aeschbacher, P. and Bose, M. (2010) *Planning and Design Strategies for Healthy Living, Parks and Recreation in the Pottstown Area*, University Park, PA: The Pennsylvania State University.

HHS (Health and Human Services) (2008) 'Physical activity guidelines for Americans', online, available at: www.health.gov/paguidelines/guidelines/chapter2.aspx (accessed 1 December 2013).

Hou, J., Francis, M. and Brightbill, N. (2005) *(Re)Constructing Communities: Design Participation in the Face of Change*, Davis, CA: Center for Design Research.

Institute at the Golden Gate (2010) *Park Prescriptions: Profiles and Resources for Good Health from the Great Outdoors*, Sausalito, CA: IGG.

Jarrott, S.E. (2007) *Tried and True: A Guide to Successful Intergenerational Activities at Shared Site Programs*, Washington, DC: Generations United.

Jarrott, S., Kaplan, M. and Steinig, S. (2011) 'Shared sites: Avenues for sharing space, place and life experience across generations', *Journal of Intergenerational Relationships*, 9(4): 343–347.

Kaplan, M. (1994) *Side-by-Side: Exploring Your Neighborhood Through Intergenerational Activities*, San Francisco, CA: MIG Communications.

Kaplan, M., Haider, J., Cohen, U. and Turner, D. (2007) 'Environmental design perspectives on intergenerational programs and practices: An emergent conceptual framework', *Journal of Intergenerational Relationships*, 5(2): 81–110.

Kent, J., Thompson, S.M. and Jalaludin, B. (2011) *Healthy Built Environments: A Review of the Literature*, Sydney: Healthy Built environments Program, City Futures Research Centre, UNSW.

Lanningham-Foster, L., Nysse, L.J. and Levine, J.A. (2003) 'Labor saved, calories lost: The energetic impact of domestic labor-saving devices', *Obesity Research*, 11(10): 1178–1181.

McCurdy, L.E., Winterbottom, K.E., Mehta, S.S. and Roberts, J.R. (2010) 'Using nature and outdoor activity to improve children's health', *Current Problems in Pediatric and Adolescent Health Care*, 40(5): 102–107.

National Center for Creative Aging (n.d.) 'Generations for a healthy community', online, available at: www.creativeaging.org/creative-aging-program/8387 (accessed 27 November 2013).

National Youth Affairs Research Scheme (2006) *Community Building Through Intergenerational Exchange Programs*, report to the National Youth Affairs Research Scheme, Canberra: Australian Government Department of Families, Community Services and Indigenous Affairs (FaCSIA).

Preiser, W. and Smith, K. (2011) *Universal Design Handbook*, second edition, New York, NY: McGraw-Hill.

Rapoport, A. (1990) *The Meaning of the Built Environment: A Nonverbal Communication Approach*, Tucson, AZ: University of Arizona Press.

Sallis, J.F., Cervero, R.B., Ascher, W., Henderson, K.A., Kraft, M.K. and Kerr, J. (2006) 'An ecological approach to creating active living communities', *Annual Review of Public Health*, 27: 297–322.

Sillito, D. (2006) 'Finns open playground to older adults', *BBC News*, 8 February, available at: http://news.bbc.co.uk/go/pr/fr/-/1/hi/world/europe/4691088.stm (accessed 2 December 2013).

Smart Growth America (2014) National Complete Streets Coalition overview, online, available at: www.smartgrowthamerica.org/complete-streets (accessed 18 February 2014).

Steinfeld, E. and Maisel, J. (eds) (2012) *Universal Design: Creating Inclusive Environments*, Hoboken, NJ: John Wiley and Sons.

Sullivan, K.J. (2002) *Catching the Age Wave: Building Schools with Senior Citizens in Mind*, Washington, DC: National Clearinghouse for Educational Facilities.

Teufel, J., Holtgrave, P., Dinman, M. and Werner, D. (2012) 'An intergenerational, volunteer-led approach to healthy eating and active living: CATCH healthy habits', *Journal of Intergenerational Relationships*, 10(2): 179–183.

Thang, L.L. and Kaplan, M. (2013) 'Intergenerational pathways for building relational spaces and places', in G.D. Rowles and M. Bernard (eds) *Environmental Gerontology: Making Meaningful Places in Old Age*, New York, NY: Springer, pp. 225–251.

van Vliet, W. (2011) 'Intergenerational cities: A framework for policies and programs', *Journal of Intergenerational Relationships*, 9(4): 348–365.

4 Intergenerational shared spaces in the UK context

Julie Melville and Alan Hatton-Yeo

Introduction

Naturally occurring opportunities for exchange and interaction between the generations are not as prevalent in contemporary society as they perhaps once were. This means that young people and older adults are now more likely to spend a significant amount of their time in age-segregated settings; children often spend their days in school and/or child-care centres, younger people with their friends, and many older adults in age-isolated facilities such as senior centres or retirement homes/communities (Johnson and Bytheway 1994). This viewpoint is reiterated in a survey of European citizens who were asked their opinions about existing relations between the young and the old. The survey found that the majority of citizens felt there were insufficient opportunities for older and younger people to meet and work together, via associations and local community initiatives (European Commission 2009).

In response to these concerns, it has been suggested that both younger people and older adults thrive when resources are used to bring the generations together rather than separate them (Generations United 2005; Jarrott and Weintraub 2007). Intergenerational Shared Spaces (IGSS) have therefore been promoted as a means of addressing some of the negative social implications of an increasingly age-segregated society and they present opportunities for frequent, structured and informal activities and have the potential to establish a more age-integrated community that can meet the diverse needs of its members (Generations United 2005; Hayes 2003).

In the context of the current social, political and economic circumstances this chapter will provide a critical review of the current literature on intergenerational shared spaces (IGSS) and relate this to the UK policy and practice context; discuss the relevance of intergenerational shared spaces to the UK with reference to current political and economic drivers and circumstance; and explore potential future actions for policy makers and practitioners.[1]

Background

The concept of intergenerational practice is not new, but is historically embedded in the familial and patriarchal relationships of different cultures (Hatton-Yeo and Ohsako 2000: 12). Many elements of intergenerational practice have been around for decades, but it was not until the 1980s that intergenerational practice was recognized as being particularly relevant to addressing a variety of social problems and issues such as drug and alcohol abuse, low self-esteem and isolation that were affecting two of the most vulnerable population groups (Beth Johnson Foundation 2011) – namely older and younger people.

From the 1990s onwards, the scope of intergenerational programmes altered to include the regeneration of local communities through various action programmes. Intergenerational programming had increased dramatically across Europe by the end of the 1990s in response to such issues as the integration of immigrants, the need to enhance social inclusion and active ageing, and the perception of the breakdown of familial solidarity (Beth Johnson Foundation 2008). By the beginning of the twenty-first century, intergenerational practice was progressively seen as a way of addressing tensions between the generations and varied projects were being established internationally (Beth Johnson Foundation 2011).

Many studies have primarily focused on how programme participants benefit from involvement in intergenerational relationships. More recently, attention has been drawn to the potential of such programmes to build cohesive communities and promote civic engagement. Kaplan *et al.* (2007) suggest that this has been accompanied by a conceptual shift in how intergenerational practice is defined and understood with a move away from a singular emphasis on structured programmes of intervention, to encompass a wider emphasis on the cultural and communal practices involved in bringing older adults and younger people together. This shift resonates with UK developments, as intergenerational practice has become increasingly well established here.

Twenty-five years ago, intergenerational practice in the UK was about younger people 'doing things to/for' older adults with minimal contact between the generations (Bernard 2006). Early projects were school-based, mainly focusing on mentoring schemes. Today, intergenerational practice is based much more on exchange and reciprocity, where younger people and older adults are brought together in a variety of settings and environments to engage in mutually rewarding activities (Bernard 2006). Accordingly, intergenerational practice is no longer limited to individual participants and how they benefit, but is increasingly applicable to intergenerational relationships within the wider community. More programmes are recognizing this shift and are concerned with such outcomes as the creation of social capital; the potential to develop the capacity of communities; the diversification of volunteering; and the greater involvement of educational institutions in their communities (4Children 2011; Springate *et al.* 2008).

However, this focus on outcomes as a way of defining intergenerational practice, whilst useful, has been criticized for failing to acknowledge that outcomes may accrue to just one generation rather than the other and that they may well be influenced by characteristics such as age, class, income, gender and so on (Mannion 2012). What is more, Springate *et al.* (2008) argue the need for greater clarity around the definition of intergenerational practice, and Mannion (2012: 388) has recently called for 'a more nuanced approach to saying what is distinctive about intergenerational practice'. Mannion (2012: 396) further makes the case for understanding intergenerational practice as being, 'all-age, reciprocal and multigenerational' but as a prelude to offering an expanded definition of intergenerational education, rather than intergenerational practice per se.

Whilst accepting that there is no single agreed definition of intergenerational practice and that it is still a contested area requiring greater attention, for the purposes of this review, we have adopted the Centre for Intergenerational Practice's definition of intergenerational practice:

> Intergenerational practice aims to bring people together in purposeful, mutually beneficial activities which promote greater understanding and respect between generations and contributes to building more cohesive communities. Intergenerational practice is inclusive, building on the positive resources that the young and old have to offer each other and those around them.
>
> (Centre for Intergenerational Practice 2014)

Intergenerational shared sites/spaces – definition and models

As intergenerational practice has evolved as a field there has been debate and discussion about how it is defined, structured and approached. Twenty years ago, the National Council on Aging in the United States defined intergenerational programmes as: 'activities or programs that increase cooperation, interaction or exchange between any two generations' (Kaplan *et al.* 2007: 83). Since then, there has been significant change and growth in the volume and diversity of programmes.

The North American model of IGSS is arguably the most prevalent (having been around the longest) and is based primarily on a physically constructed shared site, in contrast with more naturally occurring shared sites such as public spaces or parks. For example, in North America, this model typically consists of a day-care facility for children based within an older adults' long-term care facility (Kuehne and Kaplan 2001). However, the emergence and growth of IGSSs over the past 20 years has resulted in a wide variety of models being established. Some 15 years ago, the American Association of Retired People (Goyer and Zuses 1998) released the results of their survey of IGSS, which detailed the range of shared-site programme possibilities and reported the most common programmes in existence. Since then, there have been very few (systematic) international reviews of the literature available on these programmes (Melville 2009; Jarrott and Bruno 2007) and, as with intergenerational practice, understandings of what IGSS are vary across studies and in the research and practice literature.

For the most part, an early and well-known definition of an intergenerational shared site used by AARP – 'programs in which multiple generations receive ongoing services and/or programming at the same site, and generally interact through planned and/or informal intergenerational activities' is still used (Goyer and Zuses 1998: p. v). However, in a recent special issue of the *Journal of Intergenerational Relationships* on shared sites, the guest editor argues that this original definition is limited in that it does not 'necessarily apply across all countries and cultures' (Jarrott 2011: 40).

Again, whilst acknowledging that there is no agreed definition of IGSS – and because of the, as yet, limited UK work on them – we have adopted the following North American definition to anchor this chapter:

> Intergenerational shared sites are programs in which children and/or youth and older adults participate in ongoing services and/or programming concurrently at the same site (or on the same campus within close proximity), and where participants interact during regularly scheduled, planned intergenerational activities, as well as through informal encounters.
>
> (Generations United 2005: 13)

The majority of literature and examples of IGSS are from North America and as such, the literature is disproportionately shaped by a particular cultural and political context. Whilst we proactively looked for literature in the UK, the majority of examples found either involved a single generational facility, such as a school, children's centre or residential home, which engaged in specific contained and time-limited intergenerational activities or a community setting that promoted use by people of all ages but activities were not intentionally intergenerational and usually age segregated. While IGSS programmes continue to grow in the US, with over 280 documented sites now in existence (Generations United 2005), shared sites in the UK are a relatively new phenomenon. At the time of writing, few intergenerational sites/spaces with explicit intergenerational focuses were identified in the literature. It should be noted that there may be other sites/ spaces in which older adults and young people share facilities, but these facilities do not offer or actively encourage 'shared' activities (Vegeris and Campbell-Barr 2007).

Intergenerational shared sites/spaces – the research evidence

Whilst intergenerational programmes have developed steadily over the past 20 years, our knowledge and understanding of how these programmes work, and if they meet their aims and objectives, is still limited (Steinig 2006; Granville 2002; Kuehne 1999). To date, only four overviews of intergenerational practice have been conducted in the UK (Martin *et al.* 2010; Springate *et al.* 2008; Beth Johnson Foundation 2006; Granville 2002). These reviews are limited in their scope and, for the purposes of this review do not specifically focus on IGSS. However, these reviews highlight the fact that compared to the rapidly growing quantity and variability of programmes and projects, the number of documented evaluation and research studies is not keeping pace (Jarrott 2011; Melville 2009;

Kuehne 2003a; Granville 2002; Kaplan and Kuehne 2001). It should also be noted that similar concerns have been expressed about the limited amount of research being conducted in Europe and countries other than the United States (Melville 2009; Statham 2009; Granville 2002). It is important to sound a caution here that lessons learned from intergenerational practice in the US are not necessarily directly transferrable to other cultural and policy contexts, nor to countries where there may be different traditions and methods of conducting research.

That said, in her early review of intergenerational practice in the UK, Granville (2002) argued that more research was required to validate claims made by practitioners about the benefits of intergenerational practice. Pain (2005) adds to this discussion by asserting that evaluation of outcomes is a challenging task for projects, and that both 'hard' and 'soft' outcomes are difficult to quantify as they are often diffuse and long term. According to Springate *et al.* (2008), the evidence base remains weak, particularly in relation to outcomes. To date in the UK, there has been no large-scale, formal, systematic evaluation of intergenerational projects and the international base of reliable data on outcomes is also relatively small (Jarrott 2011; Jarrott and Weintraub 2007; Epstein and Boisvert 2005). Many researchers have suggested that this is due to the fact that intergenerational practice is inherently 'unsuited' to quantitative research methodologies given the often small sample sizes, the variability of projects and schemes, confusion over what aims and objectives to measure, and difficulty in obtaining a control group (HM Government 2009; Epstein and Boisvert 2005; Bowen *et al.* 2000).

Evaluation data often focus on outcomes without attention to the nature of the interactions between generations and have ignored more fundamental questions pertaining to whether, and how, participants interact (Jarrott 2008). However, understanding the *process* of intergenerational contact is central to understanding its outcomes. Moreover, experiences of many practitioners in the field have highlighted interaction (both formal and informal) as the central mechanism for achieving mutual benefit in intergenerational practices (Jarrott 2008). Linked with this, there are a limited number of documented research studies of intergenerational practice drawing on established theoretical frameworks (Kuehne 2003b; VanderVen 2011; Melville 2009). Reports that do exist are often based on anecdotal information that lacks any clear conceptual framework to begin from and only emphasizes the immediate effect of a programme.

The body of research that does exist is often based on small, non-representative samples of participants and is not subject to rigorous methodology (Jarrott and Bruno 2007; Raynes 2004; Granville 2002). However, it does permit us to examine some of the benefits and obstacles associated with the development of IGSSs and help us, as Jarrott (2008) argues, to ask critical questions such as at what expense are these benefits acquired; and what are their connected 'costs'? With that said, below is a detailed description of the 'potential' benefits of IGSS found in the literature.

Intergenerational shared sites/spaces: benefits

Existing studies of IGSS programmes have consistently indicated that they are mutually beneficial for all participants, and can yield positive outcomes for individuals, both younger and older, for staff, for the organization itself, for local communities and for the wider society (Jarrott *et al.* 2011; Beth Johnson Foundation 2011). For example, one of the overviews of intergenerational practice (Springate *et al.* 2008) found that it has great potential for changing negative perceptions of older adults and young people and increasing the health and well-being of those involved. According to the Beth Johnson Foundation (2011), there is also clear evidence that intergenerational practices can improve service design and delivery, and improve the quality of life experienced at the local community level. More specifically, IGSSs have been identified as key developments in local communities with the potential to explore solutions to conflicts over public space, contribute to regeneration projects, enhance active citizenship among generations, improve community cohesion and deliver aspects of neighbourhood renewal schemes (Pain 2005; Granville 2002; Kuehne and Kaplan 2001). For the purposes of this chapter, benefits are looked at in relation to individual participants; practical administrative and operational issues; financial considerations and the wider community.

Benefits for individual participants

IGSS administrators and staff have reported that the most successful aspects of IGSS are the positive benefits to participants and the increased frequency of positive, informal interactions among the generations (Jarrott *et al.* 2008). Martin *et al.* (2010) found that the most fundamental outcome for participants was that they enjoyed the activities. More specifically, Springate *et al.* (2008) have identified four main outcomes for both younger and older participants: increased understanding, friendship, enjoyment and confidence. Other research on 'generational attitudes' has explored whether contact was successful in fostering positive images of ageing, thereby reducing stereotypes and various forms of discrimination and suggests that young people and older adults both benefit from shared experiences and daily contact (Hayes 2003; Salari 2002).

Benefits specific to older adults relate mainly to improved health and well-being, reduced isolation and social exclusion, and a renewed sense of worth (Springate *et al.* 2008; Jarrott and Bruno 2007). Outcomes particular to young people include gaining specific skills, improved self-esteem, and greater empathy for older adults (Springate *et al.* 2008; Jarrott and Bruno 2007). Yet other studies have examined the personal and social skills of youth and younger children taking part in IGSS programmes. Results indicate that skills development – verbal, cognitive and practical, is enhanced in comparison to their counterparts in non-IGSS programmes (Rosebrook 2008). It has also been suggested that young people often thrive on the individual attention that older adults are able to provide in this type of setting (Jarrott 2008). For their part, older adult volunteers obtain a sense of satisfaction from participating in such activities and have reported increased self-esteem, an enhanced sense of belonging and increased social interaction (Jarrott 2011).

Another consistent theme highlighted in the literature is the benefits that accrue to staff, such as the ability to learn about the role and importance of other populations that they do not normally work with (Jarrott *et al.* 2008). Additionally, it has been suggested that the availability of, and access to, on-site child care can improve staff recruitment and retention. The staff in most IGSS programmes also report positive feelings about their programmes; and this, and the added benefits of on-site child care in some facilities, contributes to lower staff turnover in an area that is often plagued with high staff turnover and stress (Jarrott *et al.* 2008).

Practical and administrative benefits

Several research and evaluation studies have considered the administrative and practical dimensions of IGSSs as well as exploring the significance of the context and setting for programme objectives and outcomes. As Kuehne (2003b) states, institutional contexts can be important to IGSS programme outcomes but are often not apparent on the surface. Kuehne and Kaplan (2001) also warn us that the impact of institutional variables and the nature and quality of administrative leadership in IGSS programmes should not be underestimated. For example, administrators are more likely to provide truly intergenerational activities if they hold positive attitudes toward intergenerational exchanges in general (Kuehne and Kaplan 2001).

This raises the associated issue of the quality of staff working in IGSSs and several studies have emphasized the need for adequately trained staff: staff who possess the skills and knowledge related to meeting age-appropriate developmental needs; and who are able to develop and implement IGSS programmes (Kuehne and Kaplan 2001). The importance of having – or developing – expertise in working with both younger and older participant groups has also been highlighted by many researchers as essential in planning and facilitating developmentally appropriate programmes (Springate *et al.* 2008; Hayes 2003). Similarly, the research evidence shows that IGSS programmes work best when participants are involved from the beginning, with staff, in planning what they want to do rather than having things done to and/or for them (Kuehne and Kaplan 2001).

The 'setting' has emerged as another significant factor in the success or otherwise of IGSSs (Kaplan *et al.* 2007; Jarrott *et al.* 2006). This is supported by Salari's (2002) research examining age-appropriate environments in intergenerational interaction which concluded that variations in the physical environment, staff demeanour and activity content of programming can generate vastly different responses and outcomes. Likewise, Hayes' (2003) study and Jarrott and Bruno's (2003) work illustrate that time and frequent, regular, opportunities for intergenerational contact and interaction are the best combination for building positive relationships within a particular IGSS space.

Financial benefits

Research shows too that IGSSs are potentially cost-effective, and cost containment has been one of the main benefits identified in the literature (Jarrott *et al.* 2008). For example, it has been proposed that IGSSs have the unique ability to 'expand funding options' by attracting new grants drawing from traditional children, young people and older adults' services and/or sources (Jarrott *et al.* 2008). Similarly, Hayden's (2003) financial analysis of an IGSS programme generated a list of 'cost-effective' benefits including shared services such as child and nursing care and volunteering. Perhaps the most comprehensive and noteworthy study in this area is Generations United's (Jarrott *et al.* 2008) comparative analysis of the operational costs of IGSS. The findings show that the use of shared sites can result in a decrease in total expenditures, and a lessening of programme costs when older adult and young people's services share expenses. Furthermore, programmes with high levels of intergenerational contact cost only as much, or less, to operate than programmes without high intergenerational contact.

Benefits for the wider community

Beyond the narrow focus of benefit(s) to individual participants, several beneficial outcomes for the wider community have also been identified in the IGSS literature. These include: improved community cohesion; its ability to help address other community-related social issues; to help build social capital and develop community capacity; its impact in terms of the growth in volunteering; and on the ways in which educational institutions have become more involved in their communities (Springate *et al.* 2008).

Role of the environment in intergenerational practice

A review of the literature has clearly demonstrated that one of the critical issues emerging within the intergenerational field is a lack of attention to how the built environment plays a crucial role in influencing intergenerational interaction. As Mannion (2012: 391) states: 'interpersonal relations are always located in a place'. Therefore, the development and management of intergenerational shared spaces must consider both the environment and, specifically, people's relationships with their environments. More specifically, we would argue that any intergenerational shared space must focus on the interaction between person and environment, rather than on one or the other exclusively and explore how the design and use of a physical environment or space promotes or inhibits the interactions taking place within it.

Kaplan and Kuehne (2001) in their informative paper *Evaluation and Research on Intergenerational Shared Site Facilities and Programs* introduce the novel idea of the intergenerational quotient (IQ) where a community/shared space with a high 'IQ' would consciously ensure that such a community would be developed to be elder and child/youth-friendly and where the design would put intergenerational exchange as a high priority. Kaplan and Kuehne (2001: 5) go further and identify a number of core principles that are central to developing successful IGSSs:

1   Adopting a lifespan perspective for training staff on human development;
2   Cross training of all staff in an interdisciplinary model that ensures everyone has bought into the model is an absolute priority;
3   Partnership is central to success;
4   People need to be clear what the purpose of an IGSS is and consequently how success is defined.

Age-friendly cities and communities: a framework for the future of IGSS?

Building on the World Health Organization (WHO) programme to develop Age-Friendly Cities and Communities (WHO 2007) and the European Year for Active Ageing and Solidarity between Generations – 2012, a vision for a European society for all ages has been developed. This movement has gained visibility with an increasing number of stakeholders now interested in a collaborative effort to achieve fair and sustainable solutions that promote the best use of financial and human resources to develop age-friendly environments. As a result, there is now a vision for an All Age-Friendly Europe by 2020 (AGE 2012). AGE's Manifesto for an Age-Friendly European Union by 2020 outlines a list of benefits from creating an age-friendly European Union. While all benefits listed in the manifesto have relevance to this review, two benefits highlighted here are particularly significant to the development and fostering of IGSS:

1   Goods and services that are adapted to the needs of all highlighting the need for solutions that are based on the concept of Design-for-All and Universal Design – essentially intervention for environments with the aim that all ages can enjoy participating in the 'construction' of our society with equal access to use and understand their environment with as much independence as possible.
2   Accessible outdoor spaces, buildings and transport as well as adapted housing and physical activity facilities that promote participation in society for longer, while increasing opportunities for exchange within and across generations.

In many ways, the development of intergenerational shared spaces mirrors findings emerging from the World Health Organization on Age-Friendly Cities and Communities concept of considering the needs of all the generations in planning and allocating resources and services as an essential principle for community planning.

In an effort to achieve fair and sustainable solutions that promote the best use of financial and human resources, the development and use of existing shared spaces is even more important and pertinent. According to WHO, the physical and social environments are key determinants of whether people remain healthy and independent throughout the lifecourse. Demographic change is a major challenge facing all EU countries. As such, promoting age-friendly environments is an effective approach for responding to demographic change. Creating age-friendly environments means adapting our everyday living environment in order to empower people to promote social inclusion, active participation and maintain an autonomous and good quality of life in later life. What is more, such a solution has the potential help societies better cope with demographic ageing in a way that is fair for all generations.

Intergenerational shared spaces and 'age-friendly places': a new concept for intergenerational interaction

The North American model of IGSS, as has been illustrated, is a particular model primarily based on a physically constructed shared site. Although such initiatives have existed for over 20 years there is still a real lack, and depth, of evidence or theoretical underpinning to this work. Consequently, a much deeper understanding of how to develop successful intergenerational programmes is needed. This review has shown that the literature has primarily focused on interactions within age-segregated environments and has recognized a growing need in the field to consider older and younger adults' engagement with and attachment to age-integrated communities and spaces. As such, more recent literature has focused on the use of different public spaces in urban areas that are shared by many generations. Therefore, a growing interest in exploring public shared spaces and places that are intergenerational is not only increasing but essential as we develop new integrated models to address the aspirations of citizens in a time of significant demographic change. Moreover, our current climate of austerity means we have to develop new models for community planning and building that are both more effective and efficient and have the potential to achieve multiple outcomes across the lifecourse.

To date, the term 'site' has been consistently used in the literature and evidence. The term IGSS has come to signify a physical environment deliberately constructed or redeveloped to enable two or more groups of different ages to more readily interact in a location designed primarily to provide a service to each group separately. While much of the literature has focused on IGSS as the main option that could help to foster interaction between the generations, in the current economic climate there is unlikely to be sufficient funding for building new shared-site facilities. From the limited evidence available, we would question whether replicating such a model is necessary or desirable given the opportunities to utilize existing spaces (i.e. public libraries, local community centres, faith centres) that already promote opportunities for mutual exchange between the generations.

It is our belief that a more fundamental re-appraisal needs to be made of the purpose and outcomes of such initiatives before we start fixing the nature of the delivery model. Therefore, we propose the idea of 'shared space', rather than site, as a way of suggesting that the aims of an IGSS can be met by a broader range of environments than just co-located services. Such a strategy would be more efficient in terms of time, personnel and use of (limited) resources that already exist.

A model, based on the notion that many spaces currently being used for the sole purpose of one generation (i.e. schools, retirement homes) – or spaces utilized by various generations independently (i.e. library, community centres, sporting facilities) – could in fact be used 'intergenerationally'. What is more, this model or concept could be developed to include outside spaces such as parks, town centres and playgrounds.

From the review of existing evidence we suggest that for a shared space to be successful, it must possess a number of essential attributes:

1   People of different generations are able to enjoy regular contact, both formally and informally.
2   Participants from different generations are actively involved in the planning and running of all activities.
3   All staff are cross-trained and have an understanding of human development from a lifecourse perspective. Staff undertake ageism training and understand and promote the benefits of integrated working.
4   Staff and participants approach risk assessment as a positive opportunity to build safe relationships and partnerships.
5   However the space is constructed or defined, everyone continually questions how it can be made more 'age friendly' to facilitate interaction across the generations.
6   There is an explicit understanding of the fundamental aim to allow all generations to collaborate on positive activities of shared interest and benefit.

In addition we suggest that it is the combination of both the physical design of a space, as well as the furnishings within a space, that can contribute to potential interaction across the generations. We wish to stress that people who are considering embarking on the process of developing a shared space need to give equal weight to the activities, programmes and services – what happens within these spaces, as to various design principles and the end users involved. Moreover, it is critical that practitioners involved in this process consider how the environment and activities intersect, and that both staff and participants (end users) have an equal role to play from the outset. The need is to create or use existing spaces, facilitated by trained staff committed to thinking across the lifecourse that can become focal points and catalysts for regular intergenerational interaction. Such interaction will be promoted and achieved by removing barriers and challenging the stereotypes that too often shape our thinking.

Adapted from Kaplan's categorization of intergenerational programmes using a depth of engagement scale (Kaplan 2001), we suggest using 'How interconnected is your generational space?' which provides a diagrammatic illustration of the depth of engagement (or increased 'IQ') in different types of settings (see Table 4.1).

Table 4.1 Intergenerational Shared Spaces Engagement Model – How interconnected is your generational space?

Depth of engagement	Example of types of activity	Rationale and potential outcomes of activity
Centre involved in occasional 'one off' intergenerational (IG) activities	Annual concert with older people invited to attend	• Enjoyment of occasion • Can be starting point for greater future engagement
Centre that has underutilized space for hire to external groups. This may include some organized IG activities	Primary school with community rooms with separate access Children's centre that uses spare space to generate income Community centre rooms may be hired for IG projects for defined groups	• Centre has additional resources • Wider usage of centre by different groups • Some benefits achieved for participants as a consequence of specific activities and potential informal interaction
Site that has a programme of regular IG activities that are time limited with small-group facilitators	School that has a regular allotment project IGI mentoring project Time-limited skill-sharing project, e.g. IT skills, craft, etc.	• Specific outcomes for participants from activities • Some friendship and relationships developed • One or two intergenerational 'champions' in setting (but Intergenerational Practice (IP) seen as their responsibility) • Pool of potential volunteers and partners developed
Centre with core programme IG activity that also gives opportunity for informal contact. All staff trained to do IP and work in a cross-disciplinary way	Community library in school as focal point for whole-school activity including volunteering and skill sharing Community food co-op involving people all ages, volunteering, employability, cookery skills Environmental Garden Project involving whole community Archaeology site recruiting people of all ages to work together and learn together	• Relationships and respect between generations developed as a natural consequence of regular interaction • Generations learn to collaborate, listen to, share and support one another • Staff naturally work with partners to look for multiple outcomes and added value for activities as appropriate • The whole community is engaged and demonstrates greater confidence and capacity

We also propose that a model based around communities of interest rather than constructed around short-term projects is more sustainable with greater long-term impact. By communities of interest we refer to those whose primary focus may be such topics as the environment, food, sport or the arts, that create a space for people to come together around shared aspirations and interest. If the aim is to create spaces that build generationally connected communities, we would argue that it must be based on the aspiration of communities and their vision of success. This work will need to link to the current debates on localism, community assets and the work of organizations that promote sustainable models of community activism.

To conclude, our review of the literature on the potential of intergenerational shared spaces suggests that such models of working are relevant at both a policy and practice level. However, we caution that the planning, development and use of such models will take considerable time and understanding and be anchored by the involvement of potential users. However, the potential benefits, particularly given current scarce resources, provide a model of transformative action, to develop active and engaged places that engage all the generations. Age-friendly places can promote positive engagement and participation across the lifecourse and between all the generations.

Note

1 Much of this review is adapted from Julie Melville's doctoral thesis (Melville 2013); Alan Hatton-Yeo, in collaboration with the Beth Johnson Foundation, was associate supervisor in this partnership.

References

AGE (2012) *Roadmap Towards and Beyond the European Year for Active Ageing and Solidarity Between Generations 2012 (EY2012)*, Brussels: AGE Platform Europe.

Bernard, M. (2006) 'Keynote 1. Research, policy, practice, and theory: interrelated dimensions of a developing field', *Journal of Intergenerational Relationships*, 4(1): 5–21.

Beth Johnson Foundation (2006) *Intergenerational Practice: Active Participation Across the Generations*, Stoke-on-Trent: Beth Johnson Foundation.

Beth Johnson Foundation (2008) *Intergenerational Programmes and Practice. Briefing Paper: Intergenerational Practice Ministerial Roundtable Discussion, House of Commons*, Stoke-on Trent: Beth Johnson Foundation.

Beth Johnson Foundation (2011) *A Guide to Intergenerational Practice*, Stoke-on-Trent: Beth Johnson Foundation.

Bowen, G.L., Martin, J.A., Mancini, J.A. and Nelson, J.P. (2000) 'Community capacity: Antecedents and consequences', *Journal of Community Practice*, 8(2): 1–21.

Centre for Intergenerational Practice (2014) 'About us', online, available at: www.centre-forip.org.uk/about-us/what-is-ip (accessed 10 January 2014).

Epstein, A.S. and Boisvert, C. (2005) *Let's Do Something Together: Identifying the Effective Components of Intergenerational Programs, Final Project Report*, Ypsilanti, MI: High Scope Press.

European Commission (2009) *Intergenerational Solidarity: Analytical Report, Flash Eurobarometer 269*, Brussels: European Commission.

4Children (2011) *For All Ages: Bringing Different Generations Closer Together*, online, available at: www.4children.org.uk/Resources/Detail/For-All-Ages (accessed 1 February 2014).

Generations United (2005) *Under One Roof: A Guide to Starting and Strengthening Intergenerational Shared Site Programs*, Washington, DC: Generations United.

Goyer, A. and Zuses, R. (1998) *Intergenerational Shared-Site Project, A Study of Co-located Programs and Services for children, youth, and Older Adults: final report*, Washington, DC: AARP.

Granville, G. (2002) *A Review of Intergenerational Practice in the UK*, Stoke-on-Trent: Beth Johnson Foundation.

Hatton-Yeo, A. and Ohsako, T. (2000) *Intergenerational Programmes: Public Policy and Research Implications: An International Perspective*, Hamburg, Germany: UNESCO Institute for Education.

Hayden, C.D. (2003) *Financial Analysis & Considerations for Replication of the ONE-generation (ONE) Intergenerational Daycare Program*, Oakland, CA: National Economic Development and Law Center.

Hayes, C.L. (2003) 'An observational study in developing an intergenerational shared site program: challenges and insights', *Journal of Intergenerational Relationships*, 1(1): 113–131.

HM Government (2009) *Generations Together: A Demonstrator Programme of Intergenerational Practice*, London: Department for Children, Schools and Families.

Jarrott, S.E. (2008) 'Shared site intergenerational programs: Obstacles and opportunities', *Journal of Intergenerational Relationships*, 6(3): 384–388.

Jarrott, S.E. (2011) 'Where have we been and where are we going? Content analysis of evaluation research of intergenerational programs', *Journal of Intergenerational Relationships*, 9(1): 37–52.

Jarrott, S.E. and Bruno, K.A. (2007) 'Shared site intergenerational programs: A case study', *Journal of Applied Gerontology*, 26(3): 239–257.

Jarrott, S.E. and Weintraub, A.P. (2007) 'Intergenerational shared sites: A practical model', in M. Sánchez (ed.) *Intergenerational Programmes: Towards a Society for all Ages*, Barcelona: la caixa Foundation, pp. 125–146.

Jarrott, S.E., Gigliotti, C.M. and Smock, S.A. (2006) 'Where do we stand? Testing the foundation of a shared site intergenerational program', *Journal of Intergenerational Relationships*, 4(2): 73–92.

Jarrott, S.E., Morris, M.M., Burnett, A.J., Stauffer, D., Stremmel, A.S. and Gigliotti, C.M. (2011) 'Creating community capacity at a shared site intergenerational program: "Like a barefoot climb up a mountain"', *Journal of Intergenerational Relationships*, 9(4): 418–434.

Johnson, J. and Bytheway, B. (2004) 'Ageism: Concept and definition', in J. Johnson and R. Slater (eds) *Ageing and Later Life*, Thousand Oaks, CA: Sage Publications.

Kaplan, M. (2001) *School-Based Intergenerational Programs*, Hamburg: UNESCO Institute for Education.

Kaplan, M. and Kuehne, V. (2001) *Evaluation and Research on Intergenerational Shared Site Facilities and Programs: What We Know and What We Need to Learn*, Generations United Background Paper: Project SHARE, Washington, DC: Generations United.

Kaplan, M., Haider, J., Cohen, U. and Turner, D. (2007) 'Environmental design perspectives on intergenerational programs and practices: An emergent conceptual framework', *Journal of Intergenerational Relationships*, 5(2): 81–110.

Kuehne, V.S. (1999) 'Building intergenerational communities through research and evaluation', *Generations*, 22(4): 82–87.

Kuehne, V. (2003a) 'The state of our art: Intergenerational program research and evaluation: part one', *Journal of Intergenerational Relationships*, 1(1): 145–161.

Kuehne, V. (2003b) 'The state of our art: Intergenerational program research and evaluation: part two', *Journal of Intergenerational Relationships*, 1(2): 79–93.

Kuehne, V. and Kaplan, M. (2001) 'Evaluation and research on intergenerational shared site facilities and programs: What we know and what we need to learn', Generations United background paper, Project SHARE, Washington, DC: Generations United.

Mannion, G. (2012) 'Intergenerational education: The significance of reciprocity and place', *Journal of Intergenerational Relationships*, 10(4): 386–399.

Martin, K., Springate, I. and Atkinson, M. (2010) *Intergenerational Practice: Outcomes and Effectiveness*, LGA research report, Slough: National Foundation for Educational Research.

Melville, J. (2009) 'A critical, library-based review, of intergenerational shared sites: Promoting communication between the generations, unpublished MSc dissertation, Vrije University, the Netherlands.

Melville, J. (2013) 'Promoting communication and fostering interaction between the generations: A study of the UK's first purpose-built intergenerational centre', unpublished doctoral dissertation, Keele University.

Pain, R. (2005) *Intergenerational Relations and Practice in the Development of Sustainable Communities*, background paper for the Office of the Deputy Prime Minister, Durham: International Centre for the Regional Regeneration and Development Studies, Durham University.

Raynes, N. (2004) 'Where are we now with intergenerational developments: An English perspective', *Journal of Intergenerational Relationships*, 2(3/4): 187–195.

Rosebrook, V. (2008) 'Shared sites and their contribution to a society for all ages', *Journal of Intergenerational Relationships*, 6(3): 380–383.

Salari, S.M. (2002) 'Intergenerational partnerships in adult day centres: Importance of age-appropriate environments and behaviours', *Gerontologist*, 42(3): 321–333.

Springate, I., Atkinson, M. and Martin, K. (2008) *Intergenerational Practice: A Review of the Literature*, LGA Research Report F/SR262, Slough: NFER.

Statham, E. (2009) 'Promoting intergenerational programmes: Where is the evidence to inform policy and practice?' *Evidence and Policy*, 5(4): 471–488.

Steinig, S. (2006) *Intergenerational Shared Sites: Troubleshooting.* Washington, DC: Generations United.

VanderVen, K. (2011) 'The road to intergenerational theory is under construction: A continuing story', *Journal of Intergenerational Relationships*, 9(1): 22–36.

Vegeris, S. and Campbell-Barr, V. (2007) *Supporting an Intergenerational Centre in London: Scoping the Evidence*, Policy Studies Institute, London: London Development Agency.

World Health Organization (2007) *Global Age Friendly Cities: A Guide*, Geneva: World Health Organization.

5 'It's a really nice place to live!'

The ethnographic encounter as a space of intergenerational exchange

Helen Lomax

Introduction

This chapter draws on insights from visual sociology and human geography to consider the ethnographic encounter as a space of intergenerational exchange. Building on theoretical developments in children's geography and sociology (Fox Gotham 2003; Hendrick 2003; Valentine 2008; Vanderbeck and Dunkley 2004) and methodological insights from visual sociology and discursive psychology (Luttrell 2010; Taylor 2010; Wetherell 1998) the chapter considers the ways in which the ethnographic encounter makes visible the complex dynamics of intergenerational relationships and identities in disadvantaged neighbourhoods. The analysis draws on sequences of ethnographic film-making with young people, focusing on exchanges between the child-interviewers and their adult subjects in order to explore the ways in which residents articulate their experiences of life in low-income neighbourhoods. These include sequences in which adults emphasize the positive aspects of community life and disavow wider negative stereotypes and imaginings of poorer places (Parker and Garner 2010). The chapter suggests that these narratives can be understood as a form of place-making work through which residents seek to construct a positive identity in the context of dominant stigmatizing narratives about the negative social burden of life in low-income neighbourhoods (Geddes *et al.* 2010; Pearce 2012).

The chapter argues that these identity practices can offer an alternative perspective on intergenerational relationships to that offered in policy, practice and media representations of low-income neighbourhoods (Fink and Lomax 2014 in press; McKendrick *et al.* 2008), not least that these exchanges might serve as collective bonding work between generations and within neighbourhoods in response to external identity threats. In addition, following Valentine's (2008) proposal for an evidenced-based examination of the practices of living and Fox Gotham's (2003) emphasis on the fluid nature of social identity, the chapter considers how a focus on lived encounters can enable a more nuanced understanding of the ways in which adults and children negotiate a sense of community and inclusion in the face of significant social and economic challenges.

Place matters: inequality and the neighbourhood effect

> Communities and neighbourhoods that ensure access to basic goods, that are
> socially cohesive, that are designed to promote good physical and psycho-
> logical wellbeing, and that are protective of the natural environment are
> essential for health equity.
>
> (Marmot 2008: 4)

The starting point for this chapter is that place matters, and it matters particularly
to children and families in low-income neighbourhoods. This can be seen in
national statistics on life expectancy and health, with those living in the poorest
neighbourhoods having significantly lowered life expectancy and disability-free
years then their better-off neighbours (Office for National Statistics 2011). While,
overall, life expectancy is rising, inequalities in health are increasing for those
living in the poorest localities (Torjesen 2012). People living in low-income
neighbourhoods experience diminished access to life-enhancing resources and
greater environmental burdens in the form of poor housing, poorer air quality,
lack of access to green space, higher levels of pollution and traffic (Geddes *et al.*
2010; Marmot 2008). Children living in low-income neighbourhoods are less
likely to participate in outdoor activities due to reduced access to safe places to
play (Popay *et al.* 2003; UNICEF 2011). These data are supported by growing
evidence on the psychosocial burden of poverty, inequality and spatial stigma and
their impacts on the physical and psychological health of communities and individuals (Wilkinson and Pickett 2009; Pearce 2012; Prilleltensky 2012). At a
policy level there is increased focus on addressing these burdens through initiatives aimed at enhancing community resilience and social capital (Buck and
Gregory 2013; Cabinet Office 2013). An important theme underpinning such
approaches is the recognition of the ways in which place-based relationships are
protective of health. This is supported by sociological research, which reveals
how supportive relationships enhance quality of life and promote resilience, and
corroborated by cohort studies, which demonstrate the protective effects of neighbourhood or 'proximal' relationships for all age groups across the social gradient
(Bartley 2012; Higgs *et al.* 2003; Steptoe *et al.* 2013). Although it is important to
note that these effects follow the social gradient; good health and good relationships are more common in those who are better financially resourced (Bartley
2012). These differences are explained by poorer people channelling limited financial and social resources into basic needs, resulting in significant long-term
negative implications for health (Wilkinson and Pickett 2009).

However, whilst there have been a number of important studies of low-
income neighbourhoods there has been a tendency to examine adult and children's experiences in isolation, with little research exploring adults and
children's collective experience of place (Besten 2010; Brent 2009; Rogaly and
Taylor 2011). The analysis presented in this chapter aims to redress this gap in
the literature in order to explore residents' experience from both adult and
children's perspectives and, at the same time, to elaborate the potential of

participatory visual methods for enabling this process. It asks: how are neighbourhood spaces in low-income areas experienced by adult and child residents? And what can creative participatory methods add to our understanding of children's and adult's experiences of life in low-income neighbourhoods?

Neighbourhood and research contexts

The research was undertaken in two low-income neighbourhoods in the UK: *Springfield* and *Riverside*.[1] Both areas are established estates of predominantly social housing (approximately two-thirds of householders rent their properties from the council, a housing association or a private landlord). Built during the 1970s and comprised of low-density terraced dwellings surrounded by parkland, cycle and footways, Springfield is an estate of approximately 850 households (2,000 residents). Riverside is much smaller, home to 350 households (less than 1,000 residents). They are also two of the 10 per cent most deprived neighbourhoods in England with high levels of disadvantage as measured by the percentage of adults claiming income support. Life expectancy is the lowest in the city with residents, on average, living ten years less than those from more affluent areas of the city. Other measures which indicate disadvantage include the high number of family referrals, children in receipt of free school meals, teenage conceptions, poor exam results, neighbourhood complaints, criminal damage and emergency hospital admissions (Anonymized neighbourhood statistics 2011). Additionally both estates are stigmatized places, regularly labelled as the 'worst' places to live in local print and online media.

However, there remains a lack of understanding about how these neighbourhood statistics translate into children and adults' lived experiences. The aim of the research was to address this deficit, working with residents in order to articulate their experiences of life in each neighbourhood. To this end the research offered participants a choice of participatory visual methods including drawing, photography and film-making with which to explore their experiences. Visual methods were selected as a set of methods which, if used critically and within a participatory framework, can support people to articulate and give voice to their experiences (Luttrell 2010; Luttrell and Chalfen 2010; Piper and Frankham 2007). Here, my approach acknowledges the emerging critique from within childhood studies and visual methods, recognizing the ways in which research outputs are shaped by the social and discursive contexts in which they are produced. As Mannay (2013: 137) argues, creative research outputs: 'do not exist in a vacuum; they are based on the ... experiences and feelings of the person who made them and ... the power relations that surround them.'

Mannay's position reflects my own concern to explore the ways in which children and young people's activities as knowledge producers are shaped by the dynamic relationship between themselves and the people around them, from the adults and children whom they interview, film and photograph and with whom they co-produce the visual outputs. In this, my focus has been to explore the messy and dialogic nature of participatory work (Harper 2002) through a focus on the immediate productive and discursive contexts of fieldwork as well as to attend to the ways in which images are read by audiences (Lomax and Fink 2010; Lomax 2012). This chapter extends this analysis to consider the ways in which dominant narratives and counter-narratives (about place, neighbouring, parenting and a 'good' childhood) are brought to bear in the productive context of participatory visual research with children and young people. It considers how an understanding of these discursive contexts can illuminate children's and adults' experiences of their neighbourhoods and children's roles as knowledge producers. This includes the ways in which these data, in the form of images and narratives, can be seen to 'answer back' (Clarke *et al.* 2007) to dominant and stigmatizing narratives about life in low-income neighbourhoods. The chapter suggests that theorizing the research encounter as a space of intergenerational exchange and attention to ways in which different (adult and child) voices are brought to bear on children's knowledge production in these contexts can offer a richer understanding of place and place-making as a situated negotiation between adults and children in these contexts.

Findings

The data presented in this chapter is drawn from the participatory films (*Springfield Friends* and *I Love Riverside*) and photography walks made in each neighbourhood. Data was analysed using methods developed within discursive and narrative psychology (Taylor 2010; Wetherell 1998) and focuses on children's and adults' talk recorded during film-making, neighbourhood walks and interviews. This comprised filmed interviews in which children talked to other young people and adults, including, on occasion, their own parents. In addition, it includes some sequences of talk between Katrina, a professional production editor who visited Riverside for a brief period during fieldwork in order to interview residents for an audio podcast about neighbourhood well-being. Katrina's interviews with adults and children provide an important contrast with the children's style of interviewing which further illustrate the importance of giving critical attention to the productive contexts of interviews. As the analysis considers, Katrina is particularly persistent in her questioning about the ways in which neighbourhoods may be experienced as threatening and harmful. The adult responses, which disavow the assumptions about poorer neighbourhoods implied in Katrina's line of questioning, provide an important source of information about adults and children's place-making and the ways in which it is rhetorically shaped and recipient designed (Billig 1987; Heath *et al.* 2010).

Felt aspects of neighbourhoods

An important theme which is revealed through the children's film-making and photography is the significant attachments and pride in neighbourhood, themes which are connected to and observable in the children's pleasure in the creative process and their deployment of particular media (captions, drawings and still photography) in their film-making. This includes a potent sequence in which the children took turns to glide down a slide, proclaiming 'We love *Springfield*' and another in which they worked together to create a living diorama in which their bodies collectively spell out the estate's name (see Plate 5.1).

A further amusing sequence involved one of the children creating a voice-over in which he pretends to interview a stick. Adopting a comic voice he asks 'What do you like about Springfield?', the stick replying 'Everything!' to peals of laughter from the other children and adults. These sequences collectively and powerfully convey the significance of neighbourhood and its importance for friendship. Drawing also provided an opportunity for expressing feelings about place, the final scene in 'Springfield Friends' includes a large red heart which the children had drawn, with the words 'We love Springfield'. This choice of images and creation of captions to accompany a selection of some of the photographs the children had created suggests a similar reading. Here the children selected a range of images from their fieldwork including portraits of older adults attending a lunch club; other children and adults photographed during the period of the study and multiple scenes of the parks, green spaces and play areas on the estate. Captions such as 'I took this picture'; 'Family ... family ... family' 'Go us! We are amazing!' and '*Springfield* is not just an estate. We are a <u>TEAM</u>! suggest pride, friendship and social connectivity.

Plate 5.1 Film-makers spell out the letters of the estate's name, Springfield (photograph by anonymous interviewee, © author).

In Riverside too this was seen in the ways in which the children chose to photograph each other, posed for individual and group portraits and 'having fun' in the community garden. This is mirrored in the children's filmed interviews with each other in which they talk about the importance of friendship and playing out. Georgia, for example, in response to a question about what she likes doing best in the community garden explains that she likes to 'play and talk' while Ellie describes 'playing in the park and my front garden' as her favourite activities. The sensory experience of place was seen in the children's discussion about why they took particular images, explanations which were based on the aesthetics of the landscape ('it's beautiful' and 'I like the way that the water ripples'; Plate 5.2) and which encompassed their everyday experience of place.

For example Amy's photography of the hillock in Plate 5.3 was accompanied by her recollection of sledging there with her siblings while Paulo's photograph of the canal was accompanied by a tale he had been told about a child who was said to have drowned there. Both explanations indicate a familiarity and connection with the landscape including its joys (sledging) and risks (drowning). They also show an appreciation of the natural environment of the estate (water, green space, plants). Their comments about parks and green spaces in particular

Plate 5.2 The mound, Riverside (photograph by anonymous interviewee, © author).

Plate 5.3 Ripples on the canal, Riverside (photograph by anonymous interviewee, © author).

echoing, as the next section considers, what adult residents also claim to value about their neighbourhoods, as providing a pleasant, friendly (and safe) environment for children. Of significance is the ways that adults and children's responses here contrast noticeably with the ways in which children's use of, and visibility in public space is problematized in media and policy (Morrow and Mayall 2009; Moss 2012; Pain 2006). Rather, in both neighbourhoods, adults were unequivocal that children should be seen and heard in neighbourhood spaces. These views, the ways in which they are expressed in the intergenerational spaces of the research and how they can be seen to shape children's articulation of their experiences are considered in the following section.

Neighbouring: social connectivity and belonging

Adults and children in both neighbourhoods talked about the importance of safe spaces and opportunities for play and participation in community activities. This included parks and green space for children to play out as well as events and activities organized in the community garden and centre. This is expressed by

Jane, a Springfield parent, who in a filmed interview with her daughter, Chloe, contrasts Springfield with many such opportunities with a previous neighbourhood in which there was a lack of such provision. As Jane says, 'there's a lot more activities going on compared to what we used to have.' Similarly Laura, a Riverside parent talks about the importance of the play areas and parks as providing important opportunities for children. As she explains, in a response to a question from Katrina (the production editor mentioned earlier) about amenities on the estate, 'We use the Children's Centre. They have great activities for the little one. Don't they, Holly?' Laura and Jane's comments here are significant. Both talk about the important opportunities that each estate provides for their children but they do so in a very specific way. Jane's use of 'we' in her response to her daughter's question both includes and invites her agreement. Similarly, Laura, who on this occasion is being interviewed by Katrina in the presence of her young children, includes her ten-year-old daughter, Holly in her response, inviting her to agree with her claim about the provision of 'great activities'. This is reinforced a few moments later in Laura's comments about the estate's parks and green spaces in which she describe how her children can play out. Here she employs a similar rhetorical device, including her daughter Holly in the reply. As she says 'There's the park, my little one goes out there, she's only three but my big one can take her can't you?' As Jane does earlier, Laura contrasts her children's freedom to play out with friends; the ease with which they can 'go to the park', 'have a picnic' and 'kick a ball in the back park' with the lack of such opportunities in a previous neighbourhood, where busy roads prevented the children playing out. Of interest is the way that the children's identities, as active participants in the life of the community are mobilized in these responses to reinforce a narrative of participation and belonging. Of further interest is the way that Holly, Laura's eldest daughter who is present and included in this sequence (referred to as 'my big one') mirrors her mother's statement in her own subsequent comments to Katrina. As Holly says, 'it's nice to see kids playing out', a comment which might suggest that she is 'ventriloquizing' parental narratives (Luttrell 2010) but is perhaps better understood as a particular response to the antagonistic line of questioning by Katrina in this context.

The importance of 'getting involved' and volunteering in community activities was articulated by several adults who expressed in interviews with both Katrina and the children that they enjoyed taking part in neighbourhood events and celebrations. As Laura explains to Katrina 'we always try to take part in everything that is going on in the community'. The strong sense of social connectivity was also evident in adults' responses to children's questions about how they felt about where they lived. For example, Kelly, in response to her daughter's question 'what do you like about Riverside?' explains that 'I like the people – everyone's friendly and works together' while several adults used the phrase 'community spirit' to talk about what they liked about their neighbourhood. This importance of community is further exemplified in explanations about neighbourly responsibility expressed in the adults' interviews with Katrina. For example, Laura explains to Katrina how a neighbour had informed her when her daughter, Holly, was subjected to an unpleasant and worrying incident by an unknown adult. As she explains 'people look out for one another in Riverside.' The idea of neighbourliness as watching out for others is further suggested in Liz's (a Riverside parent) response to a question from Katrina about how safe the estate feels. Liz's response includes the assertion that while she herself has never witnessed any anti-social behaviour on the estate she would have no hesitation in intervening should the occasion arise. In this way Liz and Laura's responses disavow the idea that the estate is anything other than a safe place while displaying their credentials as good neighbours who look out for others.

Stigma: resisting negative narratives

The idea that low-income areas are stigmatizing places to live was plainly evident in participants' talk, mentioned directly by adults and children alike. For example Callum, aged 12, explains in a filmed interview with Springfield children, that 'people don't think much of Springfield', a comment which reveals his attentiveness to the ways in which the estate is stigmatized locally. Children's comments during the Riverside photography walking tour also suggest an awareness of social difference, privilege and disadvantage. For example 'the mound' (a grassy hillock which separates the executive homes from the terraced housing in Riverside in Plate 5.1) was mentioned as being a place to have fun (rolling down and sledging) whilst also that it 'separated the two classes of people'. Such comments reflect children's understanding of local narratives about the ways in which the estate is designed and landscaped to reinforce social division (large private dwellings separate from smaller council housing). These social divisions are also articulated by the adults on the Riverside estate. Laura, a Riverside parent, narrates an unpleasant encounter to Katrina in which her partner was verbally abused by a member of staff in a local shop, abuse which centred on his status as a Riverside resident. Of significance is the way in which in this comment, in common with the earlier discussion of the incident involving her daughter, blame is attributed to people from outside the estate. In this way Laura defends against the possible interpretation that these incidents are attributable to the moral character of those living on the estate. Of further methodological interest is the observation that both these comments are made out of earshot of the child-filmmakers, suggesting that there are some aspects of neighbourhood life that parents prefer to protect their children from.

   Disavowal of negative assumptions about life on the Riverside estate was also seen in adult responses to Katrina's questions about vandalism, crime and litter. Lesley, who was helping the children plant the community garden at the time of Katrina's visit, responds to persistent questioning about the impact on the garden of 'litter' and mess with an outright denial ('This area doesn't seem to suffer'). Instead, Katrina's question about the possibility of malicious damage is reformulated in a reply which constitutes 'damage' as an inevitable and natural consequence of plants and children in the same space. Lesley describes the unavoidable consequence of 'children want(ing) to play and flowers getting broken' and her planting solutions to maximize everyone's enjoyment of the garden. In so doing she rebuts the implicit charge that public spaces in low-income neighbourhoods are rife with community and intergenerational conflict. Similarly, in another example in which Kyle, a young single man who lives on Riverside, is asked about what he likes about the neighbourhood ('the community centre and activities') and what he likes less ('people riding around in their cars really fast, people hanging about on the street corners as some people do'). However when pressed about this by Katrina, who asks him 'Is it about lack of employment?' he is quick to challenge this interpretation, pointing out that 'most people have got jobs, everyone is working really hard and trying their best.' One interpretation of Kyle's response is that it is an explicit and carefully crafted defence against the moral sanction implied by a dominant narrative about low-income neighbourhoods rife with crime and worklessness.

This section has explored the ways in which adults may disavow dominant negative narratives about poorer neighbourhoods. The following section explores the ways in which both adults and children resist and silence the dominant narrative and 'imagination ... of the slum' (Rogaly and Taylor 2011: 50) through a counter-narrative of low-income neighbourhoods as flourishing.

Flourishing neighbourhoods

As I have explored elsewhere (Lomax *et al.* 2011) adults responses to children about life in low-income neighbourhoods can be seen to safeguard childhood from wider negative portrayals of life in disadvantaged neighbourhoods. For example, presenting neighbourhoods as spaces of inclusion and cohesion and glossing over troubles (such as speeding cars, petty crime and financial hardship). Data presented in an earlier paper (Lomax *et al.* 2011), for example, explored how Jane, the mother of one of the child-filmmakers, sought to present the move to the estate from a rural neighbourhood some distance away, as a positive choice in terms of the opportunities it afforded her daughter. Other residents, in interviews with the children, downplayed suggestions of anti-social behaviour and discord ('every estate has its problems') while describing the ways in which the estate is 'coming together'. Narratives from Riverside adults and parents filmed and interviewed by children are similarly positive. The estate is described as a place 'I love' (Leanne); with great parks and 'just home to me' (Kelly) and with 'a good community spirit' (Serena). This is not to suggest that adults are insincere or misrepresenting their experiences of life on the estate but rather that neighbourhoods are complex places which can be both social and anti-social, friendly and hostile. Moreover, the adults' responses to children need to be understood in the context of the geographies of poverty and wider stigmatizing discourses of low-income neighbourhoods. As the earlier section considered, both estates are disadvantaged on a range of indicators, disadvantage that children might not be explicitly aware of (Ridge 2011; Sutton 2009) but which parents and adults must negotiate and manage on children's behalf. Serena's and the other adults' comments need to be understood in the context of place-making and identity work, including the ways in which postcodes can label and stigmatize (Wacquant 2008). As Billig (1987) argues, where one lives brings its own rhetorical baggage which can threaten residents' social and moral identities as parents, neighbours and citizens. Viewed through this lens, narratives about hard work, neighbourliness and community spirit are a means of challenging dominant narratives of shirkers and skivers (Stuckler and Reeves 2013). As Clarke *et al.* (2007: 142) explain 'people are not just addressed or summoned by dominant discourses but also "answer back".' Their efforts in this regard can be viewed as an attempt to challenge the prevailing narrative of 'improper places' (Popay *et al.* 2003) and to assert an alternative or 'emergent' (Williams 1977) narrative in order to claim a legitimate social and moral identity. As Finnegan (1998: 172) argues: 'story-tellings are used in the claiming or maintenance of identity, for self-legitimation and the validation of experience.'

In this way, the employment of particular tropes such as 'community spirit' and the 'good neighbour' are an important means by which adult residents are able to claim a particular moral identity in a socio-economic and spatial context in which this is routinely threatened.

Concluding remarks

As Luttrell and Chalfen (2010: 198) argue, there is an assumption in participatory visual research that such methods 'give voice' to marginalized and less powerful groups. This includes the suggestion that creative and visual methods enable individuals to articulate their experiences in ways which are meaningful to them and to challenge decision-making and policy. However, as Walmsley and Johnson (2003: 41) argue, whose story is being told, and for whose benefit, is not always made clear. Within childhood studies this has included a call for those working participatively to evidence their claims (Gallacher and Gallagher 2008). This chapter has taken up this challenge, building on methodological insights from geography and sociology (Harper 2002; Hunleth 2011; Mannay 2010, 2013) to consider the ways in which participatory visual research is constituted in and through the complex, messy and dynamic intergenerational relationships that can characterize fieldwork. It has included a focus on broader dialogues or structures of feeling (Williams 1977) that characterize imaginings of life in low-income neighbourhoods and through which residents of all ages can be seen to be in continual dialogue.

In so doing, the chapter has articulated the ways in which adults' voices may sometimes drown out those of child participants in child-led research, elaborating the ways in which parents and other adults may unwittingly (and wittingly) seek to influence the data that children produce. However rather than seeing this as a failure of 'child-led' methods, I have suggested that a methodological focus on this process, exploring what is being said, what is silenced and what left unsaid, can enable a richer, more nuanced understanding of adults and children's lives in low-income neighbourhoods. Moreover, framing the research encounter as a space of intergenerational exchange and elaborating the potential of discursive methods for 'hearing' the visual (Piper and Frankham 2007) can enable visual researchers to get to grips with 'whose voice is whose' (Luttrell and Chalfen 2010: 198). Theorizing ethnographic spaces in this way can offer insights into place-based identities including the ways in which these are produced in and through participants' narratives while elaborating the constraints of children's and adults' knowledge production in these contexts. Understanding these processes enables us to better understand them as partial representations, like ethnographic texts, framed by wider discursive contexts and participants' understanding of what is tellable within these contexts.

In summary, visual ethnographers need to think critically about the ways in which creative and visual methods enable participants to speak for themselves. This chapter has explored how wider narratives (of poverty, parenting and neighbourhood) structure the ethnographic encounter. This includes adult responses to a growing negative rhetoric about poorer places and people. Thus, while the ethnographic encounter can provide a space through which lives in low-income neighbourhoods can be made visible it would be unwise to view this space as a neutral space for adults and children to articulate their lives. Rather, participatory visual data is best seen as the outcome of a dialogue with wider narratives and contexts, an examination of which can offer a richer account of life in poorer neighbourhoods.

Note

1 Springfield and Riverside are pseudonyms.

References

Bartley, M. (2012) *Life Gets Under Your Skin*, London: ESRC International Centre for Lifecourse Studies in Society and Health, UCL Research Department of Epidemiology and Public Health.

Besten, O. (2010) 'Local belonging and "geographies of emotions": Immigrant children's experience of their neighbourhoods in Paris and Berlin', *Childhood*, 17(2): 181–195.

Billig, M. (1987) *Arguing and Thinking: A Rhetorical Approach to Social Psychology*, Cambridge: Cambridge University Press.

Brent, J. (2009) *Searching for Community: Representation, Power and Action on an Urban Estate*, Bristol: The Policy Press.

Buck, D. and Gregory, S. (2013) *Improving the Public's Health: A Resource for Local Authorities*, London: The King's Fund.

Cabinet Office (2013) *Wellbeing: Policy and Analysis. National Wellbeing Collection, Policy Paper*, London: Government Office, online, available at: www.gov.uk/government/publications/wellbeing-policy-and-analysis.

Clarke, J., Newman, J., Smith, N., Vidler, E. and Westmarland, L. (2007) *Creating Citizen-Consumers: Changing Publics and Changing Public Services*, London: Sage.

Fink, J. and Lomax, H. (2014 in press) 'Challenging images? Dominant, residual and emergent meanings in on-line media representations of child poverty', *Journal for the Study of British Cultures*, 1(21).

Finnegan, R. (1998) *Tales of the City: A Study of Narrative and Urban Life*, Cambridge: Cambridge University Press.

Fox Gotham, K. (2003) 'Towards an understanding of the spatiality of urban poverty: The urban poor as spatial actors', *International Journal of Urban and Regional Research*, 27(3): 723–737.

Gallacher, L. and Gallagher, M. (2008) 'Methodological immaturity in childhood research? Thinking through "participatory methods"', *Childhood*, 15(4): 499–516.

Geddes, I., Allen, J., Allen, M. and Morrisey, L. (2010) *The Marmot Review: Implications for Spatial Planning*, London: University College.

Harper, D. (2002) 'Talking about pictures: A case for photo elicitation', *Visual Studies*, 17(1): 37–41.

Heath, J., Hindmarsh, J. and Luff, P. (2010) *Video in Qualitative Research: Analysing Social Interaction in Everyday Life*, London: Sage.

Hendrick, H. (2003) *Child Welfare: Historical Dimensions, Contemporary Debate*, Bristol: The Policy Press.

Higgs, P., Hyde, M., Wiggins, R. and Blane, D. (2003) 'Researching quality of life in early old age: The importance of the sociological dimension', *Social Policy and Administration*, 37(3): 239–252.

Hunleth, J. (2011) 'Beyond, on or *with*: Questioning power dynamics and knowledge production in "child-oriented" research methodology', *Childhood*, 18(1): 81–93.

Lomax, H. (2012) 'Shifting the focus: Children's image-making practices and their implications for analysis', *International Journal of Research & Method in Education*, 35(3): 227–234.

Lomax, H. and Fink, J. (2010) 'Interpreting images of motherhood: the contexts and dynamics of collective viewing', *Sociological Research Online*, 15(3): art. 2, available at: http://socresonline.org.uk/15/3/2.html.

Lomax, H., Fink, J., Singh, N. and High, C. (2011) 'The politics of performance: Methodological challenges of researching children's experiences of childhood through the lens of participatory video', *International Journal of Social Research Methodology*, 14(3): 231–243.

Luttrell, W. (2010) '"A camera is a big responsibility": A lens for analysing children's visual voices', *Visual Studies*, 25(3): 224–237.

Luttrell, W. and Chalfen, R. (2010) 'Lifting up voices in participatory visual research', *Visual Studies*, 25(3): 197–200.

Mannay, D. (2010) 'Making the familiar strange: can visual research methods render the familiar setting more perceptible?' *Qualitative Research*, 10(1): 91–111.

Mannay, D. (2013) '"Who put that on there … why why why?" Power games and participatory techniques of visual data production', *Visual Studies*, 28(2): 136–146.

Marmot, M. (2008) *Closing the Gap in a Generation: Health Equity Through Action on the Social Determinants of Health – Final Report of the Commission for the Social Determinants of Health*, Geneva: World Health Organization.

McKendrick, J., Sinclair, S., Irwin, A., O'Donnell, H., Scott, G. and Dobbie, L. (2008) *Media, Poverty and Public Opinion in The UK*, York: Joseph Rowntree Foundation.

Morrow, V. and Mayall, B. (2009) 'What is wrong with children's well-being in the UK: Questions of meaning and measurement', *Journal of Social Welfare and Family Law*, 31(3): 217–229.

Moss, P. (2012) *Natural Childhood*, Swindon: National Trust.

Office for National Statistics (2011) *Life Expectancy at Birth and at Age 65 by Local Areas in the United Kingdom, 2004–06 to 2008–10*, London: ONS.

Pain, R. (2006) 'Paranoid parenting? Rematerializing risk and fear for children', *Social & Cultural Geography*, 7(2): 221–243.

Parker, D. and Garner, C. (2010) 'Reputational geographies and urban social cohesion', *Ethnic and Racial Studies*, 33(8): 1451–1470.

Pearce, J. (2012) 'The blemish of place: Stigma, geography and health inequalities: A commentary on Tabuchi, Fukuhara and Iso', *Social Science & Medicine*, 75(11): 1921–1924.

Piper, H. and Frankham, J. (2007) 'Seeing voices and hearing pictures: Images as discourses and the framing of image-based research', *Discourse*, 28(3): 373–387.

Popay, J., Thomas, C., Williams, G., Bennett, S., Gatrell, A. and Bostock, L. (2003) 'A proper place to live: Health inequalities, agency and the normative dimensions of space', *Social Science & Medicine*, 57(1): 55–69.

Prilleltensky, I. (2012) 'Wellness as fairness', *American Journal of Community Psychology*, 49(1/2): 1–21.

Ridge, T. (2011) 'The everyday costs of poverty in childhood: A review of qualitative research exploring the lives and experiences of low-income children in the UK, *Children & Society*, 25(1): 73–84.

Rogaly, B. and Taylor, B. (2011) *Moving Histories of Class and Community: Identity, Place and Belonging in Contemporary England*, Basingstoke: Palgrave Macmillan.

Steptoe, A., Shankar, A., Demakakos, P. and Wardle, J. (2013) 'Social isolation, loneliness, and all-cause mortality in older men and women', *Proceedings of the National Academy of Sciences of the United States of America*, 110(15): 5797–5801.

Stuckler, D. and Reeves, A. (2013) 'We are told Generation Y is hard-hearted, but it's a lie', *Guardian*, Tuesday 30 July.

Sutton, L. (2009) '"They only call you a scally if you are poor": The impact of socio-economic status on children's identities', *Children's Geographies*, 7(3): 227–290.

Taylor, S. (2010) *Narratives of Identity and Place*, Abingdon: Routledge.

Torjesen, I. (2012) 'Rise in life expectancy is marred by widening inequality', *British Medical Journal*, 344: 1141.

UNICEF (2011) *Child Well-Being in the UK, Spain and Sweden: The Role of Inequality and Materialism*, Florence: UNICEF Innocenti Research Centre.

Valentine, G. (2008) 'Living with difference: Reflections on the geography of encounter', *Progress in Human Geography*, 32(3): 323–327.

Vanderbeck, R. and Dunkley, C. (2004) 'Geographies of exclusion, inclusion and belonging in young lives', *Children's Geographies*, 2(2): 177–183.

Wacquant, L.J.D. (2008) *Urban Outcasts: A Comparative Sociology of Advanced Marginality*, Cambridge: Cambridge University Press.

Walmsley, J. and Johnson, K. (2003) *Inclusive Research with People with Learning Disabilities: Past, Present and Futures*, London: Jessica Kingsley.

Wetherell, M. (1998) 'Positioning and interpretative repertoires: Conversation analysis and post-structuralism', *Dialogue Discourse Society*, 9(3): 387–412.

Williams, R. (1977) *Marxism and Literature*, Oxford: Oxford University Press.

Wilkinson, R. and Pickett, K. (2009) *The Sprit Level: Why Equality is Better for Everyone*, London: Allen Lane.

Part II

Memory and intergenerational (dis)continuities

6 Displaced encounters with the working-class city

Camping, storytelling and intergenerational relationships at the Salford Lads Club

Luke Dickens and Richard MacDonald

Introduction

In this chapter we approach the themes of memory, (dis)continuity and identity through a consideration of the practices and meanings of camping to members of the Salford Lads Club (SLC), a civic institution attended by over 20,000 young members from local families since it was established in the Ordsall area of Salford, Greater Manchester in 1903. The empirical content of this chapter derives from a year-long action research project undertaken with members of the SLC, which centred on a participatory, intergenerational storytelling process intended to coincide with the club's celebrations of its one-hundredth annual camping excursion to Wales in 2011. In particular, we want to unpack the suggestion by one SLC member, Archie Swift, that it is 'the camp [that] has held this club together'. Pursuing this remark, we examine the annual camp as a space of both routine and extraordinary shared experiences, enacted through extrafamilial intergenerational relationships forged between older officers and younger lads over the course of this 100-year history.

This chapter therefore responds to Vanderbeck's criticism that *extrafamilial* intergenerational relationships are under-researched by geographers and that 'comparatively little is known about where, when and how these relationships are formed and maintained' (2007: 209). Specifically we discuss how variously positioned accounts of male bonding at camp, both in the form of 'banter', fun and adventure, and modes of discipline and care, present an intersectional reading of their generational, working-class and masculine identities (McDowell 2003; Nayak 2003) over this history. Initially these relationships were based on a particular formation of seniority and the classed distance between the cadre of philanthropist leaders, referred to as 'masters', and the working-class lads seen requiring moral guidance and betterment (Davies and Fielding 1992), a situation with parallels to the kinds of 'good citizen' promoted in the early years of the Scouting movement (Mills 2013). However, over time the nature of these relationships shifted towards one grounded in a social proximity and shared experience between the adult officers and the lads in their care. We argue throughout that such experiences were heightened by their liminal and displaced encounters as 'city lads' in the countryside, and which therefore established a transformative social process so critical to ensuring the continuity of the Salford Lads Club itself.

Storytelling as participatory intergenerational engagement: sharing memories, performing identities

Our approach follows the recent effort by geographers to recuperate storytelling and narrative as way of connecting across disciplinary and conceptual concerns (Cameron 2012; Daniels and Lorimer 2012). Our research involved making audio recordings of camping memories by a group of current and ex-members, aged between 13 and 93, and running a series of workshops where participants were helped to animate their stories into short videos using the club's rare collections of 16 mm film and photography captured over its long history.[1] Elsewhere we have argued that such narrative exchanges served as an important premise for sustaining collective memory, and enhanced the digital preservation of these materials by using them as the basis for experimenting collectively with the production of a 'civic archive' of working class histories, supporting the SLC's collective effort to reflect on the dynamics of its continued existence in an area of Salford characterized by repeated cycles of dereliction and 'urban renewal' (Dickens and MacDonald in press; MacDonald *et al.* in press).

The 'tales' that emerged took a particular form of 'small stories' (Lorimer 2003), relaying personal memories of practices and experiences of the annual SLC camp, which collectively expressed recurring narratives of working class solidarity, discipline and hard work, fun and adventure, and encounters both familiar and strange. These stories were important acts of self-representation and authorship, revealing a very different image from those externally constructed images of working-class men as either hero at work or the victim of circumstance; a relaxed, intimate presentation of a specific community whose lives were drawn together within and beyond the Club.

This participatory research process is significant here as an example of the kinds of 'relational knowledge creation' – both across generational and academic/non-academic boundaries – that Hopkins and Pain (2007; see also Pain 2005) advocate as a means of countering a tendency within social science research to 'fetishize the margins' of age (i.e. concerning the very old or very young), or otherwise view age in discrete, 'compartmentalized' categories (see also Vanderbeck 2007). As they argue: 'Different markers of identity may intersect with age in interesting and important ways, influencing for example how different generational groups perceive and relate to each other in different settings' (Hopkins and Pain 2007: 290). In many ways this research encounter reflected the ways the club was already inclusive in terms of age because, as one of the volunteers noted, 'everything that we do is basically group based, so to sit there with the young lads as well as the older ones is just the norm' (Dennis). But it also created space for more explicit reflection on this collective practice:

They [lads] want something different to what we [club Officers] want out of it [...] but you don't know what bits have actually stuck in their mind until you hear the stories, and you might think 'oh, I wouldn't have thought of that.'

(Dennis)

The telling of these tales in a group context at the club – sitting in a circle and spoken between men and boys of different generations – raised a powerful sense of continuity and change across the shared experience of camping in a way similar to what Cameron describes as 'a mode of producing and expressing knowledges gleaned through embodied and intersubjective experience' (Cameron 2012: 584). Mutual listening across generations to the different perspectives on this shared experience constituted an effort in what Vanderbeck (2007) has described as 'practices of generational reengagement'. This engagement involved moments of intergenerational identification and learning (Bouchard Ryan *et al.* 2004; Mannion 2012; Newman and Hatton-Yeo 2008), prompted by what one young member described as witnessing 'the older ones being young' in the footage and photographs being discussed, and thus positioning their identities as 'Salford Lads' explicitly in relation to age and ageing (Spector-Mersel 2006). The intergenerational nature of these exchanges raised uncanny similarities as well as differences across generations of lads, enriching approaches that use oral histories of older people when they were younger (Andrews *et al.* 2006) as a means of understanding past social life in connection with the present.

Camping, liminality and the (dis)placing of intergenerational encounter

The stories of camping that emerged from this research are significant for the ways they provide insights into 'camp' as a liminal space, and the particularly situated intergenerational experiences that have served to construct such a place away for one week of every year for over a century (see Plate 6.1). Initially, the practice of organized camping in rural environments by working-class clubs and associations were linked to a wider discourse of rational recreation and the value of fresh air and natural, open space for factory workers. Much like other youth movements at the start of the twentieth century, the SLC adopted and recast these impulses by viewing the rural camp as a space for the moral betterment of children and 'brightening' of young lives. For example, as Mills has examined, Robert Baden-Powell's ideas for the informal training of boys in the practice of Scouting, particularly the teaching of self-discipline, self-improvement and resourcefulness were tested on an experimental camp in 1907, drawing on a wider history of viewing rural settings as important sites for 'citizenship training' for young men in particular (Mills 2011: 540–541; see also Cupers 2008; Matless 1995).

Plate 6.1 Group photo from the first SLC camp in Llandulaas, Wales, 1904; and on the 100th SLC camp in Llanbadan Fawr, Wales, 2011 (source: Salford Lads Club).

For the Lads Club movement, with its origins in the industrial urban areas of northern England, changes in the association of the rural as liminal space can be detected following the shifts away from a heavily industrialized to a de-industrialized (and depopulated) city. As these processes of economic and political change were felt sharply in cities like Salford and Manchester, the resulting atomization of community and attendant loss of communal space was countered by seeking spaces elsewhere, renewing the purpose of camping together for clubs such as the SLC.

Within these efforts to enact the SLC annual camp is an implicit understanding of both place and identity as relational concepts, held in a productive tension along axes of urban/rural and self/other. Here, Turner's (1964) seminal notion of 'liminality' has purchase for examining the SLC's periodic movement between Salford and rural Wales as a means of temporarily opening up a dialectic, dynamic space between discipline/freedom and familiar/strange in the young lads' lives. Likewise, encountering an 'other' (rural) place but one enacted through familiar relationships seems to heighten an awareness of what it is to be a 'Salford Lad'. Importantly, as we discuss in the rest of this chapter, camp in these terms serves as a key site for producing at least the fundamental conditions for an 'age integrated society' (Riley and Riley 2000), where the usually separated spheres of learning, work and leisure – and their associated lifecourse stages – are experienced across generations.

'Tales from camp': memories, (dis)continuities and identities (*c.*1904–2011)

This section outlines themes raised in the SLC members' stories about camp, though we stress that this is no substitute for watching and listening to them as a collection.[2] The following section will pick up on where these themes touch on the dynamics of the intergenerational relationships that have emerged through the interrelated spaces of 'camp' and 'club'. Indeed, Archie's suggestion that the SLC annual camp was the reason for the longevity of the club itself is a potent statement of the ways the specificities of this shared experience can forge lasting intergenerational bonds between club members. In his story, he hints at the contingencies of continuing the annual camp after the devastating effects of the Second World War on club members and their families, noting that:

> My first camp was in 1946.... They called it a taster because the Seniors, who ran everything, they wanted to know if it was gonna work ... after the war like they did before the war ... they looked after us, fed us and watered us, you know, kept us dry and shown us how to camp. And I'll never forget it, it was a marvellous experience.
>
> (Archie)

For another current Officer and ex-lad in his late seventies, Bill, this longevity could be traced across the full history of camping at the club through individual members of his own family, a situation that was not unusual for current Officers:

> There's always been about three of us, me two uncles and myself, right the way through [...] Harry was the eldest, and he went to his first camp in 1920. Then [there] was John, Bill and Harry, who played on the football team, and there was Dave and Ralph [...] this year, the hundredth [camp], there was myself, my son and my grandson. So we've always carried on.
>
> (Bill)

In other stories, lads also told of a range of relationships both familial and extra-familial across generations. For Liam, an Officer and ex-lad now in his twenties, the 'banter' between campers was 'just like my mates back home who are all in their twenties, but these guys are all in their sixties and seventies'. He felt this was central to the effective functioning of the club, which 'takes the kids and puts them on the right track and sticks with them into adulthood'. But his story also picked up on the contingencies within Archie's story, reflecting on the sadness of the death of one of the oldest Officers, 'a real character', which leads him to 'wonder about the future of the club'.

Other stories mused on the recurring sense of the 'Salford Lad' experiencing the countryside for the first time: the photos illustrating Dennis' film show the lads in their 'Sunday best' – flared suits and ties – ignoring the requests of officers not to wear their best clothes on camp. For Alan, now in his late sixties, this experience was encapsulated in his story about bats, but also contrasted the care and compassion with the banter between the older and younger boys as a lasting memory within his story:

> the only bat they've ever seen before is a cricket bat! And the other lads, the older lads, were telling us that these bats bite onto your head, suck your blood.... We were scared but it was a funny scaredness.... Fifty years later I can still remember that and I can still remember how I felt.
>
> (Alan)

Similarly, Dennis, a current Officer, recalled a trick he had played on Alan, repeatedly soaking a vest that he had left on the camp washing line: 'this went on for about two days, and he never twigged, he just kept saying: "I can't understand why this vest is not drying"!'. Other stories also pick up on the necessary chores and everyday domesticity created within the camp space: Jimmy, who joined the club at 13 and is now an Officer in his late seventies, recalled working 'in the cookhouse maybe thirty-odd years', and particularly the song the campers used to sing while peeling potatoes, noting that 'as you get older you still remember those songs!'.

Addressing this blend of discipline and domesticity with the fun and adventure in the rhythms of camp, Brian C. describes one lad falling in a river running back from the pub to the camp site, hurrying before the sound of the bugle rang out, 'because if you didn't get there by then you got what they called "Jankers", which could be anything from picking up paper or being kept in for a night'. Similarly, 93-year-old Nicholas, whose first camp was in 1939, recalled the 'competitions every day for the cleanliness of the tent' alongside the joy of swimming in the river then being 'treated to a hot drink and a sweet afterwards'; while 17-year-old Ryan enjoyed the fact that more recent camp activities like assault courses, go-karting and fishing were 'about having fun but aiming as well for the trophy at the end'. Both point to the ways the camp was run with a view of instilling discipline within the lads, but also with recognition for doing so, even as Nicholas recalls, the camp itself being a reward for his efforts in night school at the club throughout the year.

Plate 6.2 Setting off for camp outside the SLC in 1908 (above) and 2009 (source: Salford Lads Club).

The youngest member's story, from Dylan aged 13, was an account of the rituals of leaving from outside the club, a send-off and journey much documented over the years, and the anxieties and excitement he felt with all the 'mums and dads waving as the kids put their face onto the glass of the coach' (see Plate 6.2). For Niall, now an Officer in his early twenties having attended the club since he was 13, this process was something he had recently experienced, and was a key part of the ways young lads on camp were supported to become independent beyond their immediate families:

Camp is important for the young people at the club as it teaches them a lot of new skills and allows them to *gain experiences you wouldn't get from living in the city all the time* … it taught me to become independent, having to look after myself at camp without my mum or dad being there.

(Niall, emphasis added)

As Niall's emphasis suggests, the importance of the liminal time-space of camp, and the relationships it affords, were at the heart of the camp experience.

Displaced encounters with the working class city: intergenerational relationships between 'camp' and 'club'

In this section we look in further detail at the dynamics of the intergenerational relationships that have developed over the course of the 100 years the camp has been held. Here, we draw on a series of in-depth interviews with members conducted to explore the issues raised during the production phase of *Tales from Camp*.

Class transitions of paternal care

The relationships forged at camp cannot be understood without linking them to those fostered within the more everyday spaces of the Club itself. The SLC was established by local brewery owners, and much like similar lads clubs at the time (Russell 1905), was done so through localized philanthropic efforts to improve the prospects of local working-class children growing up in slum-like neighbourhoods. In particular, lads clubs of this kind sought to counter street life and local gang culture prevalent among young working class men (Davies 1999b, 2009, 2011; though see Davies 1999a). At this time there was a clear distinction between the mercantile-class 'gentlemen' who funded, managed and ran the SLC, and the working-class 'lads' they supervised. The relationship between these two groups was characterized by a level of deference, whereby senior club officers were referred to as 'Mas' (short for 'master'). This is exemplified in the suggestion by one current officer now in his late seventies, Eddie, that: 'in them days [...] people in my position now, the kids would salute us and it was "yes sir, no sir"'. Similarly, Archie, now in his eighties, recalled one club leader in the 1940s as being 'a very, *very* serious man where youth was concerned'. Despite these hierarchical social distinctions, and the clear lines of respect that the lads at the time had to obey, the accounts of current Officers who can remember this early period still maintained that their relationships with the Masters could be a caring one. For one ex-lad, Mike, who attended the club in the early1960s, such paternal care was especially valued because it was often lacking in family life:

they took great care of me. There were dysfunctional families [...] because there was no state support so you could get a family with eight, nine, ten kids whose dad didn't work and somehow managed to stay drunk, but the kids wouldn't eat and wouldn't have clothes on their backs and it would take charity to help feed and clothe them [...] you could come somewhere like this [Salford Lads' Club] and [...] it was just a load of men who cared about young working-class men and wanted to help them and that was the beginning and end of it.

(Mike)

However, in the decades following the Second World War, the management of the club gradually moved away from the philanthropic efforts of local businessmen, and was increasingly taken up by ex-lads who had grown up through the club and wanted to return to help run the weekly club nights. In terms of the intergenerational relationships at the Club (as elsewhere in British society) there appears to have been a softening of that deference and hierarchy so marked in the 1940s and 1950s, particularly as the generation of leaders that Archie refers to moved on. Later, at around the turn of the millennium, the actual management of the Club shifted symbolically from the Groves family to a committee of ex-lads and volunteers. The transition was made when the then club leader and chair of the board decided to close the club in the face of mounting debts and the pressures of wider industrial decline of the nearby Salford Docks. Bill explained that rather than close, a group of ex-lads had opted to take on the management themselves: 'There was like seventeen of us workers, we'd all been through the club, so we took it on from then'. This resulted in a discernible shift in the hierarchical relationships between the new cohort of Officers and lads, based on a shared working-class experience, with 'seniority' passing as definition of life experience rather than the more overtly classed category used in the first half of the century. For Niall, this shift redefined the ethos of paternal care that characterized the club, from the philanthropic and hierarchical to a more direct level of shared personal experience: 'they're not all qualified youth workers, but they do just as good of a job because they've been there and they've seen how it's done properly, and then they just grew up into that' (Niall).

These shared experiences, as Bill explained, translated across the various ages of Officers and lads now attending the club into a socially equal as well as proximate relationship: 'everything is equal in here whether you be thirteen or seventy. You've still got rules to abide and that's the only way, you know, one rule for all'. For the Club's project manager, Leslie, the ability of the Officers to relate to the lads throughout the weekly cycle of activities within the club itself, was also pivotal to the running of the annual camp in a similar vein:

they can organize these camps like clockwork because they've done it for like forty years. And they know every kind of experience that these kids are going to feel. [...] they still know those experiences now, and I think that's really important.

(Leslie)

Within the SLC, camp was seen as a highlight of the year and a pivotal moment in the enactment of these kinds of paternal relationships, an opportunity in 'giving the kids something to remember' (Niall), based on the tacit, long-term and collective experiences of generations of lads leaving Salford and living together as a group. As Niall explained: 'nothing's really changed [on camp], everything's still the same, it's just that everyone's growing up', drawing attention to the ways camp was both performed through the repetition of specific activities, while also being marked by new cohorts of young lads growing up in the process.

Extended family, youth transitions and the transformative practice of camp

These relationships of paternal care were increasingly kept *within* a family of sorts, or what Mike described as 'strong and familiar, "*en famille*" I think the French call it ... an extra part of your family'. As the many recollections of being cared for by more senior lads illustrate, these were deeply formative bonds that were shaped through the shared practice of camp.

Discussing the particular nature of these extrafamilial relationships as they occurred on camp today, Ryan (now a junior Officer) explained of the seniors running camp: 'they do act like your parents but [they] give a little bit of leeway where you can do your own stuff and it just gives kids more time to bond in a different place'. So, while the Club Leader, Leon, stressed that 'Camp is hard work. There's no "ifs" or "buts" about it', it was also clear that there was scope for young men to establish their own boundaries in this liminal time-space. Thus, unlike the highly disciplined, closely observed and ultimately nationalistic forms of civic embodiment within the Scout movement (Cupers 2008; Mills 2013; Robinson and Mills 2012), which also informed the initial approach of the SLC, it appeared that a socially proximate, tacit understanding of young men's experiences within the club, coupled with a sense of leeway afforded by camp, had come to characterize the ways Officers supported the transition of young lads towards adulthood. As Brian noted: 'I'd like to think we're setting them out to be their own person kind of thing, that we're [...] not showing them what to do, but helping them learn for themselves'.

Importantly, this process was supported by adult Officers who were also often fathers and uncles. As Dave, a middle-aged Officer in his fifties explained: 'my son came [...] and he was only seven and there were two more kids in the tent about the same age. Normally they'd have to be ten but with their fathers being there it was OK'. Such sentiments point to the 'fuzzy boundaries' (Hopkins and Pain 2007) of lifecourse transitions, whereby the participation of young lads in camp activities was negotiated through implicit notions of 'emerging adulthood' (Arnett 2004). Ryan, talked through this process with reference to the formative experience of domestic duties, particularly the use of the cookhouse, which signified an adult space:

INTERVIEWER: Is it like a formal thing, is it like, 'Now you're a junior?'

RYAN: No not really it's just when you start getting older [...] you don't really do things that the kids do, and the kids can't go in the kitchen and stuff and when you get older you can, and then you just start doing different jobs and helping out [...] and really you get the name junior officer from camp.

(Ryan)

For the Officers, practices of camp were also appreciated as transformative of intergenerational relationships between themselves and the lads. Reflecting on the banter of camp, Brian noted that 'What you have to remember is that you don't ever grow up really', suggesting both a genuine enjoyment in Officers' relationships with the younger lads, and a reluctance to pass judgement on their youthful behaviour, primarily because they had once been in the same situation. Significantly, as Niall explained:

the kids who haven't been to camp, they might be say naughty or cheeky in the club, they go to camp and because they've spent a full week with the Officers and Junior Officers, they come back with more respect for them because ... when we're at camp they see that we're just normal people.

(Niall)

Therefore, the result of these transformed relationships was also a means of securing the future of the camp for future generations:

we're now trying to give the kids jobs to do [...] so when they grow up they know how to take the tents down, they know how camp runs properly and stuff like that. So they can just naturally step up into the role of the officer and know exactly what happens at camp.

(Niall)

The annual camp not only established bonds within the club, however, but also extended to a range of relationships sustained over many years of the Salford Lads camping in Wales.

Encounters with the rural 'other': intergenerational relationships with Welsh people

The annual return of Salford lads to the campsite outside Aberystwyth had helped form lasting friendships between the Officers and nearby villagers. As Niall outlined, 'the locals that we speak to, the older generation have built up really, really good relationships with them [...] they'll have loads of stories about us'. The older Officers in particular had a strong appreciation of the ways the camp site was made hospitable by the efforts of local people, an extension of welcome and care, illuminating a networked geography of social relationships that has underpinned and nourished a displaced sense of belonging and connection between campers and their rural hosts (Bunnell *et al.* 2011). As Brian B. explained: 'the lady that owns the field, *our* field, we're pretty friendly with her and the family and they're very good to us, I mean we don't pay any rent or anything for the field'. Similarly, Dennis retold of a man who owned a local electricity business: 'we've known [him] for years […] and we've never paid for any gas all the years we've been camping there […] he just doesn't blink an eye because we've known him that long'.

However, it was the football match between the SLC and the local Aberystwyth team – the centrepiece of the annual camping ritual featured every year since at least the mid-1930s – 'that's the one to cement relationships' (Dennis). Brian elaborated: 'that's where the other friendships develop […] we always encourage our lads to mix with the opposition, the Welsh lads, because that's what we've done over the years'. Similarly, as Dennis made clear:

> one of the things that you try and instil in the kids [is] that you can have friends 140 miles away, it's not all about wanting to fight somebody. These relationships can last for years […] sometimes it's nice to be nice, and that's just how we are as a group of lads I think. We respect the other people, they look after us and we look after them when we can.
>
> (Dennis)

These ongoing relationships, and the mutual respect that sustained them, were therefore passed down through the generations of SLC members much like the other practices of camp.

Conclusion: voluntarism, (dis)continuities and the future of the Salford Lads Club

The hundredth anniversary of the SLC's annual camp to Wales was an opportunity to celebrate the history of this practice, and reflect on its pivotal role within the club over this time. For Niall, who at the age of 20 had recently experienced being a young lad on camp and was now taking on the role of Officer, noted that as a result of the celebrations 'I think the younger generation now have realized what camp means to Salford Lads Club'. In particular, the *Tales from Camp* videos and related research had helped him to appreciate how the bonds established at camp had produced valuable stories that lasted beyond the camp, noting that 'you can guarantee every year something funny will happen at camp and then that will be remembered for years afterwards […] the older ones then pass it on to the younger ones'.

However, such explicit reflections had also prompted some consideration of the sustainability of camp and, indeed, the future of the club itself. Vanderbeck's interest in the transmission of culture across generations (2007: 204, 208) speaks directly to these considerations among club members, who expressed both hope and concern for the ways the meaning of camp, and the relational identities bound up in being a 'Salford Lad', might be passed on to future generations without foreclosing or predefining the kinds of experiences they might seek. Interestingly, both the junior Officers and the older Officers and volunteers viewed continuity in terms of the potential openness to change. Thus for Ryan, now aged 17, this was both a process of carrying on with offering care and positive experiences to new cohorts of lads, but also of the club 'getting better' as a result of such longevity:

> the officers done it for me and they made my club night and my camp and stuff special, they made if fun, so why can't I do it for other kids and make it fun for them? And then previous people who I've helped will also do it for the younger generation and it will just make the club better and it will keep it going.
>
> (Ryan)

For Brian, now in his seventies, any secure future for the club would be based on recognizing the ongoing importance of allowing the next generation of lads the freedom to decide for themselves in the same way as his generation had been able to do:

> I sometimes wonder whether we put young people off because we're, you know, a different generation [...] I'd like to bow out gracefully eventually but come down every now and again. [...] because the way young people think now is not the same as ... they're not my way of doing things and that means we're selfish if you don't [...] let them do their own thing, you know, like we've had.
>
> (Brian)

Implicit in both comments were the tacit, localized and *voluntary* forms of social reproduction that characterized the extrafamilial intergenerational relationships established within the club. Behind them was a feeling commonly expressed by both older and younger members that the club continues to need volunteers from the Ordsall neighbourhood to keep things going in the future. The voluntary basis of the club, with only one or two paid workers, was a great strength and a source of independence in terms of the rich relationships described above, but it was also a potential vulnerability, raising questions about where volunteers will come from in the future, or how different might the club's Leader and Officer roles be if they were professionalized more formally as 'youth work'. Yet despite such anxieties, the resilience of the club's culture of voluntarism and its proven transition from a philanthropic members' club to an organization rooted in shared experience and social proximity across generations has already provided the basis, at least, for a unique and lasting form of 'generational reengagement'.

Notes

1 See http://storycircleuk.com/2012/02/13/100-camps-digital-storytelling-with-the-salford-lads-club/, http://salfordladsclub.org.uk/camping/ and http://salfordladsclub.org.uk/camping/100-camps-timeline/.
2 See www.youtube.com/playlist?list=PL8vzwBxG4B2OrGBNsuaUCiuW-6ad2z3q4 for the *Tales from Camp* videos and www.youtube.com/watch?v=OqFjU9S4QqA for accompanying *100 Camps* photographic exhibition.

References

Andrews, G.J., Kearns, R.A., Kontos, P. and Wilson, V. (2006) ' "Their finest hour": Older people, oral histories, and the historical geography of social life', *Social & Cultural Geography*, 7(2): 153–177.

Arnett, J. (2004) *Emerging Adulthood: The Winding Road From the Late Teens Through the Twenties*, Oxford: Oxford University Press.

Bouchard Ryan, E., Pearce, K.A., Anas, A.P. and Norris, J.E. (2004) 'Writing a connection: Intergenerational communication through stories', in M.W. Pratt and B.E. Fiese (eds) *Family Stories and the Lifecourse: Across Time and Generations*, Mahwah, NJ: Erlbaum, pp. 375–399.

Bunnell, T., Yea, S., Peake, L., Skelton, T. and Smith, M. (2012) 'Geographies of friendships', *Progress in Human Geography*, 36(4): 490–507.

Cameron, E. (2012) 'New geographies of story and storytelling', *Progress in Human Geography*, 36(5): 573–592.

Cupers, K. (2008) 'Governing through nature: Camps and youth movements in interwar Germany and the United States', *Cultural Geographies*, 15(2): 173–205.

Daniels, S. and Lorimer, H. (2012) 'Until the end of days: Narrating landscape and environment', *Cultural Geographies*, 19(1): 3–9.

Davies, A. (1999a) ' "These viragoes are no less cruel than the lads": Young women, gangs and violence in late Victorian Manchester and Salford', *British Journal of Criminology*, 39(1): 72–89.

Davies, A. (1999b) 'Youth gangs, masculinity and violence in late Victorian Manchester and Salford', *Journal of Social History*, 32(2): 349–369.

Davies, A. (2009) *The Gangs of Manchester: The Story of the Scuttlers, Britain's First Youth Cult*, Preston: Milo Books.

Davies, A. (2011) 'Youth gangs and late Victorian society', in G. Barry (ed.) *Youth in Crisis? 'Gangs', Territoriality and Violence*, Abingdon: Routledge.

Davies, A. and Fielding, S. (1992) *Workers' Worlds: Cultures and Communities in Manchester and Salford, 1880–1939*, Manchester: Manchester University Press.

Dickens, L. and MacDonald, R. (in press) ' "I can do things here that I can't do in my own life": The making of a civic archive at the Salford Lads Club', *ACME: An International E-Journal for Critical Geographies*.

Hopkins, P. and Pain, R. (2007) 'Geographies of age: Thinking relationally', *Area*, 39(3): 287–294.

Lorimer, H. (2003) 'Telling small stories: Spaces of knowledge and the practice of geography', *Transactions of the Institute of British Geographers*, 28(2): 197–217.

MacDonald, R., Couldry, N. and Dickens, L. (in press) 'Digitisation and materiality: Researching community memory practice today', *The Sociological Review*.

Mannion, G. (2012) 'Intergenerational education: The significance of reciprocity and place', *Journal of Intergenerational Relationships*, 10(4): 386–399.

Matless, D. (1995) 'The art of right living: Landscape and citizenship, 1918–39', in N. Thrift and S. Pile (eds) *Mapping the Subject: Geographies of Cultural Transformation*, London: Routledge, pp. 93–122.

McDowell, L. (2003) *Redundant Masculinities: Employment Change and White Working Class Youth*, Oxford: Blackwell.

Mills, S. (2011) 'Scouting for girls? Gender and the Scout movement in Britain', *Gender, Place & Culture*, 18(4): 537–556.

Mills, S. (2013) '"An instruction in good citizenship": Scouting and the historical geographies of citizenship education', *Transactions of the Institute of British Geographers*, 38(1): 120–134.

Nayak, A. (2003) 'Last of the "Real Geordies"? White masculinities and the subcultural response to deindustrialisation', *Environment and Planning D: Society and Space*, 21(1): 7–25.

Newman, S. and Hatton-Yeo, A. (2008) 'Intergenerational learning and the contributions of older people', *Ageing Horizons*, 8: 31–39.

Pain, R. (2005) *Intergenerational Relations and Practice in the Development of Sustainable Communities*, background paper for the Office of the Deputy Prime Minister, Durham: International Centre for the Regional Regeneration and Development Studies, Durham University.

Riley, M.W. and Riley, J.W. (2000) 'Age integration: Conceptual and historical background', *The Gerontologist*, 40(3): 266–270.

Robinson, J. and Mills, S. (2012) 'Being observant and observed: Embodied citizenship training in the Home Guard and the Boy Scout Movement, 1907–1945', *Journal of Historical Geography*, 38(4): 412–423.

Russell, C.E.B. (1905) *Manchester Boys: Sketches of Manchester Lads at Work and Play*, Manchester: Manchester University Press.

Spector-Mersel, G. (2006) 'Never-aging stories: Western hegemonic masculinity scripts', *Journal of Gender Studies*, 15(1): 67–82.

Turner, V.W. (1964) 'Betwixt and between: The liminal period in *Rites de Passage*', *The Proceedings of the American Ethnological Society, Symposium on New Approaches to the Study of Religion*, Washington, DC: U.S. Government Printing Office.

Vanderbeck, R.M. (2007) 'Intergenerational geographies: Age relations, segregation and re-engagements', *Geography Compass*, 1(2): 200–221.

7 Bridging the generation gap

Holidays, memory and identity in the countryside

Michael Leyshon and Tea Tverin

Introduction

Although the positive benefits of leisure activities on children's social, emotional and physical well-being have been relatively widely researched, these studies tend to overlook disadvantaged young people and the importance of holidays as forms of leisure activity. In this chapter we explore how disadvantaged young people are affected by their experiences of respite holiday care. In particular we evaluate critically children's holidays provided by CHICKS (Country Holidays for Inner City Kids), a charitable organization that runs free respite breaks for disadvantaged children and young people. CHICKS holidays aim to produce happy memories, new opportunities and experiences within a 'safe' rural environment. For children and young people whose everyday lives are characterized by poverty, abuse or responsibilities beyond their years, respite holidays are an opportunity to experience joy and fun away from everyday adversity, with many potential benefits. The relationship between kindness, care, memories and play is currently poorly understood in geography and the social sciences more widely. This chapter differs signally from previous work by illustrating how disadvantaged young people organize their thoughts and memories about their holiday experience and how 'happy memories' are created and sustained in the face of hardship. We further argue that intergenerational relationships organized in what we call, following DeLanda (2002), a *flat ontology of care* produce memories through multiple emotional geographies. In so doing we contribute new knowledges to geographies of care, love and engagement with rural/natural places.

CHICKS: placing the research

CHICKS is a charitable organization that runs free respite breaks for disadvantaged children and young people between the ages of 8–15. In its first year of operation in 1992 CHICKS enabled 25 children to have a holiday. Since then CHICKS has grown drastically, hosting nearly 1,200 children and young people in 2012. CHICKS has two centres: Moorland retreat in Brentor, Devon, and Coastal retreat in Tywardreath, Cornwall. Both centres have multiple indoor and outdoor play areas and large grounds set within rural surroundings. All the children attending a CHICKS break are referred there by a professional social worker, youth worker or teacher. The children may have experienced domestic violence, poverty, emotional/physical/sexual abuse or they are a young carer.

The CHICKS 'camps', as they are called, run from Monday to Friday March to December each year at both centres. The two age groups, 8–11 and 12–15, are at separate camps. The camps are run by trained supervisors assisted by adult volunteers. At camp, young people have the opportunity to have fun canoeing, rock-climbing and playing group games. CHICKS' ethos is to provide hope and happy memories. Additionally, the organization seeks to show the young people that there are safe adults who care for and support them. CHICKS staff send a Christmas card and a present and a birthday card to everyone who attends CHICKS camp that year and encourage the children to stay in touch.

Memory, holidays and care

Geographical studies of memory have significantly increased in recent years focusing on areas including landscape (e.g. Jones 2005; Wylie 2007), historical events (e.g. Della Dora 2008) and managed retreat (DeSilvey 2012). Although these investigations have usefully enhanced understandings of memory formation, children and young people are virtually absent from these studies (exceptions include Leyshon 2011 and Leyshon and Bull 2011). Additionally, memory is often investigated in relation to memory *processes*, whereas our aim is to illustrate how memory is mobilized into the *practice* of *active care*, both in the creation of memories as well as the recall of affect. This understanding opens the possibility of investigating memory as an intervention tool for well-being. As Jones (2011: 881) states, geographers have become increasingly interested in this 'fantastically complex entanglement of self, past spatial relations and memory in current life'. If memory has the potential or if it is to be used as a mechanism for increasing well-being then it is vital that this entanglement can be unravelled. Hence this chapter attempts to disentangle that multifaceted fusion of past, present, self, others, landscape and affect.

Holidays provide a major source of memories for many children in the global North. We argue that holidays for children in adverse family conditions are significant moments, as they can offer a change from the everyday as well as to alleviate the stresses of home conditions. This is not just a temporary condition; the benefits can extend beyond the holiday itself. As Westman and Etzion (2001) have argued, holidays are restorative for individuals by offering a break from everyday routines, alleviating job stress and improving perceived health. Further, holidays are recognized by individuals as an enjoyable event during which they forget their everyday worries and allow themselves to surrender to 'pleasurable experiences' (Thornton *et al.* 1997). However, the value and importance of holidays is not limited to enjoyment and escapism. Holidays also offer an opportunity for personal growth and development (Hunter-Jones 2003). Smith and Hughes' (1999) research on the experiences of low-income families with children on a sponsored holiday (financially assisted by a third party) illustrates that the family holiday is a welcome break from routines and a significant change for the whole family as they spend time together on outside activities.

Few studies have been conducted on the benefits of holidays for disadvantaged children. A key exception is Kearns and Collins' (2000) research on New Zealand children's health camps. Starting in the 1950s, the camps were established in rural locations and were aimed at providing holidays for children in poverty. The primary focus was to introduce those children to clean country air and teach them personal hygiene. The camps are still running today, with the explicit aim of providing underprivileged children with a respite break. In the past three decades the camps have moved away from a somewhat unreconstructed view of the healing properties of the countryside towards more social and mental-health oriented goals such as building children's self-esteem. These camps were reported to yield positive results in one to five weeks, particularly increasing life skills. However, the study does not focus on the children's own experiences.

Contemporary research that directly discusses children's experiences of therapeutic camps is sparse and usually focuses on children with life-limiting or serious illnesses (but see Morse, this volume). An exception is Dunkley's (2009) study of troubled youth at a camp in Vermont, USA, in which she argues that therapeutic places are produced through a nexus of interactions that privilege neither the social, the material nor the individual aspects of a place. For young people the therapeutic process is a combination of embodied experiences with human and non-human worlds. A similar study by Kiernan and Maclachlan (2002) on therapeutic recreation for children with life-threatening illnesses examined children's experiences of different aspects of camp life. When asked how the participants would describe the camp to their friends, the most popular answers were fun and excitement (33 per cent) and the fun and varied activities (29 per cent). All the participants found the instruction-based outdoor activities more enjoyable than the indoor 'cottage-based' activities. The key findings of Stirton *et al.*'s (1997) research on holidays for young people with a life-limiting illness (cancer) were that even a short therapeutic break increased young people's confidence, built new friendships, and supported them with the knowledge that there are people out there with shared experiences. However, the causes of these improvements were elided in their study. Below we focus on the need to examine the importance of fun in producing positive outcomes for children and young people.

Children, fun and nature

In this section we seek to make two interconnected points. First, we argue that fun has the capacity to become a more-than-embodied experience: fun is more than instantaneous, temporarily experienced and transient, but can have lasting effects beyond the moment. Second, by reconceptualizing fun in this way, it can become therapeutic, acting as an alternative or complement to more traditional therapies for young people. Fun is important because, Read *et al.* (2002: 2) state, it is 'a concept that seems to comfortably belong in a child's environment'. This is not to say that children and young people do not encounter unpleasant experiences, but rather to highlight the playfulness of children's and young people's being. To access that 'playful being', Kesby (2007) argued that by creating a play-based, supportive, imagination-enhancing and creative research environment we could get powerful insights into the lives and worlds of children and young people. Similarly, Evans (2008) discussed the need of 'youthful geographies' highlighting concepts of fun, exuberance and the excitement of new opportunities. Kesby and Evans are therefore calling for an acknowledgement of fun as a vital component of being a child or a young person (see also Woodyer 2012).

Zabelina and Robinson (2010) argue that children's lives and outlook on life are overall less serious and more driven by enjoyment and gratification than the lives of adults. Movement alone can be fun for children; twirling, spinning and swooping make children laugh (Berk 2003). Play hence connects fun and children (Harker 2005). This does not mean that children are purely and blindly driven by pleasure, but the lack of constraints of the adult world, such as responsibilities, give children more freedom to follow their hedonistic instincts. Moreover children are naturally curious (Berk 2003), enjoy exploring and tend to be more spontaneous than adults (Rothbart and Bates 2006). Blythe and Hassenzahl (2003: 99) write, '[D]uring fun the senses must be engaged, there must be spectacle'. Also playing seldom occurs without an affective component (Harker 2005), but comes with a variety of emotions such that it would be impossible to separate the play from fun. In this light, Jones (2005) discusses his boyhood experiences of exploring and playing in the countryside. He describes roads, trees, creeks and smells very vividly, evoking enjoyment, a sense of adventure and strong emotions. Ward (1978, 1988) similarly described children playing in cornfields or in long grass or (in cities) amongst buildings and cars, carefree with smiles on their faces, exploring their surroundings. Ward (1978) argues that children are able to produce fun, joy and happiness by creating their own *happy habitat* by having the child's capability to always find something to do, to explore and to detach from the other realties of their lives. Drawing on this knowledge play therapy has emerged as a psychological tool through which the child can work through unsolved emotional issues. It is suggested that play therapy, playing and having fun, helps young people put something that was lost back into their lives, thus alleviating psychological pain and helping to build a coherent sense of self (Alvarez and Philips 1998).

Schaffer (2004) argues that play in its traditional sense is a joyous activity through which children build social, emotional and verbal skills. He discusses children and fantasy play, and how fantasy and escapism help to develop and enhance imagination. Playing is a creative act through which one develops not only a sense of self but a sense of others as well (Alvarez and Phillips 1998). In this way *successful* play has been characterized as being free, unrestrained, unregulated and imaginative (Blythe and Hassenzahl 2003). However, the causal link between children's play and social development is not straight-forward, as definitions of play should not be reduced to playing activities (Harker 2005) that are homogenously experienced. In other words not all playing looks like playing, for example, teenager's play takes on different forms than younger children's. Hanging out, for example on street corners (Leyshon 2002), and drinking alcohol (Kuntsche *et al.* 2005; Leyshon 2008) can all be characterized as young people's play.

We argue that when exploring childhood, fun and holidays it is more useful to conceptualize fun in terms of a process of producing happiness that has material, embodied and cognitive effects. Happiness is both a bundle of positive feelings and emotions, as well as a judgement of those sensations (Nettle 2005). This means that positive evaluations of fun can therefore produce happiness. Happiness and the pursuit of happiness have been explored extensively else-where (Nettle 2005) and are beyond the scope of this chapter. However, as Read *et al.* (2002) discovered, happiness is linked to anticipation and expected fun. Also they suggest that experiences of fun produce stronger memories than less-fun experiences. In this light, fun in the lives of children almost becomes some-thing more important than a giggle or an enjoyable moment, but something capable of intensifying experiences and something that combines affect with cognition. This obviously has implications for all kinds of applied settings, such as designing learning environments and materials, and possibly even using fun as a therapeutic tool in a creation of strong happy memories, as in the work of CHICKS, discussed below.

Methodology

This project was conducted using ethnographic methods that were both iterative and inductive in nature. They were inductive in the sense of approaching the research arena with an open mind rather than with set hypotheses, and iterative in manner which ideas, data gathering and analysis were all fluidly intercon-nected. Our approach was designed to investigate how young people experience, live and interpret their lives through events and happenings. Furthermore we have attempted to contextualize young people's subjective experiences by placing emphasis on the human contact by recognizing the irreducibility of human experience.

Although we used a suite of methods to conduct the research, in this chapter we focus particularly on interviews with key actors (staff and young people) and participant observations made of children on holidays at CHICKS. These were supplemented with art and photo-elicitation projects, a social media platform and online questionnaires. Interviews were conducted with 26 young people, 14 staff members, 29 volunteers and 24 referral agents. The main methodological chal-lenges within the project were ethical issues, power relations and positionality of both the research subject and researchers. Whenever researching young people, ethical dilemmas and concerns are everyday operational issues, such as provid-ing anonymity to the interviewees, to finding 'safe' spaces in which to conduct the research. These practical challenges affected the subject position of the researchers and the multiple roles that an ethnographer often takes when carrying out the research and further to how one acts and interacts within these roles. Leyshon (2002) discusses becoming a hybrid persona of a researcher-youth worker when carrying out research with rural youth, taking on multiple dynamic roles. Similarly this research was carried out in a hybrid manner by the researchers constantly changing their subject positions. This helped alleviate the power dynamics while opening up an observational research space where one could simultaneously observe, sense and be part of the affective experiences. The different physical spaces invited different types of interactions and communication modes. For example on the trampoline and basketball courts the researcher would often adapt the stance of an incompetent adult where the young people had a chance to teach and advise. We followed Langevang's (2007) advice that sharing and moving between the spaces that young people occupy allows the researcher to build rapport and relationships, both vital for discovering the hidden nuances of young people's experiences.

Creating memories through care

Hampl (1995 in Jones 2005: 211) has posited the notion of 'remembering through the heart', a view that captures the qualitative difference between the experiential memory and what that memory might represent. For example, we might have a limited recollection of a particular childhood playtime with friends. We might not remember exactly who was present or exactly what was said, but we are likely to remember the smiles, the laughter, the sense of belonging and togetherness. We might even be able to recollect the smells, textures and tastes. These types of emotional geographies are pervasive and can be carried through to adulthood and beyond (e.g. Wylie 2007). However, there are many children and young people in the UK with less than pleasant memories who have not been afforded a carefree childhood of innocence and enjoyment, and often have substantial care responsibilities. Organizations like CHICKS attempt to counteract negative memories by offering free respite breaks to disadvantaged children and young people. At CHICKS the children get to experience time away from 'home' with adults and peers in a safe environment going climbing, kayaking, spending a fun day in a theme park and engaging in lots of other activities. Ally (14 years old) describes her memories nearly a year after her trip to the beach whilst on a CHICKS holiday.

> Body boarding, that was amazing. I loved like being on top of the wave and the supervisor got this amazing picture of me ... I was on top of the wave and he was in front of me and it is an amazing photo 'cos there is the wave and then me.

> (Personal interview)

Here Ally is remembering an affective experience with both the landscape and her holiday supervisor. She eloquently recalled her sense of connection to the spaces of the beach as well as her affective performance on a body board. Wylie (2007) uses the term 'landscaping' to explain the human connections beyond representation that are sensory, emotive as well as remembered. Ally does not experience the landscape alone, indeed as this quote demonstrates there is an entanglement of self, other and landscape in a way that is meaningful to her. The quote also illustrates how for Ally the practice of memory has two distinct levels. First, she is remembering an engagement in an activity that occupied her body; essentially she is creating a memory through this activity. Second, she is and has actively practised that creation or memory by looking at the photo and recalling the event months after it occurred. In other words she is telling herself a story about herself. Through this process of self-narration or storytelling, she is constructing her lifecourse autobiography (Leyshon and Bull 2011). Yet her story is not just that of an urban youth experiencing the countryside and enjoying a multitude of connections to the natural environment. Her story has other people in it, adults and children alike, who co-produce with her caring networks.

At CHICKS there is seldom any human interaction where care is not present in one form or another. Care is present not only in actions but words, empathy, touch and expressions. Throughout people's lifecourses they both receive and give care and would not thrive without it (Milligan and Wiles 2010). Geographers have examined care, for example, in relation to care-workers (Meintel *et al.* 2006) and teachers (Hargreaves 2001). These are care professions where lesser emotional investment is required in which care systems are invariably organized instrumentally in a top-down fashion. Policies are constructed that produce regulatory systems of practice and monitoring that are delivered by care-workers to those deemed in need of care. This leaves the impression that health-care systems are strongly embedded in somewhat impersonal wheels of national and international bureaucracy and decision-making at the highest political level. An alternative to this is a flat ontology of care. Drawing on the research of DeLanda (2002), a flat ontology refers to interconnected entities or agents where the spatial–temporal value of these connections may be different but the ontological status is the same. Hence, in this research we interpret a flat ontology of care to mean that there are multiple centres of care, that reproduce over time through interactions with a variety of experiences, emotions, spaces, practices etc. rather than clear lines of hierarchy and the delivery of specifically determined 'care'. To demonstrate this point the quote below is from a volunteer in her early twenties who has volunteered on three CHICKS camps.

> This is a bit going off track but I have suffered from eating disorder so I go to CHICKS and I know I have to eat the food to show the kids and it is so good to leave all that behind and just be at CHICKS and it sounds bizarre … it sounds a bit selfish cause I do go there for the kids (laughs) … I feel a bit … not a different person but you see it in the kids as well that you get to be a different person … they feel at home. So I can see how the kids can see different parts of themselves.
>
> (Personal interview with volunteer Anna)

Being able to 'leave behind' oneself and become someone different is what Anna is describing. For her, volunteering at CHICKS is a progressive act of forgetting her adult self such that she can position herself in relation to children. This enables her to fulfil her personal commitment to helping young people enjoy their holiday through producing a fun and safe environment that is culturally credible to her and them. It is not unusual for caregivers to emotionally benefit from caring (Milligan and Wiles 2010) but what Anna is describing here is more than that, by being at CHICKS she is simultaneously the object as well as the subject of care. In other words there are multiple interconnected centres within the care processes at CHICKS. This way the flat ontology of care allows for the dissolving of professional distance. This is illustrated below as Laura (15 years old) describes how she perceived the adults on her CHICKS break.

> I expected them kind of to be more like teachers ... but they were just happy and fair and you could get away with doing stuff but not like too much stupid stuff. I see you like you are really like at our level. Like if someone would have a problem they wouldn't really hesitate to go and speak to one of you. Seeing more like friend rather than a figure of authority.
>
> (Personal interview with Laura)

Laura is describing the outcome of a flat ontology of care. Importantly the perception of power is absent but sameness is observed – a shared understanding of reciprocal care between adult and child. Such is the emotional investment of both adults and children on these camps that temporal and spatial geographies of love and care are visible. Wylie (2007) describes geographies of love as the polar opposite of segregation, division and separation. This can be seen from one of the researchers diary extracts below:

> Sitting all together in the lounge, very relaxed, no worries, no rush, everyone chilled, smiles on their faces, adults kids all as one. It's quiet. Stuart (volunteer) and his guitar, plays a few notes, someone requests something. A song begins, a few adults join in, then the kids, everyone singing, smiling, chilling, enjoying. Fun, relaxed.
>
> (Research diary entry 28 August 2010)

Harrison (2007) suggests that geographies of love represent an affective bonding between individuals as they 'reach out' or extend care across space(s) to each other. Reaching or extending ourselves across space is an intriguing notion that can be understood as an emotional reach as well as, in this research, a literal bridging of the generational gap. Although Vanderbeck (2007) has cautioned that this process can often be a hindrance in intergenerational interaction as young people are rightly circumspect about adult–child relations, at CHICKS young people apparently see adult volunteers differently. Emma (13 years old), for example, suggests that volunteers do not belong to a different generational group or that they are simple present to 'care' for the children but rather she focuses on describing the sameness between the young people and the adults stating 'volunteers act like kids. Volunteers are like big kids. It's funny'.

The flat ontology of care is therefore reciprocally produced and its outcome can be seen in multiple geographies of love and affection that are constantly being made, negotiated and remade. Fun, place and care are central to this process, and they work in an interconnected way that enables young people and adults alike to formulate memories. The next section will examine how the memories are produced and used to promote emotional advantages.

Memory as an intervention

Memory and memories play a central role in the processes of becoming (Lee 2001) and are vital for the development of self and autobiography. However, we would like to suggest that reconceptualizing memory as an intervention and practice shifts the emphasis away from the memory of memory to one of memory as performance. Jones (2011) discusses memory as a non-linear trajectory in which a host of different affective states become mobilized depending on what is being remembered. This opens the possibility for using memory as a tool for well-being. If memories are produced with the purpose of therapeutic/happy interventions into the lives of young people, we need to question which types of memories produce positive affects and which memories are sustained in autobiography.

As discussed above, both a flat ontology of care and geo-spatialites of love can produce pervasive positive emotional geographies. This research tentatively demonstrates that emotional geographies can be enhanced through the concept of fun. Fun has the capability to become memoried, embodied and hence is more than temporary and transient, but can also have lasting effects exceeding beyond the moment (Harker 2005). In relation to children and young people fun is important because, as Read *et al.* (2002: 2) put it 'fun is a concept that seems to comfortably belong in a child's environment'. Although research is limited, we suggest that play therapy can help children produce happy memories because, through play, fun and enjoyment become central to a child's life. In effect fun helps replace something that was lost and, in this way, fun alleviates psychological pain and helps to build a coherent sense of self (Alvarez and Philips 1998). This is how Jason (12 years old) and Carmen (16 years old) describe how being on a CHICKS holiday differs from everyday life.

> JASON: Everyone doesn't really ... cause it is like no one really lives in Cornwall. No one lives there ... weird people ... no offence ... (laughs.) So like you don't have any regrets cause you are not gonna go back there so you don't have to meet the people and stuff. You will get to know everyone so quickly so that you can just be so open to everyone and everyone is like all crazy so ... it is like the atmosphere the main thing just like makes you feel a bit crazy...
>
> CARMEN: Yeah like freedom really. Freedom to just dance around in a middle of a shopping centre if you like.

Being 'crazy' and having the 'freedom' to dance around are described as aspects of the carefree, joyful and fun atmosphere that characterizes CHICKS experience. These types of positive emotional geographies do not cancel out prior, less positive memories but they enable the young people, albeit temporarily, to forget through creating a (liminal) time-space in which happier memories can be made. For example, Reece (15 years old) explains what CHICKS means to him.

TT: What does it mean to you to go to CHICKS?

REECE: Break from all your worries and enjoying the life and relax and people.

TT: So you don't think about your worries…?

REECE: Yeah, you do, but you do the activities and just forget about that for the whole week and it makes it lot easier.

As Lee and Yeoh explain (2006) remembering and forgetting seem to be woven together. In other words in order to remember one first needs to forget. Care, love and fun all aid in this process of forgetting by creating an interconnected dynamic loop. These happy, positive and good memories then build up an emotional and resilient tool-box giving the young people new skills with which to cope with past and potential future challenges. The acquisition of life skills by young people is a significant area of research beyond the reach of this chapter but we would like to emphasize that CHICKS also enables children to acquire different skills while they are on holiday and that this is evidenced in the quotes below from Jake (age 13), Dylan (age 10) and Becky (age 14):

TT: Feel different at CHICKS than in everyday life?

JAKE: Yeah, cause you don't know people but you get used to them and know what they are doing and everything and you get your confidence up every day.

TT: What is it about CHICKS that builds your confidence?

JAKE: Everybody is welcoming you all the time and making you feel safe. Everybody is nice to you, yeah.

DYLAN: People are easy to make laugh. Instead of thinking people gonna be friends, instead of, like, always thinking people hate you, you just join in.

BECKY: [It's] good to get away but good to come back and realize that you do love your family and especially if you have, like, fallen out and then you go, like, I hate my mum and then you go, no I do like them, so it kinda opens your eyes. It's like … I did self-harm pretty bad and I got help now so there are all these other sides to it. So if you are looking at the right side. Other people's stories and how they have coped, like

Jake, Dylan and Becky are all alluding to how they produce coping strategies predicated on memories of CHICKS. Their enlarged sense of well-being often grows in increments (Hart *et al.* 2007) through increased confidence, seeing themselves (perhaps for the first time) as someone who can make friends, and self-realization that there other people experiencing similar (traumatic) events. These contribute to resilience, positivity or happiness that can be a key for increased well-being.

Conclusion

In this chapter we set out to examine memory as practice and a potential tool for an emotional intervention to enhance the well-being of young people. We have argued that memory can be used as an effective intervention in the care of disadvantaged young people but there are caveats on this understanding. First, it is vital that care practices are set out in a manner of a flat ontology. Through this understanding we propose decentring the power relations between carer and cared-for, by recognizing that care is not one-directional and is reciprocal. Hence, caring and being cared for is co-constituted and relational. Furthermore, we discussed how a flat ontology of care produces care that has multiple centres, care moving and bouncing from one to another, shifting in direction and intensity transforming caregivers into care receivers.

Second, the experience of fun by children and carers alike allows for forgetting to take place as well as positive memories to emerge along a memory trajectory. Additionally this chapter argued that power executed within the landscape of fun promotes cooperation and shared agency and is a useful tool when working with and caring for (disadvantaged) young people. These positive memories can then give birth to new set of emotional and social skills or realizations of one's potential, hence making memory a potentially effective form of intervention for people who have experienced adversity.

The affective qualities of fun are predicated on a reciprocity of experience between adults and young people, creating multiple moments of laughter, humour and smiles that gain in power and effect over time. These momentary glimpses of happiness never fully arrive or become exhausted or simply forgotten, they produce memories that become written on the body, that travel with the individual to create new paths and directions. The young people and adults are co-creating a network of affect that is embedded in care and has the capacity to transform lives. That formation is rooted first and foremost in the landscapes of fun, e.g. the play spaces of holidays, that enable both adults and young people alike to use power in an affective and caring way rather than as a function of control. The concept of fun and care we argue is therefore intertwined into reciprocated networks of care that are non-hierarchical and decentred as a flat ontology.

The environment provided opportunities for discovery, magic and interaction. These spaces of care that the adults created with the young people were created and manifested through movement. This movement becomes landscaping, an active tactile engagement not only with the natural environment but also with those others in it. And although the natural environments have been widely reported to be beneficial for children and young people, at CHICKS the emphasis is on the nature of care. In other words the affective connections, embedded in fun and love, between people seemed to produce stronger memories than purely interactions with the natural environment.

References

Alvarez, A. and Phillips, A. (1998) 'The importance of play: A child psychotherapist's view', *Child and Adolescent Mental Health*, 3(3): 99–103.

Berk, S. (2003) *Child Development*, New York, NY: Alyn and Bacon.

Blythe, M. and Hassenzahl, M. (2003) 'The semantics of fun: Differentiating enjoyable experiences', in M. Blythe, K. Overbeeke, A. Monk and P. Wright (eds) *Funology: From Usability to Enjoyment*, Dordrecht: Kluwer Academic Press.

DeLanda, M. (2002) *Intensive Science and Virtual Philosophy*, New York, NY: Continuum.

Della Dora, V. (2008) 'Mountains and memory: Embodied visions of ancient peaks in the nineteenth-century Aegean', *Transactions of the Institute of British Geographers*, 33(20): 217–232.

DeSilvey, C. (2012) 'Making sense of transience: An anticipatory history', *Cultural Geographies*, 19(1): 31–54.

Dunkley, C.M. (2009) 'A therapeutic taskscape: Theorizing place-making, discipline and care at a camp for troubled youth', *Health and Place*, 15(1): 88–96.

Evans, B. (2008) 'Geographies of youth/young people', *Geography Compass*, 2(5): 1659–1680.

Hargreaves, A. (2001) 'Teaching in a box: Emotional geographies of teaching', in C. Sugrue and C. Day (eds) *Developing Teachers and Teaching: International Research Perspectives*, London: Falmer Press.

Harker, C. (2005) 'Playing and affective time-spaces', *Children's Geographies*, 3(1): 47–62.

Harrison, P. (2007) 'The space between us: Opening remarks on the concept of dwelling', *Environment and Planning D: Society and Space*, 25(4): 625–647.

Hart, A., Blincow, D. and Thomas, H. (2007) *Resilient Therapy with Children and Families*, London: Brunner Routledge.

Hunter-Jones, P. (2003) 'Managing cancer: The role of holiday taking', *Journal of Travel Medicine*, 10(3): 170–176.

Jones, O. (2005) 'An ecology of emotion, memory, self and landscape', in J. Davidson, L. Bondi and M. Smith (eds) *Emotional Geographies*, Aldershot: Ashgate Publishing Limited.

Jones, O. (2011) 'Geography, memory and non-representational geographies', *Geography Compass*, 5(12): 875–885.

Kearns, R. and Collins, D. (2000) 'New Zealand children's health camps: Therapeutic landscapes meet the contract state', *Social Science & Medicine*, 51(7): 1047–1059.

Kesby, M. (2007) 'Methodological insights on and from children's geographies', *Children's Geographies*, 5(3): 193–205.

Kiernan, G. and Maclachlan, M. (2002) 'Children's perspectives of therapeutic recreation data from the "Barretstown Studies"', *Journal of Health Psychology*, 7(5): 599–614.

Kuntsche, E., Knibbe, R., Gmel, G. and Engels, R. (2005) 'Why do young people drink? A review of drinking motives', *Clinical Psychology Review*, 25(7): 841–861.

Langevang, T. (2007) 'Movements in time and space: Using multiple methods to research with young people in Accra, Ghana', *Children's Geographies*, 5(3): 267–282.

Lee, N. (2001) *Childhood and Society: Growing Up in an Age of Uncertainty*, Buckingham: Open University Press.

Lee, Y. and Yeoh, B. (2006) 'Introduction: Globalisation and the politics of forgetting', in Y. Lee and B. Yeoh (eds) *Globalisation and the Politics of Forgetting*, Abingdon: Routledge.

Leyshon, M. (2002) 'On being "in the field": Practice, progress and problems in research with young people in rural areas', *Journal of Rural Studies*, 18(2): 179–191.

Leyshon, M. (2008) ' "We're stuck in the corner": Young women, embodiment and drinking in the countryside', *Drugs: Education, Prevention and Policy*, 15(3): 267–289.

Leyshon, M. (2011) 'The struggle to belong: Young people on the move in the countryside', *Population, Space and Place*, 17(4): 304–325.

Leyshon, M. and Bull, J. (2011) 'The bricolage of the here: Young people's narratives of identity in the countryside', *Social and Cultural Geography*, 12(2): 159–180.

Meintel, D., Fortin, S. and Cognet, M. (2006) 'On the road and on their own: Autonomy and giving in home health care in Quebec', *Gender, Place & Culture*, 13(5): 563–580.

Milligan, C. and Wiles, J. (2010) 'Landscapes of care', *Progress in Human Geography*, 34(6): 1–19.

Nettle, D. (2005) *Happiness: The Science Behind Your Smile*, Oxford: Oxford University Press.

Read, J., MacFarlane, S. and Casey, C. (2002) *Endurability, Engagement and Expectations: Measuring Children's Fun*, Eindhoven: Shaker Publishing.

Rothbart, M.K. and Bates, J.E. (2006) 'Temperament', in W. Damon, R. Lerner and N. Eisenberg (eds) *Handbook of Child Psychology: Vol. 3. Social, Emotional, and Personality Development*, New York, NY: Wiley.

Schaffer, R. (2004) *Introducing Child Psychology*, Oxford: Blackwell Publishing Ltd.

Smith, V. and Hughes, H. (1999) 'Disadvantaged families and the meaning of the holiday', *International Journal of Tourism Research*, 1(2): 123–133.

Stirton, J., Pownall, J. and Carroon, B. (1997) 'A holiday break for adolescents provided by the Malcolm Sargent cancer fund for children', *European Journal of Cancer Care*, 6(2): 154–155.

Thornton, P., Shaw, G. and Williams, A. (1997) 'Tourist group holiday decision-making and behaviour: The influence of children', *Tourism Management*, 18(5): 287–297.

Vanderbeck, R.M. (2007) 'Intergenerational geographies: Age relations, segregation and re-engagements', *Geography Compass*, 1(2): 200–221.

Ward, C. (1978) *The Child in the City*, London: The Architectural Press.

Ward, C. (1988) *The Child in the Country*, London: Robert Hale.

Westman, M. and Etzion, D. (2001) 'The impact of vacation and job stress on burnout and absenteeism', *Psychology and Health*, 16(5): 595–606.

Woodyer, Tara (2012) 'Ludic geographies: Not merely child's play', *Geography Compass*, 6(6): 313–326.

Wylie, J. (2007) *Landscape*, Abingdon: Routledge.

Zabelina, D. and Robinson, M. (2010) 'Child's play: Facilitating the originality of creative output by a priming manipulation', *Psychology of Aesthetics, Creativity, and the Arts*, 4(1): 57–65.

8 Mother and daughter 'homebirds' and possible selves

Generational (dis)connections to locality and spatial identity in south Wales

Dawn Mannay

Introduction

This chapter builds on previous intergenerational studies of social mobility within the mother–daughter relationship by highlighting the importance of place. Place has become more centralized within recent psychosocial accounts; for example, Walkerdine and Jimenez's (2012) sensitive analysis of the spatial and temporal aspects of a Welsh de-industrialized town and the affective relationships of this marginalized community. However, in order to extend contemporary understandings and inform policy, it remains useful to revisit the mother–daughter relationship and seek to combine spatial identities with psychosocial accounts of everyday lives; couched within the maternal relationship.

This chapter is drawn from a four-year Economic and Social Research Council-funded project that took place in a marginalized housing area in urbanized south Wales, United Kingdom. The project employed visual techniques, including collage, mapping and photo-elicitation to explore the everyday lives of mothers and their daughters (Mannay 2010, 2013a). However, in order to move beyond the everyday and examine intergenerational continuities and discontinuities it was important to develop an approach that allowed for retrospective engagement. This chapter discusses the method of positive and negative 'possible selves' in which mothers wrote future narratives about the selves that they wanted to become and feared becoming in the future and retrospective narratives about the self that they had wanted to become and feared becoming, from the point of view of being their daughters' age.

The method provided an insight into intergenerational spaces of working-class femininity and raised issues that were unexpected, sensitive and previously unspoken (Mannay 2011, 2013b, 2013c). This chapter presents data from these accounts to demonstrate how employing a 'possible selves' approach can engender new understandings about the creative and imaginative practice of being an individual, and the intergenerational legacies that both limit agency and also instigate processes of change. In this way the chapter argues that it remains necessary to continue to engage with social lives at the level of the home, the local, the cultural and the everyday, but also at the level of the psychological: acknowledging the ways in which the past shadows both present and intergenerational futures.

The study

Research was conducted in Hystryd, a predominantly white urban area, which ranks as one of the most deprived communities in Wales (Welsh Assembly Government 2008). Nine mothers and their daughters participated in the project. I previously lived in the area and this shared sense of geography positioned me as 'experience near' (Anderson 2002: 23). Consequently, it was important to address my position as an indigenous researcher and make a deliberate cognitive effort to question my taken-for-granted assumptions of that which I had thought familiar (Delamont and Atkinson 1995).

Participant-directed visual data production techniques of photo-elicitation, mapping and collage were selected to limit the propensity for participants' accounts to be overshadowed by the enclosed, self-contained world of common understanding. Participants took photographs, drew maps and made collages depicting meaningful places, spaces and activities (for a full discussion of these techniques see Mannay 2010), and then discussed them with me in tape-recorded interviews to ensure that I understood what they intended to communicate (Rose 2001).

The majority of the images and talk focusing on place and space were based on the 'here and now' but some participants linked the past with the present and offered an insight into their imagined futures. This temporal dimension was useful in elucidating discourses of gender identity, generation, social reproduction and processes of identification. In order to move beyond the everyday and examine intergenerational spatial continuities and discontinuities it was important to develop an approach that engendered temporality.

Initially, I was drawn to the work of Markus and Nurius (1986: 954) who attempt to provide a conceptual link between cognition and motivation by exploring individuals' possible selves; their 'ideas of what they might become, what they would like to become, and what they are afraid of becoming'. Although influential and frequently employed (Chang *et al.* 2006; Guimond *et al.* 2007), this early work has been criticized for the forced choices presented in the questionnaire, which limited participants' responses to the categories offered by the researchers.

A more compatible approach was discovered in the work of Lobenstine *et al.* (2004) who worked with mothers and their daughters, employing mapping exercises and elicitation interviews to examine positive and negative conceptions of 'possible selves'. More recent qualitative work has demonstrated how orientation to the future occurs within a certain social, geographical, cultural and historical context that acts as a 'straitjacket' on our ability to envision and realize future possibilities (King and Hicks 2007; Fletcher 2007; Casey 2008; Susinos *et al.* 2009).

Everyday life then, represents a constant and continual state of fluidity in an ever-changing landscape of who we might have been and who we might still become. A continuous fluidity that Ringrose and Renold (2011) argue is intensified in the lives of girls and women; therefore, contradictions arise between past, future and present selves (Mannay 2013b). The following sections explore these intergenerational spaces, focusing on the home and utilizing possible selves narratives and a photographic image to examine the imagined futures and lost pasts of mothers and daughters on the margins of contemporary Wales.

Same place, different space

Previously, I revisited Diana Barker's (1972) seminal paper 'Keeping close and spoiling in a south Wales town', by drawing on the data produced with one mother and daughter dyad, who were given the pseudonyms Mary and Adele (Mannay 2013c). Barker focused on geographical closeness and the strategies employed by parents to keep their children living at home, rather than sending them to university; in my response I explored the contradictory nature of remaining geographically close, living within the family home and commuting to a local university.

I argued that this spatial arrangement cannot negate the psychological separation: rather the maternal relationship can be threatened by an 'ambiguous loss', the loss of a loved one who is physically present but psychologically absent. In this chapter, I return the lens of focus to the psychological strategies employed by Mary and Adele to engender closeness and maintain continuity as well as drawing on Bourdieu and the concept of habitus; however, rather than centring the educational experience, I will explore their relationship with their immediate locality, an area I have called 'Hystryd'.

As Fink (2012) argues, statistics diminish people's lives by treating them as figures and this research was interested in the individual within the shadow of similarities; focusing on the locale and the intergenerational. For this reason, the pseudonym, Hystryd, was created to reflect this interest in the everyday salience of home by drawing on the Welsh word for the street, *y stryd. Y stryd* appreciated the focus at this level of the local but the study was specifically interested in the lives of women and girls, of mothers and their daughters; and to incorporate this lens a feminization was engendered drawing from the Welsh word for she, *hi*, and taking the first letter to form the amalgamated 'Hystryd'.

Place is an arena where people are actively engaged in a process of constructing themselves through the complexities of difference and similarity, and as the chapter will explore, Mary and Adele experience the geographical site of Hystryd differently; this difference is intimately tied to their unique biographies. The youngest of seven children, Mary was born in 1967. Along with a substantial Catholic population of families who relocated from older, crowded, urban centre areas, Mary's parents moved to Hystryd as children in the 1930s, moving into one of the 4,000 newly built council homes. Once married, they moved into their own council home in the heart of Hystryd, where Mary was born and where her mother and father remain. Mary and all but one of her six siblings have remained in Hystryd and brought their own children up in the area.

Mary's fifth child, Adele, was born in Hystryd in an ex-council house. Hystryd is dominated by council housing but council areas are not homogenous.

Adele was born on one of the council estates in a street that is sought after by council residents. Here many of the houses display well-maintained gardens, symbolizing and evoking a distinction between other streets in council housing areas; when Adele was 12 her family moved to a private estate situated on the outskirts of Hystryd. The private estate is in the boundaries of Hystryd but it is a clearly marked enclave of home ownership and upward social mobility with an absence of the markers of deprivation such as graffiti and disrepair that characterize many social housing estates.

There are close ties between spatial and social distinctions, and this biographical distinction is reflected in the different relationship that mother and daughter have with their local area. Mary conveys a potent connectedness, while Adele, despite identifying as a 'homebird', has a much more ambivalent attachment to Hystryd.

Situating the self

In Hystryd there are close ties between spatial and social distinctions, both actual and ideological; thus, in order to preclude 'place-centrism' (Jessop *et al.* 2008), it was important to explore notions of territory by comparing the accounts of those residing in council-owned properties in the heart of the estate with those on the borderlines of Hystryd in privately owned homes. Many of these private and public homes are within the boundaries of the stigmatized area, but some residents invest in belonging to other constituencies, even though the land registry denies this premise. The position of these houses then is debatable as Hystryd was in existence before the council housing arrived. Hystryd began as a locale with privately owned homes lining some of the busiest roads and dispersed randomly in small pockets and later the council housing programme changed the landscape. These homes were then joined by housing associations, providing low-cost rental accommodation, and private housing estates built along Hystryd's boundaries.

Mary identifies as a 'homebird' and has lived in Hystryd all her life apart from a short time when she was first married and moved to a Welsh valley town, where houses could be purchased at an affordable price. In Mary's narrative she only refers to crying twice and they both relate to her connectedness to Hystryd: 'I've only moved out of Hystryd for two years my whole married life … I got married on the Saturday, I was crying on the Wednesday I was'. After two unhappy years away from Hystryd, Mary returned and moved in to an ex-council house, a few minutes' walk from her parent's home, where the family remained for 16 years. This home was located in the heart of Hystryd's largest council estate but it was in a desirable location where many people have purchased their properties from the council. It was the decision to move from this street to a house on a newer private estate on the periphery of Hystryd, nine years ago because, as Mary explains, their house was too small to hold a family of eight, which is the one other incident she describes as characterized by crying: 'When I left the (old street) my neighbours cried … and I said to the one, I been crying since I come up here'.

Newer housing estates on the edge of Hystryd, are often populated by residents who have moved from council areas in a bid for social mobility through the 'right to buy' scheme, a government initiative that provides an opportunity for eligible council tenants the *right to buy* their tenancy property from their council at a discount. The move can illustrate a 'geographical imagination' where individuals born and bred in Hystryd now describe their residence by the name attached loosely to the original housing development at the time of construction or to neighbouring areas, despite the fact that their postcode marks them as a resident of Hystryd. In this way residents on bordering, newer housing estates are, explicitly and implicitly, constantly proliferating distinctions between their spatial positioning and that of their neighbours (Skeggs 1997).

Mary describes herself as 'Hystryd born and bred' and finds the neighbour's premise of living elsewhere ridiculous. What Mary finds laughable is the fact that most of her new neighbours were at one time living in council housing in Hystryd. Hystryd is not a place that attracts many private residents from outside the immediate area and so the pattern of social mobility often means that residents simply move to a different street rather than leave Hystryd.

However, the move to a new street is often accompanied by the 'idea' of living somewhere else, and this geographical imagination was recently captured by the parish priest at the funeral of a 'local' man who had always said he never lived in Hystryd. As the priest commented, more housing is built in Hystryd every year, yet paradoxically, every year Hystryd appears to get smaller. This complexity is also illustrated in relation to the schooling experience of Mary's daughter Adele:

> No one calls this Hystryd, Dawn (laughs) (both laugh) Adele used to go to school with the girl next door, called Jo, they were in the same class, Jo lived in (area just outside), Adele lives in Hystryd. The teacher used to say how come if you live next door to her, Jo lives in (area just outside) (laughs) and you live in Hystryd? (laughs)
>
> (Mary)

For Mary, and perhaps for her neighbours, it is the presence of this 'outside within' (Massey 1994: 170) that in part helps to construct their specificity to the local place: Hystryd and a denied Hystryd. Mary will not engage in the geographical imagination that the neighbours maintain and she identifies her current home as being part of Hystryd rather than disguising her address. As most of her current neighbours moved from Hystryd's council housing and like Mary are 'Hystryd born and bred', she views them as actively rejecting their culture. As place is tightly bound up with identity, in many ways this creation of distance can be seen as a rejection of the self (Walkerdine *et al.* 2001); and it is not a rejection with which Mary can align: 'Yeah they are snotty they are, Dawn, *snotty*, they think they live somewhere else honestly, and they're snotty'.

This defence is perhaps more prominent as four of Mary's five siblings have remained in Hystryd, and three of these live in council housing areas.

Importantly the family relationship within Hystryd can be described as one of 'joint organization' (Bott 1957: 87), featuring considerable visiting and exchange of services. This means that Hystryd is not simply a geographical space but the site of a close-knit, intergenerational, kinship network. Drawing from Bourdieu (1990), Mary's habitus is characterized by a value system in which Hystryd is central; therefore, when members of the community, in this case, residents of Hystryd, begin to deny Hystryd, she is forced to defend her sense of belonging: 'You know, I'm not ashamed to say I'm from Hystryd, Dawn, you know, and if people look at you and stereotype you as being from Hystryd, that's their business'. As Scourfield *et al.* (2006) maintain, childhood experience is the foundation upon which self-identity is built and Mary's upbringing, family and friendship networks have been consolidated in the council housing that makes up the heart of Hystryd. The quotation illustrates the way in which Mary invests in her identity as being part of Hystryd, whilst acknowledging the associated stigma. However, Mary does not see these judgements made by outsiders as a threat to her sense of self-worth. It is the ways in which those living on the outskirts, but within Hystryd, continually create distance to establish distinction that angers Mary: and these outsiders within are her neighbours.

In Bott's seminal work *Families and Social Network* (1971 [1957]) she found that families who had had to move to find suitable housing, like Mary needing a larger home, tended to idealize their old areas, only remembering the friendliness. Similarly, Mary minimizes the negative aspects of Hystryd associated with its vast social housing estates, whilst emphasizing the positive features of both the place and those who embrace their local identity. As Massey (1994) argues, 'mother' often personifies a place that does not change and we tend to invest in a place where we once belonged, which belonged to us and where we could safely locate our identities. The positive and negative affiliations of place then are not just based on our everyday experience but on memories, hopes and imaginings.

Mary is unable to move away from Hystryd either geographically or socially. The only reference to tears is to those shed in the tangible and conceptual departure from Hystryd. I did not ask Mary about moving away, but in discussions of possible future selves, I commented that one day it will be just Mary and her husband in the family home. It is at this point that Mary tells me of a desire that she has beyond the boundaries of Hystryd, but even dreams for the future are shattered before they have even been described:

> I always wanted to go to (seaside town), Dawn, and (other seaside town) and things, I always wanted to … I've always wanted, it was always my thing, you know, to re … Retire me and (husband) but to be honest (pause) I don't know if we could.

Mary is comfortable with Hystryd, for this geographical and social space is the bedrock of her sense of who she is and where she belongs; she identifies as a 'homebird', a term that remains central to ideas of acceptable working-class femininity (Barker 1972; Mannay 2013d). In this way, for Mary, Hystryd is not the site of ignominy but a continual confirmation of the securities of the self. It is also, and perhaps more importantly, the home of family, friends, Mary's six children and now grandchildren, the first of whom was born when her youngest child was only ten years old. As Mary explains: 'I've had grandchildren quicker than I thought I was going to have them, you know'.

Mary talks about having 'no time in between' and the role of grandmother has come to represent another significant tie to Hystryd. Leaving now, when she is needed to look after her grandson, would challenge the legitimacy of her motherhood/grandmotherhood and also the institution of the family, both of which are paramount for Mary. As Bourdieu (1984) maintains, habitus refers not only to values, disposition and lifestyle but to our expectations of and for self; creating a distance from Hystryd, geographically or psychologically, is then a betrayal to self and others that Mary feels unwilling and unable to initiate. This responsibility can in part explain why although Mary 'always wanted to go' the desire is countered with the statement 'I don't know if we could'.

It is in a conversation about Mary's social life, which revolves around her family, that she indirectly reverts back to the topic of a seaside retirement and tells me 'Like, really, I couldn't up and go now because I wouldn't be able to have the grandkids every day'. Interestingly, she adds 'you know you're still sort of, your choices are not your own choices as you get older' supporting the contention that the present identity is strongly dependent on the past identity (Wetherell 1996; Erikson 1980). It is the strength of the past identity, tightly bound up in caring for others that means that the present is a place where 'your choices are not your own choices'; rather, they become part of an indelible and inescapable legacy and the present self is built upon both past and future 'lost possible selves' (King and Hicks 2007).

Un-situating the self

In preparation for our first interview Adele took a series of photographs to represent her everyday life; all but one of these images included family members. The photographs were used as an elicitation base but I left all copies of this recognizable visual data with Adele. The prevalence of the family in these photographs and the accompanying talk situated Adele as a 'homebird'; extending a habitus of acceptable working-class femininity. However, the photograph that Adele describes as 'gone wrong' opens up different conversations, drawing on a habitus that is not bound within the value system of the home (see Plate 8.1).

The photograph has 'gone wrong' as it is not aesthetically pleasing and its subject matter escapes from the lens; however, it retains its usefulness as an elicitation tool. Adele had attempted to take a photograph of her car from her living room window but the separating glass pane acted to reframe the subject so that the viewer is presented with Adele's hands and the camera inside her home. The camera is disposable so Adele does not see the image, until I deliver the unopened package of photographs to her, so there is no attempt to retake the

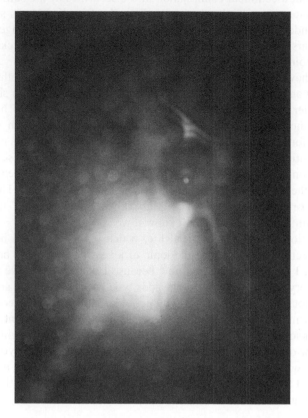

Plate 8.1 'Gone wrong' (photograph by anonymous interviewee 'Adele'[1]).

shot. The car has been bought with a university grant and is symbolic of the journey to a different and conflicting world (Mannay 2013d); however, the reflective image places Adele firmly back 'home'. This positioning of inbetweenness, where the nexus of biographical, social and local factors conjoin possibilities with the contexts and conditions of lived practices and future expectations, characterizes Adele's talk.

Earlier in the chapter, I introduced a quotation to illustrate how Mary finds her neighbour's premise of living elsewhere, when the boundaries of their ward place them firmly within Hystryd, both ridiculous and treacherous. The quotation suggests that while her classmate, Jo, denied her Hystryd residency, according to Mary, Adele was comfortable with the association. As Bourdieu (1990) comments, familial habitus results in a tendency for young people to acquire expectations that are adjusted to what is acceptable 'for people like us'; however, as Adele has matured there has been a departure from Mary's adage 'I'm not ashamed to say I'm from Hystryd', which had previously been a stance taken in unison. However, 'not ashamed' does not demonstrate the same strength of allegiance as 'proud of' and throughout Mary's account the demonstrative use of the negative 'I'm not' rather than the positive 'I am' indicates an uncertainty, which manifests itself in Adele's account.

> Up until like quite recently, if people, like, took the mick out of Hystryd I'd stick up for it and say I'm from Hystryd, I'm not a gyppo [*sic*] or … my parents are working, but now I think I am really just seeing what it is like

Rather than feeling that she should defend Hystryd, Adele has shifted her perspective to acknowledge that perhaps Hystryd, or at least some areas of Hystryd, are indefensible, and Hystryd can be experienced as a site of ignominy. Adele expands on this position of 'seeing what it is like' in the following quotation:

> I think just as I've got older I'm seeing it for what it really is like … where my brother lives, like, we go there sometimes to get food and it's, like, every time you go in there there's a little yobbo in there, like, abusing the man working there, and outside the shops causing trouble…. And I just thinks, I wouldn't want my kids, if I had kids, I wouldn't want them to turn out like that.

Mary continually made comparisons between herself and her neighbours on the private housing estate, creating distances and establishing distinctions between the *snotty, horrible, funny* and those who maintained their allegiance to Hystryd. However, Adele does not engage in the processes of pseudo-specification avowed by her mother but instead provides a set of distinctions inside Mary's idyll. It is in the heart of the council estate that Adele raises distinctions between herself and the person she would want her children to become, in comparison with the 'rough, yobbo and troublemaker'.

Returning to Scourfield *et al.*'s (2006) premise that childhood experience is the foundation upon which self-identity is built, Adele invokes her move from a council area to private housing at the age of 12 to account for the way her views of the council estate have changed. In the next excerpt, Adele reflects on her views of Hystryd when living on her old street in an area of social housing as compared to living at her current address, and being a visitor to Hystryd's council housing. Adele explains how 'she never noticed anything' when she was living on the council estate because it formed part of her everyday life. As Skeggs (2004: 41) maintains, 'perspective is always premised upon access to knowledge' and from her perspective as a resident on the council estate, Adele was unable to notice familiar, everyday situations which are often taken for granted until she 'come out of it'.

> ADELE: And there wasn't no private houses but I think when you come out of it there's a difference when you go back like, when we lived there I never noticed anything

INTERVIEWER: Mmm

ADELE: 'Cause I was used to that but when you move from there I think you go back and you think

As academics then we often strive to 'make strange the social context that we assume to understand by virtue of taken for granted cultural competence' (Atkinson *et al.* 2003: 47) and employ techniques that allow the researcher 'to make the familiar strange and interesting again' (Erikson 1986: 121). It is Adele's move from the council estate combined with her frequent visits to family and friends on the council estate, which have acted as a technique of 'defamiliarization' (Gurevitch 1998), forcing Adele to slow down her perception, to linger and to notice or in Adele's words 'I think you go back and you think'. Accompanying this 'defamiliarization' is a narrative of investment in a perceived class fraction and a complex process of position-taking that is articulated through the socio-spatial distinctions of Hystryd. The following extract explores these different ways of knowing and understanding:

ADELE: I don't know, I think it is like society's getting worse, I think, but I don't know if I've noticed it because I've come out of it a bit…. And where you go back in, you see it then, I don't know … and I wouldn't want to live here when I'm older and bring my kids up in here

INTERVIEWER: Yeah (pause) you was happy when you was a kid though, growing up in the (council estate) it didn't bother you

ADELE: No, it didn't bother me whatsoever but I think up here I think you are a bit out of it…. Then I never minded living on the (council estate) and I liked living there, like I like living here but, but if I had kids, I don't think I'd want to bring them up here, if I could afford to live somewhere else I would

Adele uses the phrase 'society's getting worse' that suggests a universal nature of change. However, the remainder of the account constructs a set of clear socio-spatial distinctions that geographically situate the perceived neighbourhood decline. Adele positions herself as '*out of it*' twice in this quotation making a clear divide between council and private areas and talks about her 'visits' to areas of social housing as going '*back in*'. For Adele, this process of (dis)identification (Skeggs 1997) is fragile because although she may be '*out of it*', a short walk takes her '*back in*'. Adele's tool of 'defamiliarization' has provided a gateway to destinations that lay beyond her repertoire of preconceived understandings and changed her view of Hystryd. Being notionally 'out of it' in an enclave of private housing is not enough. Adele desires a future where being 'out of it' is no longer tenuous but certain, distinct and geographically tenable; 'if I had kids, I don't think I'd want to bring them up here, if I could afford to live somewhere else I would'.

This realignment of Adele's perspective has resulted in an incompatibility in regard to mother's and daughter's sense of place. Spatial narratives are intimately linked to individual biography and crossing the state/private housing divide at a life stage where people are actively engaged in a process of constructing themselves, rather than as an adult, appears to provoke a different interpretation of the social relations of space and also erodes certainties of the self. For this reason, it is important to think carefully about locality and appreciate that it is the micro level of specific streets in combination with the social relations of the wider area that construct individual habitus.

In her possible selves narrative and her interview talk, Adele stresses that she would not want to live here when she is older, that she would not want to bring her children up in Hystryd and that if she could afford to live somewhere else she would. All of these statements form an opposition not only to Mary's position of 'allegiance to her manor' (Evans 2007) but also to Adele's and Mary's strong identification with the concept of the 'homebird'. As discussed earlier, Mary constructs a social value out of territoriality and in order to negotiate some consensus with her mother, Adele is required to create an alternative set of distinctions.

> ADELE: It's because of the people here, like yesterday I was coming home from uni and it was, like, ten past two and you know where they are building the wall by (shop)
> INTERVIEWER: Yeah
> ADELE: Men was stood there drinking cans of beer at two o'clock.... It's, like, get a life get a job

As Evans (2007: 28) contends, social classes in Britain endure via a 'segregation that is emotionally structured through mutual disdain'. Where Mary engaged in processes of pseudo-specification to celebrate allegiance to the council estate she grew up in and to denigrate the manners of her 'turncoat' neighbours in private housing, Adele employs different distinctions to set residents apart within the council estate. Perhaps it is useful here to draw again on Bourdieu (1984) because Adele actively relegates another social group to an inferior status through the appropriation of a different habitus in terms of working-class respectability. The term working-class is important here for work is the tool of differentiation, those who are 'drinking cans of beer at two o'clock' operate within a different habitus and are both without a job and without a life, a theme that is developed in the next quotation.

> It depends what the person was like, they might fit in (laughs) (both laugh) you know, I'm not saying everyone in (the council estate) is, like, benefit person but the people who are spoil it for the people who aren't.... Like, people spoil it for ... the good people who live in Hystryd.

The 'benefit person' fails to engage with working-class respectability and comes to represent the destruction of the community, 'spoil it for the people'. By utilizing discourses of 'them and us' Adele is able to identify with Mary's idyll of

Hystryd, for the space can exist in this 'good' form at a time when it was unspoilt, Hystryd, or more importantly Mary's connectedness to Hystryd, can be justified and supported by Adele as long as the bad can be placed with individual others. Adele has further stories of the 'bad' that are living off the state, drinking, taking drugs and buying stolen goods that resonate with wider stereotypical mediated images of the working-class (Hayward and Yar 2006).

Importantly, for Adele the 'benefit person' is one constructed in a contemporary age of neoliberalism, competitiveness and individualism. As Charlesworth (2000) argues, in marginalized areas of de-industrialization, those over 40 have a coherent sense of unemployment as the result of changes in the wider political and economic realm. However, for Charlesworth (2000: 2), as one moves through the generations the narrative of the social transforms to a narrative of the individual; so that 'coherence of a spoken common understanding based on mutual respect and shared sources of value, become more infrequent until, among the very young, understanding and value seem impossible'. For Adele, understandings of place, and the people within spaces, is intimately tied to a biography and temporality that inform her perspective and create a different Hystryd from the one experienced by her mother,

Conclusion

Representations of the self are always formed within wider social imaginaries of place (Anderson 1983), and maintaining a liveable sense of self demands the creative and imaginative practice of spatial situating and un-situating. Employing a possible selves framework offers an opportunity to explore how the past is implicated in the present so that relations with space are always informed by our individual biographies. In this way intergenerational legacies shadow both the present and the future so that even those in the same geographical locality (and home) find themselves in very different spaces. Spaces beyond the stereotypical and the homogenous, spaces for the multitude of details and intergenerational diversity of urban experience that represent the everyday experiences of mothers and daughters residing on the margins of contemporary Wales.

Acknowledgements

I would like to acknowledge the participants, who made this chapter possible, and also Professor John Fitz, Dr Emma Renold and Dr Bella Dicks, for supervising this research project. I would also like to thank Dr Robert Vanderbeck and Dr Nancy Worth at the University of Leeds for inviting me to present at their conference, Intergenerational Geographies: Spaces, Identities, Relationships, Encounters, where I began to consolidate some of the ideas presented in this chapter. The doctoral research project from which this chapter is drawn was titled 'Mothers and daughters on the margins: gender, generation and education' and was funded by the Economic and Social Research Council.

References

Anderson, B. (1983) *Imagined Communities: Reflections on the Origin and Spread of Nationalism*, London: Verso.

Anderson, G.L. (2002) 'Reflecting on research for doctoral students in education', *Education Researcher*, 31(7): 22–25.

Atkinson, P., Coffey, A. and Delamont, S. (2003) *Key Themes in Qualitative Research*, Walnut Creek, CA: Alta Mira Press.

Barker, D. (1972) 'Keeping close and spoiling in a south Wales town', *Sociological Review*, 20(4): 569–590.

Bott, E. (1971 [1957]) *Family and Social Network: Roles, Norms, and External Relationships in Ordinary Urban Families*, London: Tavistock.

Bourdieu, P. (1984) *Distinction: A Social Critique of the Judgement of Taste*, London: Routledge & Kegan Paul.

Bourdieu, P. (1990) *The Logic of Practice*, Cambridge: Polity Press.

Casey, E. (2008) 'Working class women, gambling and the dream of happiness', *Feminist Review*, 89: 122–137.

Chang, E.S., Chen, C., Greenberger, E., Dooley, D. and Heckhausen, J. (2006) 'What do they want in life? The life goals of multi-ethnic, multi-generational sample of high school seniors', *Journal of Youth and Adolescence*, 35(3): 321–331.

Charlesworth, S. (2000) *A Phenomenology of Working Class Experience*, Cambridge: Cambridge University Press.

Delamont, S. and Atkinson, P. (1995) *Fighting Familiarity: Essays on Education and Ethnography*, Cresskill, NJ: Hampton Press.

Erikson, E. (1980) *Identity and the Life Cycle*, London: W.W. Norton.

Erikson, F. (1986) 'Qualitative methods in research on teaching', in M.C. Wittrock (ed.) *Handbook of Research on Teaching*, New York, NY: Macmillan, pp. 119–158.

Evans, G. (2007) *Educational Failure and Working Class White Children in Britain*, London: Palgrave Macmillan.

Fink, J. (2012) 'Walking the neighbourhood, seeing the small details of community life: Reflections from a photography walking tour', *Critical Social Policy*, 32(1): 31–50.

Fletcher, S. (2007) 'Mentoring adult learners: Realizing possible selves', *New Directions for Adult and Continuing Education*, 114: 75–86.

Guimond, S., Branscombe, N.R., Brunot, S., Buunk, A.P., Chatard, A., Desert, M., Gracia, D.M., Haque, S., Martinot, D. and Yzerbyt, V. (2007) 'Culture, gender and the self: Variations and impact of social comparison processes', *Journal of Personality and Social Psychology*, 92(6): 1118–1134.

Gurevitch, Z.D. (1998) 'The other side of dialogue: On making the other strange and the experience of otherness', *American Journal of Sociology*, 93(5): 1179–1199.

Hayward, K. and Yar, M. (2006) 'The "chav" phenomenon: Consumption, media and the construction of a new underclass', *Crime, Media and Culture*, 2(9): 9–28.

Jessop, B., Brenner, N. and Jones, M. (2008) 'Theorizing socio-spatial relations', *Environment and Planning D: Society and Space*, 26(3): 389–401.

King, L.A. and Hicks, J.A. (2007) 'Lost and found possible selves: Goals, development and well-being', *New Directions for Adult and Continuing Education*, 114: 27–37.

Lobenstine, L., Pereira, Y., Whitley, J., Robles, J., Soto, Y., Sergeant, J., Jimenez, D., Jimenez, E., Ortiz, J. and Cirino, S. (2004) 'Possible selves and pastels: How a group of mothers and daughters took a London conference by storm', in A. Harris (ed.) *All About the Girl: Culture, Power, and Identity*, New York, NY: Routledge, pp. 255–264.

Mannay, D. (2010) 'Making the familiar strange: Can visual research methods render the familiar setting more perceptible?' *Qualitative Research*, 10(1): 91–111.

Mannay, D. (2011) 'Taking refuge in the branches of a guava tree: The difficulty of retaining consenting and non-consenting participants' confidentiality as an indigenous researcher', *Qualitative Inquiry*, 17(10): 962–964.

Mannay, D. (2013a) '"Who put that on there ... why why why?": Power games and participatory techniques of visual data production', *Visual Studies*, 28(2): 136–146.

Mannay, D. (2013b) 'The permeating presence of past domestic and familial violence in the present: "So like I'd never let anyone hit me but I've hit them, and I shouldn't have done"', in J. Ribbens McCarthy, C. Hooper and V. Gillies (eds) *Family Troubles? Exploring Changes and Challenges in the Family Lives of Children and Young People*, Bristol: Policy Press.

Mannay, D. (2013c) '"I like rough pubs": Exploring places of safety and danger in violent and abusive relationships', *Families, Relationships and Societies*, 2(1): 131–137.

Mannay, D. (2013d) '"Keeping close and spoiling" revisited: Exploring the significance of "home" for family relationships and educational trajectories in a marginalised estate in urban south Wales', *Gender and Education*, 25(1): 91–107.

Markus, H. and Nurius, P. (1986) 'Possible selves', *American Psychologist*, 41(9): 954–969.

Massey, D. (1994) *Class, Place and Gender*, Cambridge: Polity Press.

Ringrose, J. and Renold, E. (2011) 'Teen girls, working class femininity and resistance: Re-theorizing fantasy and desire in educational contexts of heterosexualized violence', *International Journal of Inclusive Education*, 16(4): 461–477.

Rose, G. (2001) *Visual Methodologies*, London: Sage.

Scourfield, J., Dicks, B., Drakeford, M. and Davies, A. (2006) *Children, Place and Identity*, London: Routledge.

Skeggs, B. (1997) *Formations of Class and Gender*, London: Sage.

Skeggs, B. (2004) *Class, Self and Culture*, London: Routledge.

Susinos, T., Calvo, A. and Rojas, S. (2009) 'Becoming a woman: The construction of female subjectivities and its relationship with school', *Gender and Education*, 21(1): 97–110.

Walkerdine, V. and Jimenez, L. (2012) *Gender Work and Community After De-industrialisation: A Psychosocial Approach to Affect*, Basingstoke: Palgrave Macmillan.

Walkerdine, V., Lucey, H. and Melody, J. (2001) *Growing Up Girl*, Basingstoke: Palgrave.

Welsh Assembly Government (2008) *Welsh Index of Multiple Deprivation 2008: Summary Report*, Cardiff: Welsh Assembly Government.

Wetherell, M. (1996) *Identities Groups and Social Issues*, London: Sage.

9 Children's engagement with intergenerational war stories

Dorothy Moss

Introduction

This chapter focuses on the intergenerational communication of memories of war, including those concerning children's need for shelter and protection, enemies and friends, domestic life and the different expectations of men, women and children. Research into childhood and war usually focuses on those directly caught up in events, considering the effects of trauma, children's perspectives and experiences, their agency and contribution (Boyden and Berry 2004; Watson 2008). However, in addition to this, it is important to explore how children relate to the war experiences of earlier generations because of the impact this has on their development and well-being. Children are affected by former wars and wars at a distance, whether or not they are directly involved. Intergenerational memories of war are selective and convey events associated with humour, fear, sadness, care and particular allegiances. This chapter analyses selective silences and uses of humour to understand how childhood memories of war are bound up in themes of nationalism, parenting and everyday life.

The original research on which the chapter is based explored how children in the latter half of the twentieth century engaged with wider social change (Moss 2010). Echoing Ansell (2009), this research arose from some frustration with approaches that focused on the micro relations of childhood but gave insufficient attention to the impact of wider social change. Watson (2006: 242) points out that research has been hampered by 'public–private' distinctions that have confined understanding of childhood to gendered and 'highly kindered' spaces. The aim was to broaden the canvas, including and moving beyond the spaces of home, school, day care, play and so forth. A number of aspects of social change were considered, including war, migration, religion, employment, technology and consumption. The 'war stories' discussed in this chapter are based on adult memories of significant social events in childhood (Moss 2013). Some respondents had direct experiences of war, but all shared stories related to wars of which they had had no first-hand experience. Here, their direct and indirect experiences are drawn together in order to explore intergenerational influences. The wars referred to include the Second World War, 1939–1945; two Indo-Pakistani wars, 1965 and 1971; war in Northern Ireland, 1968–1998; the Nigerian Civil War, 1967–1970; and conflict related to apartheid in South Africa.

Ideas from literature studies are drawn on to explore the sharing of war memories with children. These include a study of war writing for children (Goodenough and Immel 2008) where contributors discuss propaganda stories, stories that actively encourage children to think through the moral issues, and the 'minutely circumstantial realism', horror and humour associated with war and childhood (Heberle 2008: 135). The sociologies of social memory, time and space also provide different overlapping perspectives related to shared social memories. Halbwachs (1992 [1925]), a student of Durkheim, argued that collective memories underpin societal organization and that experience can only be remembered through the social frameworks that exist at the time of remembering. Hobsbawm (1983) highlighted the bias in the production of social memory, including, for example, the way powerful interests are reinforced and much is selectively forgotten. Ideas about children and families influence both the selection and the evaluation of memory. Thus memories of childhood involve dialogues between past/present and adult/child perspectives, 'Through memory, a new relation between the past and present is created' (Widerberg 2011: 330; Brannen 2004).

Particular theoretical perspectives on time enrich understanding of how former wars continue to inform everyday social relations, for example, drawing on biological concepts, Adam (1990: 66) conceptualizes human beings as 'practising centres of action rather than perpetrators of fixed behaviour'. Children must negotiate events from different time periods and distant places through their engagement with various forms of social memory, including those related to war. In a similar vein, the human geographer, Massey (2005: 119) argues that, 'Arriving in a new place means joining up with, somehow linking into, the collection of interwoven stories of which that place is made'. The passage of time and children's geographical movement inevitably involve them in engagement with different, sometimes conflicting war stories from different times and places.

These ideas are complex, overlapping and sometimes difficult to work with. Nevertheless they provide research opportunities to look both 'within and beyond' children's local environments and intimate relationships in order to reveal more of the impact of war on children's lives (Ansell 2009: 204). The former times of war should not be simply considered background to children's experience. Selective memories of war 'make their way into children's spaces', becoming part of children's experience (ibid.). As well as generating continuities with the past, this intergenerational transmission involves silences, distortions and discontinuities. In this chapter I argue for the importance of locating indirect experiences of war alongside personal experience to fully understand children's social worlds. This involves widening the lens on childhood to explore the intergenerational narratives concerning war that children engage with and in particular the uses of silence and humour within these narratives.

Researching memories of war and childhood

The research features 16 respondents born between 1950 and 1978; their childhood spanned the second half of the twentieth century. Potential respondents were initially approached from within the researcher's professional networks, using a snowball sample. These were people who worked in fields related to childhood and youth who might be willing to share some quite personal childhood experiences in order to deepen understanding of children and social change. The sample was developed to enable a rich tapestry of childhood experiences to be explored, while also attending to sensitivities to do with the subject matter. This involved ensuring there were respondents who as children had different experiences of social class, ethnicity, faith, disability, sexual orientation and gender. It is not the aim of this research to generalize about the experiences of other children growing up in the same period; instead this study offers insight into some of the complex ways children experienced events such as war, and to explore their engagement with memories of war passed through generations.

Three research tools were used. The first was a semi-structured interview where respondents shared memories of wider social change and childhood, including war. Questions were not typical of those asked about childhood and stretched the parameters for remembering, the assumption being that all children have experience, whether or not they are directly caught up in events. Respondents were told they could share first- or second-hand experience that stood out in memory. In a second interview, respondents shared recollections of daily life, including play, learning, journeys, relationships, domestic routines and so forth. The third research tool was a questionnaire where they were asked about social position, social transitions and heritage (going back three generations). These three tools facilitated triangulation and issues related to war and conflict were significant in all of them but only explicitly asked for in the first.

When analysing memories in research it is important to explore the role of emotion, the uses of humour, the selectivity of memory and the links between personal and social memory. Personal memories are drawn from clusters of events and there are aspects that are faulty. Events associated with strong emotion may be recalled easily – or may be buried (Misztal 2003). The emotion related to war may involve dark humour. This may be a distancing mechanism, for example, turning 'violence into slapstick' (Higonnet 2008: 116). However, humour is more than protective; Heberle (2008) argues it involves surrealism, absurdity, parody and irony which particularly arise when juxtaposing ideas and experience from the spheres of childhood and war. Swart (2010) argues that humour helps in the management of incongruous experience and the formation of group identity. The selectivity in all forms of memory means there are silences to consider. In the case of communications between Kinder Transport children and their own children there was considerable silence, 'Both parents and children had feared that such discussion might set off feelings in the other that would be unbearable, so they avoided the subject' (Harris 2008: 228). Family memory is transmitted orally and also contained in archives, photographs, film, letters and other memorabilia which anchor events and people in particular ways (Misztal 2003). Sometimes it is hard for research respondents to disentangle their own personal from familial memories. Such memories are 'not only of a series of individual images of the past. They are at the same time models, examples, and elements of teaching' (Halbwachs 1992 [1925]: 59). The links between personal, familial and social memory are complex, reflecting the complex relationship between everyday human experience, intergenerational relations and wider social change. Erics-son and Simonsen (2008) explore the role of national memory in relation to the experiences of children born to Norwegian mothers and German occupier fathers in the Second World War. They discuss national shame, efforts to forget, social and personal silences and the impact on children. They argue that anxiety and depression are transmitted through silences, 'Silence speaks of the unspeakable and carries powerful emotions ... insecurity, inferiority, guilt, and above all shame' (ibid.: 398). Selective memories of war are institu-tionalized in many forms within children's organizations, such as schools. Preparation for war informs the formal transitions to adulthood in most soci-eties and a thread within war writing involves the explicit mobilization of chil-dren's allegiance. Higonnet (2008: 121) argues that this is achieved by constructing 'heroic models of national childhood' and 'social archetypes of vulnerable children'.

I now move on to discuss the research findings, drawing together direct mem-ories of war as well as those transmitted through the generations. I consider the relationship between different forms of memory and the complex tapestry of war-related experience that is generated in children's lives.

Remembering war, childhood and related stories

This part of the chapter examines respondents' first- and second-hand memories of war and childhood, considering how events are altered in the memory with the passage of time. Selective silences are examined, as well as the uses of emotion, in particular humour. Wider social memories of war may be highly selective and emotive. This includes 'national' memories that are institutional-ized in everyday practices related to childhood, such as holidays and schooling. Selectivity and humour inform the transmission of familial war memories, including the remembered roles of mothers and fathers in war. Silence and emotion influence the ways that children engage with different threads of war memory in their everyday lives.

Only three of the 16 respondents remember direct experience of war. Rehana twice experienced the shelling of her home in Pakistan; James grew up during war in Northern Ireland and Richard lived in apartheid South Africa between the ages of three and 16. Although Apara experienced some of the Nigerian civil war (1967–1970) whilst a baby; she had not yet learned to talk and most of her 'memories' come from her mother. However, all 16 respondents, when asked about war and childhood, spoke with feeling about indirect experiences including family experiences that had been shared with them and what they had learned through media and school.

Nationalism, childhood and war

Wider social memories of war and childhood inform children's everyday lives. Nationalist stories that mobilize children intensify as war approaches (Johnson 2008). In the face of war, the enemy is constructed as evil. During the Indo-Pakistani wars 1965 and 1971, Rehana said there were, 'National songs, chanting, slogans against the enemy force.... "They are bad, because they are the enemy, they'll come and take us away or kill us."' For James, in Northern Ireland in the midst of conflict, there were many long-established militarized practices, 'Uniforms ... allegiances to things like the boys brigade ... badges and emblems associated with certain things, which had enormous potency ... signalled a particular allegiance to one side of the divide.' James had to tread carefully; when you met strangers, you needed to 'be aware of the cues' that signalled which half of the divide they were on, for your own safety. Leonard (2007: 492) points out that schooling in Northern Ireland at this time involved 'fictitious images', presenting 'Irish history through the lens of British colonialism'.

The impact of struggles related to Ireland informed the experience of many generations living in the UK and Ireland. George, whose family had been settled in England for three generations, recounts a repeated tale about his Irish great-grandmother. She, 'used to have to be boarded into her room ... would go out and do violence against the Catholics ... emptied the chamber pot over the crowd when they came past'. Such family stories about his departed kin entertained him but also introduced ideas about where he socially belonged, 'the family group is accustomed to retrieving or reconstructing all its other memories following a logic of its own' (Halbwachs 1992 [1925]: 52).

Prior to the Second World War, Nazi texts used ABCs, rhymes and illustrations to teach fascism, racial hierarchy and hatred of Jews to children (Johnson 2008). During and after the war, in the UK, Germans were portrayed with contempt (O'Sullivan 2008). One respondent, Claudia, had to deal with the social hierarchies related to the Second World War, even though she was born in England nine years after it ended. Her father was Romanian and her mother German. They had come to England as prisoners of war. She remembered hearing stories of an uncle who had been sent to Siberia and of soldiers in her father's unit who had been shot by the Russian army, 'There was a very negative vibe around what communism was all about and how cruel it was and unjust'. Claudia's mother also shared ideas about 'race' and country heritage that she had been taught many years before in Nazi Germany, for example, she considered Claudia's father's nationality as inferior, 'Coming from Romania and being lazy'. Claudia's understanding of war was also informed by silences, 'sort of indirect things round the fringes of life, you didn't have to hear.' She remembered as a very little girl, 'Being conscious of being different, and people calling me names'. Later, 'I was certainly teased about being German, and having lost the war'. At home, 'there were anxieties about things ... concerns about family'. Claudia's parents clearly sought to spare her the brutal details of war, 'You're hearing from under the table, so to speak. It's hidden by the table cloth'. She had to tread carefully between the conflicting versions of war in home and school. The silences and inconsistencies in the different threads of social memory transmitted family anxiety, some shame and exclusion. Claudia grew to avoid discussion which she found upsetting.

Long social memories of war transmitted to children involve coupling child and adult worlds, they caricature events in very selective ways and particular archetypes inform their construction. These include the innocence of the child in the face of war, the child as brave martyr, rather than combatant, and the call to arms to protect the child (Higonnet 2008). In South Africa, during apartheid, 'In school text books, the themes of White–Black conflict and Black barbarism were pervasive' (Dawes and Finchilescu 2002: 148). From age 11, one respondent, Richard, who was white, of English heritage and growing up in South Africa, had to join the cadets at school and wear a uniform, 'We were taught to shoot guns.... The phantom enemy that we were supposed to be shooting at were always what were called the "Kafers" which was the derogatory word for Black people'. The mobilization of white pro-apartheid forces involved the celebration of stories of Boer child martyrs from over a century before. Richard remembered two such stories that informed his schooling during apartheid:

> Rachel De Beer ... this little girl ... found herself with a younger brother ... this is supposed to be a true story, and it was freezing cold, and to protect her little brother, she shed most of her clothes, wrapped him up and then put her own body over him, and of course she perished, but he lived ... Dirky Uys ... he sacrificed himself to the Zulus ... there was a strong theme of sacrifice ... that was a very powerful message.

These stories conceal the colonization of South Africa by North Europeans, the violence against the black indigenous communities and their resistance. Here we can see how children caught up in war must navigate both direct and intergenerationally transmitted versions of war. In this case, these national, selective memories transmit constructs of heroic and vulnerable Boer children to new generations of white children in order to win their allegiance to the apartheid regime (Higonnet 2008). However, Richard had alternative versions of war to draw on. He remembered watching a film at school when he was six or seven, about a white family living in the countryside and being attacked by some black people, and the teacher saying at the end of the film 'That film was obviously made by a bunch of Nationalists'. At secondary school, there were also a few teachers who would openly share similar views, 'Not many, but some would'. At home there was criticism of apartheid, and his sister 'started to develop friendships with quite a radical whites who were really working and putting their lives in danger, to change the system.' Richard had many different stories about apartheid to 'think with' and question the dominant narrative (Tatar 2008: 242, citing Harries 2000). The very juxtaposition of competing stories involved thinking things through carefully, 'this was only one version of the truth … you have to question what's put in front of you' (Richard).

Fathers and mothers at war

The selectivity of memory in relation to combat is very evident in familial memories about fathers transmitted to children. These lacked detail and sometimes felt contradictory. The very juxtaposition of fathering and war seemed to generate particular silences. MacCallum (2011: 137) points out that silences concerning combat are both official and private. During the Second World War, men were 'discouraged from discussing wartime experiences and wives were advised not to probe'. Martin's father was a military doctor in the Second World War. Martin said, 'I don't think he had a particularly harsh war'. However later in the interview he seemed to contradict this and said that his father, 'was actually quite lucky to survive because the boat he was travelling on back from Africa was torpedoed.' Julie's stories of her father contained a similar ambiguity. Julie said, 'My father was away for seven years during the war and he had a really good war. Apparently he had a really nice time.' But later in the interview she remembered being poorly as a child and her father comforting her, 'Stroking my brow … he was saying that when he was really ill with this fever [in south-east Asia] … thought he was going to die'. These intergenerational memories convey the father's 'good' war as well as their evident near-death experiences. However Cottee (2011) points out that pleasure and pain are not binary opposites in the experience of combat, but part of a spectrum of experience. A selectively edited account of combat is intergenerationally transmitted. Although there is some truth in the narrative of father's good war, in war memories shared with children the archetype of masculinity and the 'good war' is reinforced. It is no coincidence that this informs campaigns to recruit young people into armies. Memories of fathers' roles in war were therefore quite tentative but also significant. Widerberg (2011: 334) argues in relation to her own research into remembering fathers, it is important to re-read stories, 'to grasp this father figure, I was particularly struck by the presence of his absence'.

The juxtaposition of mothering and war generates another kind of war story, also selectively edited but containing dark humour. Official propaganda stories during the Second World War in the UK emphasized women's ordinary suffering, their valour in supporting brave men, their important role in the work force and their creativity in keeping the children safe, fed and clothed throughout rationing (British Pathé 1940). Family memories related to women and war contain similar themes, but with a different slant and some grim humour. They convey some of the realities of war, the social expectations of women in relation to children and maternal guilt. During the Nigerian civil war, when Apara was a baby, 'The only thing that would send me to sleep was being driven round in [mum's] car … despite … a curfew and you'd be shot on sight, she would just put me in the car and drive round Lagos.' Cathy was told that during the Blitz, 'My Gran couldn't be bothered to get [the children] out of bed! So they were just left there ... always there was the noise, and they could hear it coming down, but it was fatalistic by then, "Oh well, whatever"'. Pamela was told that during the Blitz her mother was evacuated and her grandmother, 'made the decision to bring all the children back.... If they were going to get bombed, then they might as well get bombed at home'. These intergenerational memories convey a hint of criticism at maternal negligence, mothers' carelessness in war. Segal (2008) discusses the very real difficulties of protecting children in life-threatening situations and these intergenerational memories also convey fatigue and fatalism. They are recounted with humour and caricatures of mothers' 'extreme' behaviour. When the stories are transmitted outside of the contexts of war, the (self-)perceived failures of mothers are perhaps reinforced.

War in children's everyday lives

Memories of war and childhood, whether first- or second-hand, involve the juxtaposition of the paraphernalia of childhood with the paraphernalia of war. Lara recalled a three-day trip to Northern Ireland to Catholic relatives where she mistook army tanks for ice cream vans, 'You didn't get ice cream vans hanging around, because we saw a tank go past, and I thought it was an ice cream van'. She draws on the familiar furniture of childhood which provided benchmarks for her to remember and understand events. This incongruous juxtaposition generates humour and pathos. Rehana's memory of being shelled involves a similar coupling of childhood and war, but generates no humour and is more intense. The Indo-Pakistani War of 1965 lasted five weeks and she was four or five:

> Dad had an army background ... took over precaution measures ... taken into ... the store room ... eat by the candle ... Dad would say, 'It's coming' ... Grandma, my Dad ... my second Mum ... we'd bury our heads in their laps and we'd hear this big cannon going off and then there'd be silence ... I think my Dad could see a reflection of light before it happened.... He'd say 'It's coming' and then the burying of the head, and then we'd hear this big cannon.... That continued all night ... I just remember being petrified.

This memory conveys her fear at the time and the detailed domestic scene – represented by the protective roles of adult relatives (saving the children's hearing). Lara's and Rehana's direct experiences, albeit selectively remembered and reconstructed in stories, give some visibility to children's social worlds at the time and in the place remembered. They provide a lens on the intimate relations of childhood (the layout of homes, the relationships, the food and treats) and the effects of war. This is the minute realism of childhood and war that Heberle (2008) refers to.

In addition to such direct experiences, the intergenerational transmission of war stories means that earlier wars continue to inform children's everyday lives.

Children hear many repeated selective tales: for example, Pamela was born ten years after the Second World War ended, yet vividly 'remembers' her mother's story of sheltering during the Blitz in London,

> An L-shaped Anderson type shelter ... teachers standing in the corner ... one class down one side and another class down the other... huge wide piping with duck boarding ... knitting ... spelling bees ... mental arithmetic.... Dropping things down through the duck boarding and not being able to get them back 'cause it was all wet and muddy and the darkness and the smell.

This indirect memory contains convincing descriptive details and retains the multisensory/emotional quality of a direct experience. Some of the emotional significance has been transmitted through the generations, albeit in an edited version of events. In the analysis of data, it became increasingly apparent that many such intergenerational war stories contained humour, as did the tales of mothers in war situations discussed previously. Cathy repeats a family story from the Second World War about a gas mask, generated nearly 20 years before her birth,

> He came round ... hammered on the door and, I don't know how many times I heard this tale. My Gran opened the front door, and saw him standing there with his gas mask and everything and said something like 'Jesus, Joseph and Mary, they've landed!'

George, born 13 years after the war ended, repeats family stories about a nose and a mouse,

> [Mum] would go down to the munitions factory.... She used to feed the mouse that used to come out through the grill.... After the munitions factory, she'd have to walk back in blackout and there was a phone box at the start of the alleyway and she walked into the phone box and broke her nose!

The humour arises from the incongruity of war in the 'ordinary' domestic, street and work settings (Swart 2010). The juxtaposition of gas masks, blackouts, care for a mouse and a grandmother's curses involves the surrealism, absurdity, parody and irony that Heberle (2008) refers to. Not only is humour a means of enduring war, the interruptions of war challenge the normative social relations of peace, turning things on their head in an almost burlesque way. The humour, generated in the times of war in order to manage and understand difficult events, has created playful war stories, shareable with children across the generations. Such stories link children to particular kin, amuse them, soften war a little and involve family self-parody. They also inform children of some of what war involves. The chaos, fear and other complex emotions of the times of war are less directly visible in these transmitted memories but these stories have now become important parts of the family archives, to be transmitted on in the future; as Cathy says, 'I don't know how many times I heard this tale'.

Richard and Claudia (discussed previously) provide good examples of the continuing influence of former conflict in children's everyday lives, whether they are caught up in war themselves or not. In Richard's case the apartheid struggle and associated wars continued through his childhood, 'We were aware of things like the changing mood, things like the murder of Biko'. The archetypal stories of Boer martyr children from an earlier conflict were juxtaposed with many conflicting narratives, 'that was something else we used to do as a family, talk.' For Claudia, the Second World War was over long before her birth. Nevertheless, she had to navigate contradictory war stories at home and school. This influenced her feelings about her own family heritage, 'I suppose deep down, I find it very difficult to believe that people didn't know what was going on ... I do find it hard, and I don't want to know what they think really.'

The repetition of selective, humorous and sad war tales from earlier times and distant places has a significant influence on the everyday lives of children, 'the present generation becomes conscious of itself in counterposing its present to its own constructed past' (Coser in Halbwachs 1992 [1925]: 24). Such tales teach them about other childhoods, tie them to particular places and communities, and position them at a distance from others. They challenge and reinforce expectations about gender and family role. As is the case with written war stories for children (Myers 2008), stories shared with children in family and community may soften the brutalities of war, may help to prepare children for future war and may accentuate, sustain and contest conflict. These stories are threads of social, familial and personal memory from different times, people and places. They contain significant silences and strong emotion and their juxtaposition raises moral questions that children must navigate.

Conclusion

The chapter has drawn on research based on memories to explore war and childhood. It has drawn together direct and indirect experiences, focusing in particular on the intergenerational transmission of memory. The priority for research must be the experience of those children directly caught up in war. However, it is also important to examine additional layers of experience generated by wars in previous generations. This informs the lives of children, whether they are directly caught up or not. Analysis of memories deepens understanding of social practices related to war and childhood at the time remembered and the time of remembering. Consideration of the connections between personal, familial and social memory demonstrates how the intergenerational transmission of memory may lock children, such as James, into particular solidarities, 'You're either one or the other ... and there's a security in that'. The transmission of memory may reproduce war-related hierarchies, but also contest them. As they navigate competing perspectives on war, participants such as Claudia and Richard question and relate what they learn to their everyday childhood. Events at the time of remembering shape the selection of memory. Where conflicts continue over many generations, selective social memory serves partisan interests. The story of the Boer martyr children, which Richard shared, 'is supposed to be a true story'. It was transmitted at a time of active civil opposition to apartheid, building the allegiances of white children, whilst concealing the deaths of black people whose homelands had been invaded.

Intergenerational war stories contain elements of silence, fear, sadness, humour and horror. Such stories may express the emotional complexities of war through caricature and parody. The images of an alien in a gas mask, a mother making munitions and feeding a mouse, and a nose broken by a phone box in the blackout are examples. Humour is a means of enduring war and is generated directly from the incongruous aspects of war and childhood. The passage of time may also change some of the horror to humour, particularly in stories shared with children. The uses of humour and caricature and the ways children navigate different stories of war are both fruitful areas of analysis.

Personal, familial and wider social memories all have multisensory elements and contain intimate details from childhood. Rehana's personal memory of being shelled and burying her head in someone's lap; Pamela's familial memory of her mother 'dropping things down through the duck boarding' in the Anderson shelter and Richard's social memory of two Boer children in the freezing cold. The passage of time and the power of selectivity alter events and the silences are revealing. Attempts by adults to conceal what is considered inappropriate for children may be thwarted by children such as Claudia. Silences generate selective distortion related to fathering, mothering, childhood, ethnicity, nationality and place. Some of the memories discussed here conveyed the respondents struggle to remember accurately but also perhaps their confusions from childhood. Their expectations of parents sit uneasily in the memory when violence may be required behaviour (of fathers in particular) or when there is a failure to protect children (by mothers in particular). The caricatures of the slightly negligent mothers and fathers' 'good' wars stand out.

The research discussed in this chapter has been strengthened by themes from the study of war writing for children and by ideas from the sociologies of social memory, space and time. The overlap between personal, familial and social memory generates research difficulties, for example, in relation to childhood, it is difficult to disentangle different forms of memory, past/present and adult/child perspectives. These challenges become acute for forensic research which requires accurate evidence of events (for example, for litigation). Nevertheless, such challenges also generate research opportunities. Richard's memories of schooling during apartheid reveal allegedly 'true' war stories from a century earlier that were transmitted to white children in order to harness their emotions to the apartheid 'cause'. Examining the overlap between personal and social memory it is possible to see how events are selectively altered with the passage of time and how collective memory becomes selectively institutionalized in different childhood settings.

Ansell (2009: 198) argues that children's geographers have difficulties in, 'looking beyond children's immediate environments to things that affect children's lives'. Research into the intergenerational transmission of memory is one way forward. In this approach, former wars are not considered merely background to understanding childhood; children's everyday engagement involves navigating multiple time spans and this includes many different threads of memory and experience related to war (Adam 1990). Rather than viewing the overlap between personal, family and social memory as an impediment to research, MacCallum (2011: 131, citing Thomson 2010) argues that 'the process of remembering could be a *key* to exploring the personal or subjective meanings of lived experience.' In the feminist memory work developed by Haug (1992) and later, Widerberg (2011: 331) to explore the 'doings of gender', 'the aim was not to look for personal explanations but rather to look for social explanations (social relations and patterns) of what the stories could teach us'. Following from this, it is argued here, that the analysis of social as well as personal memories should add to understanding of the 'doings' of childhood.

References

Adam, B. (1990) *Time and Social Theory*, Cambridge: Polity.

Ansell, N. (2009) 'Childhood and the politics of scale: Descaling children's geographies?' *Progress in Human Geography*, 33(2): 190–209.

Boyden, J. and Berry, J. de (2004) *Children and Youth on the Front Line: Ethnography, Armed Conflict and Displacement*, Oxford: Berghahn Books.

Brannen, J. (2004) 'Childhoods across the generations storied from women in four-generation English families', *Childhood*, 11(4): 409–428.

British Pathé (1940) *Women and the War*, online, available at: www.britishpathe.com/video/women-and-the-war (accessed 13 May 2013).

Cottee, S. (2011) 'Fear, boredom and joy: Sebastian Junger's piercing phenomenology of war', *Studies in Conflict and Terrorism*, 34(5): 439–459.

Dawes, A. and Finchilescu, G. (2002) 'What's changed? The racial orientations of South African adolescents during rapid political change', *Childhood*, 9(2): 147–165.

Ericsson, K. and Simonsen, E. (2008) 'On the border: The contested children of the Second World War', *Childhood*, 15(3): 397–414.

Goodenough, E. and Immel, A. (eds) (2008) *Under Fire: Childhood in the Shadow of War*, Detroit, MI: Wayne State University Press.

Halbwachs, M. (1992 [1925]) *On Collective Memory*, Chicago, IL: University of Chicago Press.

Harris, M.J. (2008) 'Breaking the cycle', in E. Goodenough and A. Immel (eds) *Under Fire: Childhood in the Shadow of War*, Detroit, MI: Wayne State University Press, pp. 227–230.

Haug, F. (1992) *Beyond Female Masochism: Memory-work and Politics*, London: Verso.

Heberle, M. (2008) 'The shadow of war: Tolkien, trauma, childhood, fantasy', in E. Goodenough and A. Immel (eds) *Under Fire: Childhood in the Shadow of War*, Detroit, MI: Wayne State University Press, pp. 129–142.

Higonnet, M.R. (2008) 'Picturing trauma in the Great War', in E. Goodenough and A. Immel (eds) *Under Fire: Childhood in the Shadow of War*, Detroit, MI: Wayne State University Press, pp. 115–128.

Hobsbawm, E. (1983) 'Introduction: Inventing traditions', in E. Hobsbawm and T.

Ranger (eds) *The Invention of Tradition*, Cambridge: Cambridge University Press, pp. 1–14.

Johnson, E. (2008) 'Under ideological fire: Illustrated wartime propaganda for children', in E. Goodenough and A. Immel (eds) *Under Fire: Childhood in the Shadow of War*, Detroit, MI: Wayne State University Press, pp. 59–76.

Leonard, M. (2007) 'Children's citizenship education in politically sensitive societies', *Childhood*, 14(4): 487–503.

MacCallum, A. (2011) 'The Australian children of servicemen from World War Two: An intergenerational study', *The Intergenerational Journal of Interdisciplinary Studies*, 5(9): 129–143.

Massey, D. (2005) *For Space*, London: Sage.

Misztal, B.A. (2003) *Theories of Social Remembering*, Maidenhead: Open University Press.

Moss, D. (2010) 'Memory, space and time: Researching children's lives', *Childhood*, 17(4): 530–544.

Moss, D. (2013) *Children and Social Change: Memories of Diverse Childhoods*, London: Bloomsbury Academic.

Myers, M. (2008) 'Storying war: An overview', in E. Goodenough and A. Immel (eds) *Under Fire: Childhood in the Shadow of War*, Detroit, MI: Wayne State University Press, pp. 19–28.

O'Sullivan, E. (2008) 'Shifting images: Germans in post war British children's fiction', in E. Goodenough and A. Immel (eds) *Under Fire: Childhood in the Shadow of War*, Detroit, MI: Wayne State University Press, pp. 77–89.

Segal, L. (2008) 'Baby terrors', in E. Goodenough and A. Immel (eds) *Under Fire: Childhood in the Shadow of War*, Detroit, MI: Wayne State University Press, pp. 93–96.

Swart, S. (2010) 'The terrible laughter of the Afrikaner: Towards a social history of humour', *Journal of Social History*, 42(4): 861–887.

Tatar, M. (2008) 'Appointed journeys: Growing up with war stories', in E. Goodenough and A. Immel (eds) *Under Fire: Childhood in the Shadow of War*, Detroit, MI: Wayne State University Press, pp. 237–250.

Watson, A.M.S. (2006) 'Children and international relations: A new site of knowledge', *Review of International Studies*, 32(2): 237–250.

Watson, A.M.S. (2008) 'Can there be a "kindered" peace?' *Ethics and International Affairs*, 22(1): 35–42.

Widerberg, K. (2011) 'Memory work: Exploring family life and expanding the scope of family research', *Journal of Comparative Family Studies*, 42(3): 329–337.

Part III

The negotiation of values, beliefs and politics

10 Intergenerational recognition as political practice

Kirsi Pauliina Kallio

Introduction

This chapter introduces the concept of intergenerational recognition in the context of childhood and youth, from a relational spatial political perspective. It is approached as a dynamic social practice and a force constitutive of political agency, as unfolding in children and young people's everyday lives. Drawing from an ethnographic analysis, intergenerational recognition is suggested as a useful concept for the analysis of political communities where living together forms the major challenge as well as the promise of change.

Contextual recognition

'Recognition' is a philosophically grounded, open-ended social theoretical concept that has become established in interdisciplinary debates over the past couple of decades. On the whole, recognition refers to an ongoing social process whereby people constitute their own and other people's identities by meeting each other and seeking 'intersubjective regard' (Fraser and Honneth 2003: 1). It works on different levels of awareness and intent, gets enacted though articulations, acts, attitudes and embodied expressions, and may take individual and collective forms. Another general ideal, involving a normative undertone, is that human life can be made more ethical through recognition, while there is also always the potential for damaging misrecognitions. This ethical note has come to imply that all practices supporting recognition or leading toward misrecognition are politically consequential (Deranty and Renault 2007).

The idea of recognition draws from the thought of both Hegel (1977 [1807]) and Mead (1934) which have been developed in various theoretical directions by Charles Taylor (1994), Axel Honneth (1995) and Nancy Fraser (2000). These ideas have been taken yet further by recent scholarship that has given rise to a massive body of work on political identities (e.g. Markell 2003; Deranty and Renault 2007; Warnke 2007; McNay 2008; Staeheli 2008; Noble 2009; Häkli and Kallio 2014). Although these ideas have had influence in new disciplinary areas like childhood studies (e.g. Houston and Dolan 2008; Thomas 2012), theories of recognition have also been subject to critique. In particular, critics have called attention to the limits and dangers of identity-based recognition. The scholars cited above, amongst others, have noticed the contradictions embedded in identity categories that mark people according to certain characteristics (racial, gendered, ethnic, socio-economic, etc.). Even if useful in some policy contexts, categorical identity markers rarely fully encompass people's experiences of the self – their subjective ways of relating to the lived world.

To move beyond traditional identity categories, some critics have called for theorizations that acknowledge *contextuality* as a starting point of recognition. One such attempt is by Patchen Markell (2003), who suggests 'acknowledgement' as an alternative to identity-based recognition, emphasizing spatio-temporally situated sociability based on 'multiple and provisional subjectivities located in particular circuits of recognition which can only be captured in an ethnography of the encounter', as paraphrased by Noble (2009: 876; for contextual theorizations see also Deranty and Renault 2007; Warnke 2007; McNay 2008). These approaches are in line with the long tradition of feminist critique of universalism and the generalized other, with clear resonances with aspects of postcolonial and queer theories. They thus also suit well empirically grounded, critical spatial-theoretical research that appreciates contextuality (e.g. Robson 2004; Evans 2011).

Yet relatively few examples of this kind of work exist. Some social and political geographers have made recent contributions to these debates over the concept of recognition. These include Staeheli's (2008) introduction to the dynamics of recognition that involve state agents and institutions; Walker's (2009) study that reveals how people and places are associated in processes that produce environmental injustice through misrecognition; Noble's (2009) analysis identifying legitimacy, respect and competence as important matters to some young Sydney men's everyday existence, as compared to masculinity and ethnicity; Barnett's (2012) consideration of Honneth's idea on the relationship between experiences and articulations as illuminative of the phenomenologies of political life; and Koefoed and Simonsen's (2012) use of recognition as a counterpart to estrangement and identification in the development of political belonging and spatial attachments. In my own work, I have engaged with theories of recognition in tracing the idea of political subjectivity with a restored reading of Mead's thought, together with Jouni Häkli (Häkli and Kallio 2014).

While recognition theory is not widely deployed in spatial-theoretical research, notions of intergenerationality have been largely absent from theoretical debates over recognition. Whereas various intergenerational forms of intersubjective recognition have gained attention in empirical analyses (e.g. Houston and Dolan 2008; Reiter 2010; Thomas 2012), and the present discussions on intergenerationality implicitly involve ideas parallel to recognition (e.g. Tarrant 2010; Hopkins *et al.* 2011; Binnie and Klesse 2013), the concept of intergenerational recognition has not been developed in a theoretical sense. One important exception is by Somogy Varga (2011) who sets it forth in the study of biotechnological interventions related to prenatal birth control. He suggests intergenerational recognition as instructive in thinking about the ethics and morality of such medical technologies and the related institutional practices. As Honneth's student, Varga has identified the analytic power of intergenerational recognition. He sees the individual actors as partisans in a political game that spreads far past their everyday activities, and notices the process as influential beyond the place and time where the very acts take place.

In a similar spirit, I approach intergenerational recognition as a social dynamic at play in the mundane politics where people from different generational positions meet. Joining the contextual recognition scholars, I deem it a *dynamic social practice* which people enter from different spatio-temporal positions and a *constitutive force* that is particularly influential in the generation of youthful political subjects. In this interpretation, I disengage from the normative stance emphasized by many recognition theorists. I agree that recognition can be *used* as a technique that helps make life more ethical in certain circumstances, but I think that it also *unfolds* in everyday life in ways that are not easily estimable on the ethical continuum. This approach stems from my theoretical orientation, stressing political multiplicity and subjective experience.

In this chapter, I examine the practice and effects of intergenerational recognition in some everyday environments where Finnish children and young people lead their lives. I place the analysis in the experiential – or phenomenal – world of politics (cf. Barnett 2012; Simonsen 2013), which I enter through the biographical place-based narratives constituted in an ethnographic study. Even if tempting I will not consider intergenerational recognition in the formal politics of states and institutions (cf. Staeheli 2008). This is a practical framing, and not analytical, as I disengage from the perception where 'politics' and 'Politics' divide into distinct spheres of life (Kallio 2009, 2012; Häkli and Kallio 2014). The next section provides a short overview on the political theoretical orientation that guides my thought, hopefully informative enough to those not familiar with 'the political' in phenomenological terms. This understanding is imperative to the following empirical analysis of intergenerational recognition.

The world of plurality and its political subjects

Recent engagements with the work of Hannah Arendt (1958, 2005) have stressed the phenomenological orientation of her political theory (e.g. Barnett 2012; Dikeç 2013; Häkli and Kallio 2014). These analysts emphasize that Arendt's project was not about building ontological grounds for political thought but rather developing durable ideas concerning human political life. As an active debater on pressing societal issues, she surely did place her ideas in particular empirical settings, thus giving them seemingly context-specific forms (e.g. Arendt 1959; cf. Nakata 2008). Yet, as a political philosopher whose work built on the long tradition of democratic theory, with specific affection for Socrateian thought (Arendt 2005: 5–39), the purpose of her topical arguments was hardly to fix theorization to certain times or spatial organization of the polity. Therefore, Arendt's work contains opportune potential for theoretical rethinking and contextualization (Kallio 2009). What follows is a brief introduction to my reading of her political theory, informed by the current debates on relational space, politics and subjectivity (e.g. Cavarero 2002; Elden 2005; Marshall 2010; Vacchelli 2011; Barnett 2012; Secor 2012; Dikeç 2013).

In the Arendtian political world, every human being owns the capacity for political agency through action led by thought, from the moment of birth till the final demise. In this conceptual pair 'thought' stands for thoughtful human existence in general, including implicit understanding and internalized awareness, as well as explicit conceptions and articulate knowledge. 'Action', instead, refers to making appearances in a polity in which the given performance is significant. Here polity should not be understood in any particular scalar dimension. It is the name of the people who find themselves members of a shared world of plurality, namely the polis, which is a spatially finite yet not necessarily territorially bound political realm (Häkli and Kallio 2014). In addition to national citizens belonging to the same nation state, the polity may therefore consist of, for instance, a transnational network of people worried about the climate change, forming an 'issue public' (Kim 2012), or a fluid committed gaming community that comes together and recognizes each other in various internet-based role-playing platforms and game spaces (Ondrejka 2007). Instead of particular scalar or formal organization, definitive to the polity and the polis is hence 'living together'.

The specific term that Arendt uses for political agency is 'public speech'. This concept consists of two relative parts. 'Public' marks the space of appearance that actualizes as people meet and 'speak', and 'speech' refers to thoughtful action that may take any form and be presented by anyone as long as it is noticeable within the polity. The capacity and power of public speech, as compared to thinking devoid of action and action disengaged from thought, lies in the production of beginnings that keep the polity alive (Arendt 1958: 9). Should they cease to appear, the polis will become stagnant and the relations between people fixed, which denotes the end of freedom. This leads to an impoverished political life since to Arendt (2005: 108) 'the meaning of politics is freedom'.

Arendt states that 'Politics is based on the fact of human plurality, [and it] deals with the coexistence and association of *different* men [...] with a view to their *relative* equality and in contradistinction to their *relative* differences' (2005: 93, 96, emphases in original). 'Relative equality' refers to the idea of unique subjectivity as a basis of political autonomy (*who* we are); 'relative difference', instead, points to mutually identified categories through which dissimilarity may be articulately identified (*what* we are) (Arendt 1958: 181). 'Plurality is the condition of human action because we are all the same, that is, human, in such a way that nobody is ever the same as anyone else who lived, lives, or will live', Arendt (1958: 8) explains, thus conjoining subjectivity and plurality as the basis of living together.

These are the central concepts in Arendt's theory: thought and action, public speech, appearance and beginning, polity and polis, relative equality and difference, subject as who and what, plurality and living together, association and co-existence. To summarize, the polis is (1) a spatially finite yet scalarly open-ended experienced world; (2) formed by a polity consisting of people who are distinct both subjectively and categorically; (3) characterized by human plurality, which those longing to be a part of it must appreciate; (4) based on the principle of freedom that stands for fluid relations of equality and difference, prone to transformation and reorganization; and (5) maintained and developed through thoughtful action that is capable of creating beginnings. This phenomenological political frame informs the following analysis where intergenerational recognition is identified as political practice.

Intergenerational recognition in the world of relative equality and difference

In what follows, I provide two excerpts from the analysis where I have examined the practices and effects of intergenerational recognition in familial, institutional and public everyday settings.[1] It draws upon my ongoing research where administrative conceptions of youthful agency (formal politics of states and institutions) are contrasted and paralleled with children and young people's experiences and practices (mundane politics of everyday life) (see Kallio *et al.* 2015; Bäcklund *et al.* 2014). The ethnographic fieldwork was carried out in 2012 in Tampere and Helsinki, Finland, following principles of critical documentary ethnography (Ortner 2002). We worked with four classes of fifth graders (*n*=74) and three classes of ninth graders (*n*=55) across three different schools.[2]

Our data – collected in the form of maps, interviews, written stories and drawings – provide thick descriptions of experienced and practiced political realities of differently located and situated children and young people. I term these biographical place-based narratives. I have analysed them in an Arendtian spirit as portrayals of the 'life in polis', as conceived, experienced and practiced by our youthful participants. I present the results in the particular empirical settings of the study because, in line with Evans (2011: 344) and other critical ethnographers who have worked with families, I believe that both familial and extrafamilial intergenerationality are extremely context specific (some details that will help non-Finnish readers to understand the context are provided in the endnotes). Rather than make general arguments, my aim is to demonstrate how the dynamics of intergenerational recognition may be observed in the ebb and flow of everyday life, reflecting on what these dynamics can tell us about the mundane politics at play.

Reciprocal recognition in familial relations

Age is a contextual concept. It has not only been categorized differently over time and space, but the meanings of age differences fluctuate also within societies. As differently aged people encounter one another, their generational relations arrange distinctly, depending on the site. For instance, encounters between six- and 60-year-old persons might be understood through child–adult relations in institutional settings (e.g. a doctor's office), grand-generational positions in familial environments (e.g. religious services) and non-adult status in public commercial space (e.g. a ticket office). Respectively, as a practice of living together, intergenerational recognition takes multiple forms and modes, with variable levels of awareness, intensity, intentionality, moral judgement, normativity and reciprocity.

Moreover, even in familial contexts intergenerationality is not restricted to child–adult relations. This became evident as our research participants located their 'ordinary complexity of kinship' (Mason and Tipper 2008: 443), involving relations and encounters between parents, older and younger siblings, cousins, grandparents, aunts and uncles, neighbours and friend families, pets, and ex-spouses, step-children and other something-like-relatives. Whereas some only mentioned their family houses and cottages as a self-evident element of spatial belonging,[3] others used plenty of time and effort to express their specific familial relationships. There were labels like 'the best cousins', 'the dearest cottage', 'Sammy-granpa's ex-ladyfriend's place', to pinpoint locations of particularly affective attachments. Interestingly, such could exist also devoid of any visible permanence. For instance a ski resort that the family had visited a couple of times with a 'friend family' (a term I used to denote friendship that exists between families rather than just individuals) during the winter holiday[4] was afforded a specific label, thus transforming any cottage where this important familial life takes place into 'the cottage'.

Two affective ways of relating with differently aged people in familial life became strikingly visible: care and fun. Both belong to the sphere of socially based and organized everyday practice, underpinned by formal and informal institutions (Sevenhuijsen 1998: 21). Sometimes these two elements of intergenerationality interweave, like when taking care of younger family members by playing with them. Yet at other times caring is not fun and fun is not caring. A common case of 'unfun caring' relates to grandparents who need not be entertaining or even nice to be considered worthy of a visit or kind words. The following episode, presented by a fifth grader, makes a case in point.

> INTERVIEWEE: 'Cat-granny' [a nickname reflecting her ownership of a cat] has been a bit foolish and mum and dad can't bear visiting her anymore, and then she thought that they have abandoned her. Sometimes when me and my little sister go there she goes like 'now you are also abandoning me and your parents are running the show....' And then she trashes everyone for leaving her alone...
>
> INTERVIEWER: When you have been there, and there are these things, do you tell about them to your dad or...?
>
> INTERVIEWEE: Sure, and to mum as well, and they say that 'now it's enough, she's getting so mad that soon we will not let you there'.
>
> INTERVIEWER: Do you still want to go there?
>
> INTERVIEWEE: Yes I do sometimes [she continues by recounting an episode about how troubling these visits can be]
>
> INTERVIEWER: Who decides that you go there, overall?
>
> INTERVIEWEE: We can decide for ourselves whether we go or not.

This caring relation comes close to 'hidden care work' that has been identified as one form of intergenerationality where children's active agency plays a politically significant part (e.g. Robson 2004; Evans 2011). Yet ostensibly, visiting her grandmother is totally voluntary to this participant, to the extent that her parents are close to preventing it. She has the right to decide whether she goes to see her or not and, while visiting, may (and does) leave any moment she likes. The episode therefore hardly portrays a child obliged to caring work in her family but, rather, an attentive associate who wishes to acknowledge her grandmother's unconditional value as a family member, which is not dependent upon her characteristics or behaviour. Similar relations were brought up by many of our participants, involving differing levels of joy *about* their grandparents but great amounts of care *for* them. This subjectively grounded and socio-culturally embedded, responsible and altruistic practice is one form of youthful political agency. By caring for their grandparents, children and young people work to maintain, continue and repair their worlds (Tronto 1993: 61; Bartos 2013).

Taking quite a different shape, intergenerational recognition is practiced and political agency developed also by means of 'non-caring fun'. As one example, baking was mentioned by many of our participants as something they enjoyed doing with differently aged relatives and friends in familial settings. Baking was found rewarding for instance in terms of sociability, creativity, helping, learning, developing an appreciation for food and skill-performance. Other such seemingly ordinary activities that were brought up as 'special' on certain occasions include shopping, swimming, chatting, walking the dog, watching television, jogging and playing games. Without exception, these activities were related to clearly identifiable individuals. When talking about these relationships our participants could not often reason why a particular cousin, aunt, parent, friend family, or other important person or group of people, was so special. Instead, they simply stated that 'she is so fun to be with', 'I can be totally myself with her', 'I really like the way she is', and so on (see Bartos 2013 on friendship and environment; Marshall 2013 on beauty). In an Arendtian frame these expressions can be understood as proof of 'inarticulate humanity' – subjective uniqueness:

> The manifestation of who the speaker and doer unexchangeably is, though it is plainly visible, retains a curious intangibility that confounds all efforts toward unequivocal verbal expression. The moment we want to say *who* somebody is, our very vocabulary leads us astray into saying *what* he is; we get entangled in a description of qualities he necessarily shares with others like him; we begin to describe a type or a 'character' in the old meaning of the word, with the result that his specific uniqueness escapes us.
>
> (Arendt 1958: 181, emphasis in the original)

By emphasizing uniqueness rather than certain generally identifiable characteristics, our participants came to state appreciation of *who* and not *what* these people are, finding them*selves* similarly recognized. Gender and age, for instance, ceased to hold the same structuring power that they have in less intimate relationships. These processes of recognition thus differ notably from the caring ones where familial status is defining.

Relationships characterized by mutual enjoyment are prone to produce positive correct recognition because, to continue, they must be rewarding to the people involved. 'Having fun' does not just happen but needs to be *desired*, *accomplished* and *witnessed* by active agency each time. Simultaneously, these relationships are markedly sensitive to misrecognition, due to the high level of attentiveness, which became evident in narratives describing close relationships that had waned or come to an end. Also in these cases our participants found it hard to put a finger on 'what happened' – they just 'didn't get around to going there anymore', 'couldn't find the time to meet', 'didn't remember to call', or the like. Something that once was 'unique' had become 'common', rendering the *best* cousins *merely* cousins, among other relatives.

Therefore, contrary to what one might think, when enjoying themselves and having fun with each other people are engaged in the processes of contextual recognition in the most serious and demanding sense. It is these relationships, involving profound emotions of all kinds, that make the power of (mis)recognition most apparent. Whereas caring is about maintaining, continuing and repairing the world where one is embedded, mutual enjoyment entails creativity, sensitivity and courage by which to acknowledge the inarticulate uniqueness beyond the over-articulate identities.

Identifying plurality in the school polity

My second excerpt comes from children and young people's formal institutional lives. After the home, the most extensive framework for their daily living is the school. In Finland compulsory education incorporates basically all people between six and 16 years of age, including non-citizens, asylum seekers, and other non-permanent residents. In addition to the compulsory period, most children participate in public kindergarten programmes from an early age and continue their studies in upper secondary and vocational schools for three to four years. Even if seemingly voluntary, the choice of further schooling is practically limited to the choice of the school and not participation per se, as the second degree certificate is assumed by all educational institutes and is necessary for most occupations. Rather than nine years, the common scope of basic schooling thus comes closer to 13–19 years.

Taking that in all levels of education (including the nursery), teachers are trained in national university programmes, and the units follow a national pedagogical and curricular frame in the organization of their activities, the Finnish school system provides a somewhat consistent state-based growth environment for the first 20 years of life. As a national institution of socialization it has two major objectives. First, the *educative* intention strives for the advancement of thought and skill in substantial matters, aiming at high-level know-how that affords good starting points for further education and on-the-job learning.[5] Second, following a *pedagogical* rationale, the school aims at creating proper citizens with a shared moral mindset and respect for basic values – internalization of 'the Finnish way'. The associations between pupils and teachers[6] build respectively two-dimensionally, as student–educator and citizen–pedagogue relationships (cf. Pykett 2009a). Recognition by which the differently positioned actors identify and acknowledge each other forms an important element in the constitution and negotiation of these roles, through which individuals enter and enact the school polity.

The student–educator and citizen–pedagogue relationships are hard to distinguish in practice as the persons performing the roles are the same. Yet it is possible to separate them analytically from experience-based biographical narratives that echo the effects of (mis)recognition. I found the attitudes towards school presented by our participants' instructive on this account. They could be divided into four categories. The first group feels positively about the school: they like to study and learn, describe their teachers as good people and skilled educators, and enjoy the overall spirit of the school. The second group – nearly non-existent in our study – detests the school *tout court*, providing a contrary assessment to the previous. Also the third group basically 'hates the school' but when asked what they hate about it, it is only the *schoolwork* they dislike. The fourth group, instead, respects the school as a workplace but identifies defects in school *life*. It is the last two categories that are particularly illuminating for my purposes.

When school is disliked with respect to schoolwork, students and educators do not meet satisfyingly. The process of mutual recognition does not work at the educative level, for one reason or another, which arouses feelings of frustration, failure and lack of interest on the students' side, often leading to long-term lack of educational motivation, evasion of schoolwork and shirking classes.[7] When specifically asked during interviews, our participants would typically label one or two subjects as 'okay', but otherwise schoolwork was described as 'just too tiring' or 'boring'. Yet when asked about school life in general they had few complaints. They liked being at school together with their friends, even so much that they would sometimes stay at the yard after school, playing soccer or catch or just hanging around. The teachers were also portrayed as 'mostly okay', 'nice' and 'good people'. What these findings together imply is that this group of pupils is content with their position as *community members* in the school polity, and do not find themselves misrecognized in terms of morality, values or norms. Instead, they are uncomfortable about their position as *students* who are expected to fulfil certain curricular requirements that do not seem inviting or rewarding to them.

In contrast, the group of pupils who acknowledge the importance of studying but downgrade the school for other reasons are more conflicted about the realities of their 'lived citizenship' (Kallio *et al.* 2015, see also Lister 2007; Staeheli 2011). The negative attitude usually pertains to the staff 'who don't understand us', either as individuals or as a group, if not to bullying that also involves both the pupils and the teachers as members of the school community. One typical issue of misunderstanding concerned the 'spirit' of the school class, as portrayed by one of the ninth graders.

INTERVIEWEE: In my opinion, in our class we have trust, and a good community spirit [...] you can laugh about people and they don't take it like personally, but then you realize if something, *some thing* is like a sore spot with someone, then you stop talking about it.

INTERVIEWER: Okay. Like there's at play a kind of a mutual...?

INTERVIEWEE: Yeah, right – and then all the teachers go like 'you can't laugh at each other and talk however you like, someone may have their feelings hurt'. But we do realize if someone gets hurt and then...

INTERVIEWER: Have you talked about it in the class, with your classmates, that this is like 'our way'?

INTERVIEWEE: We have, we have, but the teachers then again don't understand it.

INTERVIEWER: You mean they have not internalized your way of being?

INTERVIEWEE: Yeah.

INTERVIEWER: Could you say that your class, as a whole, takes care of each other?

INTERVIEWEE: Yes. We are like such a, as there's only sixteen of us, we are like really close and such that ... like pretty surely we'll have low feelings now that high school finishes, we need to leave and our class breaks up completely.

When talking about their teachers, this interviewee does not discuss them as educators, but as people of the school. In these cases, the (dis-)connection hence lies between the differently positioned members of the school community, the *youthful citizens* and their *pedagogical authorities*. Along with a number of other respondents who identified problems with the social life of the school, this respondent cites encounters where teachers overlook or contradict their conceptions about the state of affairs in school, and judge their agency on the 'wrong grounds'. Such misrecognition is conducive of school life where peer cultural features become emphasized at the cost of the school community as a whole, with negative implications for schoolwork as well.

Together these findings confirm the well-known fact that, regardless of their acknowledged academic drive that has led to success in international comparisons (see IEA 2012; PISA 2012), negative attitudes towards school are common among youthful Finns. Yet my analysis suggests that the dislikes students articulate bear many connotations. When analysed in a phenomenological political frame, the message that these voices deliver does not appear merely negative. In fact, I interpret them as indications of a relatively lively mundane political life, allowing pupils active roles in the school polity. This interpretation fits with the recent development of the Finnish school where authoritarian forms of government have been substituted by deliberative ideals, leaving space also to agonistic forms of participation (Simola 2002). I will explain this interpretation using another brief example.

In cases where the school is a place where young people are happy as citizens but not as students, the pupils refuse to fill the role of 'entrepreneurial self': in other words, the student who drives for the best possible educational results in competition with his or her peers, constantly pushing their own limits ever higher (Peters 2001; Bragg 2007).[8] Rather, they navigate through their studies in ways that are merely 'satisfactory', investing in other things at school – for instance the peer life that involves different kinds of activities where skill, wit and sociability are needed. By thus rejecting the subject position proposed by the school institution and adopting one enabled by the school community, they complicate the *educators'* task to recognize them by the general label through which difference is articulately identified in school, as *students*.

This agency is subjectively grounded, set against the institutional order on a fundamental level, and has both micro- and macro-scale effects. The pupils following this path will not internalize the current neoliberal regime similarly to those who seek to please the institutional order by fulfilling the proposed 'studentship' to the full, which has obvious corollaries to their political development. Moreover, this proportion of students also proposes a fair challenge to the Finnish Government's (2011: 50) ambitious educative intention

> to make Finland the most competent nation in the world by 2020 [to] be ranked among the leading group of OECD countries in key comparisons of competencies of young people and adults, in lack of early school-leaving, and in the proportion of young people and other people of working age with a higher education degree.

In Rancière's terms, they are in active *dis-agreement* with this policy line, thus acting politically also in the present (Kallio 2012; Dikeç 2013).

In the latter case, the pupils find their own and their teachers' understandings of school life contradictory, which contests the citizen–pedagogue associations. In the current child and youth policy spirit, such expressions are typically interpreted as indicative of the lack of prospects for 'voice', calling for participatory interventions and structures.[9] Yet if understood as public speech in the school polity, the opposite seems the case. The struggle over the right to define 'good communal spirit' and the subsequent way of school life is an example *par excellence* of democratic practice where everyone may enter the space of appearance from their own stance. To be successful, these appearances need not lead to mutual consensus because, as agonistic and deliberative political theories have long professed, disagreement and dissent are important elements in a functioning democracy (e.g. Bäcklund and Mäntysalo 2010).

Thus understood, the lack of enjoyment expressed by the latter group can be read as indication of a vivid school polity where they have found their place as active partisans. This interpretation gets support from the school's overall progressive attitude.[10] Young people's agency is enabled and constricted by the plurality of the school community where equality and difference are relative, in the Arendtian sense, between pupils and teachers, but also amongst differently positioned pupils (cf. Pykett 2009b). This last notion points to another empirical finding that I did not raise above. In contrast to those views discussed, other pupils from this class expressed discomfort towards the prevailing peer cultural order, which they found cliquey and fraught. From their perspective, the introduced struggle gets yet another tone.

To summarize, with this analysis I have sought to point out that there are various processes of recognition simultaneously at play, which reveals the plurality of mundane political life. In school, differently aged actors do not practice their agency merely as teachers and pupils with singular existences, but as students and educators *associated through schoolwork*, and as citizens and pedagogues *related in school life*. On the basis of their particular relations (which are socially constituted also through peer, collegial and parental recognition), all members have the potential of both individual and collective agency, through which to partake in the school polity. This web of intersecting, cross-cutting and diverging relations provides each actor a particular stance for creating beginnings that are influential in the 'life of the polis', as well as constitutive to their own and other people's political development.

Living together by intergenerational recognition

This chapter has introduced intergenerational recognition as a dynamic practice and force that holds a central place in mundane political life. Approaching politics phenomenologically from youthful agents' perspectives, I have shown that children and young people are both objects and subjects of recognition, with potential to action led by thought and public speech yet always constricted by their particular stances in their everyday environs. By acknowledging, moulding and challenging the relations of equality and difference, the processes of intergenerational recognition establish and refigure political communities on a fundamental level. Moreover, these relationships and practices are specifically influential to youthful political development, as childhood and youth are formative moments for political agency (Kallio and Häkli 2011; Kallio 2014).

Empirically, I have introduced familial agencies related to 'care' and 'fun', and pupil–teacher relationships taking place in school, thus portraying different sides of mundane political life. Whereas the school is discussed as a field of intergenerational struggles and strains between its differently positioned members, the family appears as a network of attentive relationships. This portrayal does not suggest that these spheres of youthful living differ along those lines – relations of care and fun are formed in school as well, and family life is often fraught with power games as their members are contextually situated and constricted. Rather, I have sought to display that intergenerational recognition works in multiple directions and is enacted in numerous ways by both younger and older people. When approached from varying positions, the effects of these processes are hard to assess normatively, as they have different meanings to different people. Therefore, contextual in-depth empirical analysis is imperative for evading categorical explanations concerning intergenerationality, making sense of the meanings of these relations to those involved, and their effects in the given political reality.

I wish to conclude with an allegory where the occasions of intergenerational recognition are framed as *agreements*. In making these agreements the contracting parties consent to certain *parameters*, from the stances where they are positioned. These again place them in *relation* to each other. These relations may be oppressive, enabling or anything in between because – in the phenomenal political world – no relationship is neutral or essentially good or bad. This fundamental openness creates the prospect for *beginning* something new, for maintaining important matters, repairing ruptured relations, promoting institutional change, acknowledging personal features or just trying out a new approach in getting along. As such, intergenerational recognition appears as a *political practice* of living together in the world of plurality where fixed relations denote the end of freedom, and the space of appearance is available for anyone capable of public speech.

Notes

1 The study is based at the University of Tampere, Space and Political Agency Research Group (SPARG), Academy of Finland Centre of Excellence in Research on the Relational and Territorial Politics of Bordering, Identities and Transnationalization (RELATE), and funded by the Academy of Finland (projects SA134949, SA133521, SA258341). The theoretical work results from cooperation with Jouni Häkli and the ethnographic fieldwork was carried out with Elina Stenvall.
2 In Finland basic education takes nine years, beginning at the age of seven, and the actual school work is organized variably in units of different size and scope. Fifth graders are hence about 11–12 and ninth graders 15–16 years old.
3 In Finland nearly every family has a cottage of some kind that they can use and, because these are often passed on in the family, it is not rare to acquire the right to use various 'family cottages', which can be visited occasionally by a number of people. Marking down many cottages is hence not necessarily an indication of wealth. In terms of belonging cottages are comparable – if not more significant than – homes, as they comprise permanence, family history, sharing and closeness.
4 A national week-long school holiday in the middle of winter, traditionally 'the skiing holiday', when many families travel to ski resorts that sell and rent cottages and caravan parking space. These need not be luxurious holidays – there is a price range to fit everyone.
5 Studying is strongly state-promoted in Finland by free education, a broad network of public institutes and various subsidies. The level of educational attainment is thus high. Currently more than 70 per cent of the adult population has a vocational or higher education degree and most people build up their competence through vocational adult education in working life. The national aim is to further elevate this level to 90 per cent, by the early 2020s (see the Finnish Government Programme 2011).
6 I only discuss the relationships between children, youth and their teachers, while I realize that other people working at schools, and the parents, also partake in this intergenerational web.
7 Teachers, too, are frustrated in these relationships, feeling themselves unsuccessful educators and failing in their work. This came up in informal discussion with them but, as our fieldwork was limited to the study of youthful experiences, the analysis does not involve teacher perspectives.
8 The latest amendments to the school curricula emphasizing such studentship can be found at the Finnish National Board of Education webpages.
9 Following the international child's rights discourse and the related policy trends, Finnish policy strongly emphasizes the role of participation and democratic practice in all institutional settings (see Finnish Government Child and Youth Policy Programme 2012–2015, 2012).

10 This is evident at all levels, beginning from the school building, pedagogical and educative objectives, collegial collaboration, and further.

References

Arendt, H. (1958) *The Human Condition*, Chicago, IL: University of Chicago Press.

Arendt, H. (1959) 'Reflections on Little Rock', *Dissent*, 6(1): 45–56.

Arendt, H. (2005) *The Promise of Politics*, ed. J. Krohn, New York, NY: Schocken Books.

Bäcklund, P. and Mäntysalo, R. (2010) 'Agonism and institutional ambiguity: Ideas on democracy and the role of participation in the development of planning theory and practice – the case of Finland', *Planning Theory*, 9(4): 333–350.

Bäcklund, P., Kallio, K.P. and Häkli, J. (2014) 'Residents, customers or citizens? Tracing the idea of youthful participation in the context of administrative reforms in Finnish public administration', *Planning Theory and Practice*, 15(3).

Barnett, C. (2012) 'Situating the geographies of injustice in democratic theory', *Geoforum*, 43(4): 677–686.

Bartos, A.E. (2013) 'Friendship and environmental politics in childhood', *Space and Polity*, 17(1): 17–32.

Binnie, J. and Klesse, C. (2013) 'The politics of age, temporality and intergenerationality in transnational lesbian, gay, bisexual, transgender and queer activist networks', *Sociology*, 47(3): 580–595.

Bragg, S. (2007) '"Student voice" and governmentality: The production of enterprising subjects?' *Discourse: Studies in the Cultural Politics of Education*, 28(3): 343–358.

Cavarero, A. (2002) 'Politicizing theory', *Political Theory*, 30(4): 506–532.

Deranty, J.-P. and Renault, E. (2007) 'Politicizing Honneth's ethics of recognition', *Thesis Eleven*, 88(1): 92–111.

Dikeç, M. (2013) 'Beginners and equals: Political subjectivity in Arendt and Rancière', *Transactions of the Institute of British Geographers*, 38(1): 78–90.

Elden, S. (2005) 'The place of the polis: Political blindness in Judith Butler's Antigone's claim', *Theory & Event*, 8(1), DOI 10.1353/tae.2005.0008.

Evans, R. (2011) 'Young caregiving and HIV in the UK: Caring relationships and mobilities in African migrant families', *Population, Space and Place*, 17(4): 338–360.

Finnish Government Child and Youth Policy Programme 2012–2015 (2012) Publications of the Ministry of Education and Culture 2012:8, online, available at: www.minedu.fi/export/sites/default/OPM/Julkaisut/2012/liitteet/OKM8.pdf?lang=en (accessed 17 February 2014).

Finnish Government Programme (2011) online, available at: http://valtioneuvosto.fi/hallitus/hallitusohjelma/pdf/en334743.pdf (accessed 17 February 2014).

Fraser, N. (2000) 'Rethinking recognition', *New Left Review*, 3: 107–120.

Fraser, N. and Honneth, A. (2003) *Redistribution or Recognition? A Political–Philosophical Exchange*, London: Verso.

Häkli, J. and Kallio, K.P. (2014) 'Subject, action and polis: Theorizing political agency', *Progress in Human Geography*, 38(2): 181–200.

Hegel, G.W.F. (1977 [1807]) *Phenomenology of Spirit* [Original Phänomenologie des Geistes] trans. A.V. Miller, with analysis of the text and foreword by J.N. Findlay, Oxford: Oxford University Press.

Honneth, A. (1995) *The Struggle for Recognition: The Moral Grammar of Social Conflicts*, Cambridge: Polity Press.

Hopkins, P., Olson, E., Pain, R. and Vincett, G. (2011) 'Mapping intergenerationalities: The formation of youthful religiosities', *Transactions of the Institute of British Geographers*, 36(2): 314–327.

Houston, S. and Dolan, P. (2008) 'Conceptualizing child and family support: The contribution of Honneth's critical theory of recognition', *Children & Society*, 22(6): 458–469.

IEA (The International Association for the Evaluation of Educational Achievement) (2012) 'About IEA', online, available at: www.iea.nl/ (accessed 17 February 2014).

Kallio, K.P. (2009) 'Between social and political: Children as political selves', *Childhoods Today*, 3(2), online, available at: www.childhoodstoday.org/article.php?id=43.

Kallio, K.P. (2012) 'Political presence and politics of noise', *Space and Polity*, 16(3): 287–302.

Kallio, K.P. (2014) 'Rethinking spatial socialization as a dynamic and relational process of political becoming', *Global Studies of Childhood*, 4(3).

Kallio, K.P. and Häkli, J. (2011) 'Are there politics in childhood?' *Space and Polity*, 15(1): 21–34.

Kallio, K.P., Häkli, J. and Bäcklund, P. (2015) 'Lived citizenship as the locus of political agency in participatory policy', *Citizenship Studies*, 19(3–4).

Kim, Y.M. (2012) 'The shifting sands of citizenship: Toward a model of the citizenry in life politics', *The Annals of the American Academy of Political and Social Science*, 644(1): 147–158.

Koefoed, L. and Simonsen, K. (2012) '(Re)scaling identities: Embodied Others and alternative spaces of identification', *Ethnicities*, 12(5): 623–642.

Lister, R. (2007) 'Inclusive citizenship: Realizing the potential', *Citizenship Studies*, 11(1): 49–61.

Markell, P. (2003) *Bound by Recognition*, Princeton NJ: Princeton University Press.

Marshall, D.L. (2010) 'The polis and its analogues in the thought of Hannah Arendt', *Modern Intellectual History*, 7(1): 123–149.

Marshall, S. (2013) 'All the beautiful things: Trauma, aesthetics and the politics of Palestinian childhood', *Space and Polity*, 17(1): 53–75.

Mason, J. and Tipper, B. (2008) 'Being related: How children define and create kinship', *Childhood*, 15(4): 441–460.

McNay, L. (2008) *Against Recognition*, Cambridge: Polity Press.

Mead, G.H. (1934) *Mind, Self, and Society*, ed. C.W. Morris, Chicago, IL: University of Chicago Press.

Nakata, S. (2008) 'Elizabeth Eckford's appearance at Little Rock: The possibility of children's political agency', *Politics*, 28(1): 19–25.

Noble, G. (2009) '"Countless acts of recognition": Young men, ethnicity and the messiness of identities in everyday life', *Social and Cultural Geography*, 10(8): 875–891.

Ondrejka, C. (2007) 'Collapsing geography: Second Life, innovation, and the future of national power', *Innovations: Technology, Governance, Globalization*, 2(3): 27–54.

Ortner, S. (2002) 'Subjects and capital: A fragment of a documentary ethnography', *Journal of Anthropology*, 67(1): 9–32.

Peters, M. (2001) 'Education, enterprise culture and the entrepreneurial self: A Foucauldian perspective', *Journal of Educational Enquiry*, 2(2): 58–71.

PISA (Programme for International Student Assessment) (2012) PISA home page online, available at: www.oecd.org/pisa/ (accessed 17 February 2014).

Pykett, J. (2009a) 'Pedagogical power: Lessons from school spaces', *Education, Citizenship and Social Justice*, 4(2): 102–116.

Pykett, J. (2009b) 'Making citizens in the classroom: An urban geography of citizenship education?' *Urban Studies*, 46(4): 803–823.

Reiter, H. (2010) 'Context, experience, expectation, and action: Towards an empirically grounded, general model for analyzing biographical uncertainty', *Forum Qualitative Sozialforschung/Forum: Qualitative Social Research*, 11(1): art. 2, online, available at: www.qualitative-research.net/index.php/fqs/article/view/ 1422 (accessed 26 June 2014).

Robson, E. (2004) 'Hidden child workers: Young carers in Zimbabwe', *Antipode*, 36(2): 227–248.

Secor, A. (2012) 'Topological city', *Urban Geography*, 34(4): 430–444.

Sevenhuijsen, S. (1998) *Citizenship and the Ethics of Care: Feminist Considerations on Justice, Morality and Politics*, trans. Liz Savage, London: Routledge.

Simola, H. (2002) 'From exclusion to self-selection: Examination of behaviour in Finnish primary and comprehensive schooling from the 1860s to the 1990s', *History of Education*, 31(3): 207–226.

Simonsen, K. (2013) 'In quest of a new humanism: Embodiment, experience and phenomenology as critical geography', *Progress in Human Geography*, 37(1): 10–26.

Staeheli, L. (2008) 'Political geography: Difference, recognition, and the contested terrains of political claims-making', *Progress in Human Geography*, 32(4): 561–570.

Staeheli, L. (2011) 'Political geography: Where's citizenship?' *Progress in Human Geography*, 35(3): 393–400.

Tarrant, A. (2010) 'Constructing a social geography of grandparenthood: A new focus for intergenerationality', *Area*, 42(2): 190–197.

Taylor, C. (1994) 'The politics of recognition', in A. Gutmann (ed.) *Multiculturalism: Examining the Politics of Recognition*, Princeton, NJ: Princeton University Press, pp. 25–73.

Thomas, N. (2012) 'Love, rights and solidarity: Studying children's participation using Honneth's theory of recognition', *Childhood*, 19(4): 453–466.

Tronto, J.C. (1993) *Moral Boundaries: A Political Argument for an Ethic of Care*, New York, NY/London: Routledge.

Vacchelli, E. (2011) 'Geographies of subjectivity: Locating feminist political subjects in Milan', *Gender, Place & Culture*, 18(6): 768–785.

Varga, S. (2011) 'Primary goods, contingency, and the moral challenge of genetic enhancement', *Journal of Value Inquiry*, 45(3): 279–291.

Walker, G. (2009) 'Beyond distribution and proximity: Exploring the multiple spatialities of environmental justice', *Antipode*, 41(4): 614–636.

Warnke, G. (2007) *After Identity*, Cambridge: Cambridge University Press.

11 Intergenerationality and prejudice

Gill Valentine

In 2011 a European report, *Intolerance, Prejudice and Discrimination*, was published (Zick *et al.* 2011). Outlining the results of a telephone survey of 8,000 people from eight different European countries[1] (*n* = 1,000 per country), the report identified that rates of prejudice are higher in older than younger age cohorts. The older the respondents the stronger the expression of racism, sexism, homophobia, anti-immigrant attitudes, Islamophobia and anti-Semitism, with the oldest respondents, the over-65 cohort, demonstrating statistically significant higher levels of prejudice than all other cohorts (Zick *et al.* 2011). These findings echo generational patterns of prejudice identified by other studies in specific national contexts (e.g. Pettigrew and Meertens 1995; Gijsberts *et al.* 2004). For example, a UK study of racism using data from the British Social Attitudes surveys (a series of national surveys carried out in most years since 1983) found that

> [g]enerations brought up in an ethnically homogenous Britain express high levels of prejudice, while those who have come up since mass migration began [referring to black and Asian immigration to the UK in the 1950s] express progressively more tolerant attitudes.
>
> (Ford 2008: 620)

Indeed, the grumpy retired but loveable character given to expressing quite shocking bigotry is a familiar figure from British and American sitcoms such as *Till Death Us Do Part* and *All in the Family*. The presumption is that people born in different generations carry with them different knowledge, assumptions and experience acquired during life. In this sense, what is normative for each cohort generation differs, with older people frequently characterized as the product of less enlightened times.

In a seminal essay the sociologist Mannheim (1952) linked the formation of generations to social change. He argued that generations grow out of age groups (cohorts) under specific circumstances. People are born or located in a particular social time where they are exposed to the same set of historical, cultural or political events. Through this shared experience they can develop a common consciousness or identity which is evident in terms of their way of seeing the world or the social and political attitudes of their age group. These views and attitudes, he suggested, tend to persist over the lifecourse of cohort members.

Since the Second World War each new generation in Europe and North America has grown up in a wealthier society with increasing opportunities for education and travel (Rother and Medrano 2006). In particular, in the context of new modernity, processes of detraditionalization, globalization and accelerating social and geographical mobility mean that contemporary young people are now exposed to a much wider range of lifestyles and competing values and attitudes (both positive and negative), and are freer from social constraints to develop more individualized ways of living and to define their own personal values, than all previous generations (Beck and Beck-Gernsheim 2002). As such, the contemporary youth generation has probably been exposed to greater social and cultural diversity than any preceding generation.

Yet to date there has been relatively little research that has attempted to explore the extent to which attitudes towards social differences vary between generations, and the reasons for any patterns – whether older generations have had less exposure to otherness than younger generations in society; and whether older cohorts do view social difference through a generational lens.

This chapter addresses this neglect by drawing on empirical research conducted as part of a European Research Council-funded study, Living with Difference, to explore how individuals understand and live processes of social differentiation. Here, the focus is on multiple forms of difference (gender, age, race, class, sexual orientation, disability, religion and belief etc.) in contrast to the literature around prejudice/encounters which has primarily viewed these issues through the lens of race and racism. The research involved a survey of social attitudes conducted between February and April 2012 in Leeds, UK ($n=1,522$). The survey asked a series of questions about the respondents' encounters with people who are different from themselves in many kinds of sites. On the basis of the survey findings, 60 individuals were recruited for a qualitative case study phase of the research. Each case comprised a timeline, a life story interview, an audio-diary of everyday encounters, a semi-structured interview about attitudes towards difference, and an interview reflecting on the emerging findings ($n=120$ interviews). The participants were sampled to include those from a range of social backgrounds (in terms of socio-economic status, occupation, gender, ethnicity, religious/belief, sexual orientation and (dis)ability); whose personal circumstances and lifestyle affords them a range of opportunities for/experiences of encountering 'difference'; and to reflect the range of responses to the prejudice survey. All the quotations included here from the case study interviews are verbatim. Three ellipsis dots are used to indicate minor edits have been made to clarify the readability of quotations. The phrase [edit] is used to signify a significant section of text has been removed. All the names attributed to speakers are pseudonyms.

There is no consensus on what constitutes a generation. In this chapter generation is defined by chronological age and life experience, recognizing after

Mannheim (1952), that the shared experience of living in a particular time, with mutual reference points, is likely to produce commonalities in moral disposition (e.g. in terms of shared understandings of how people should live; who or what type of behaviours are good or bad; how people should treat 'others' and expect to be treated by them) (Valentine and Sadgrove 2013). Specifically, the chapter focuses on perceptions of the moral embeddedness of the 'older generation' defined as those aged 65 and over (until recent legal changes, the standard retirement age in the UK). In doing so, the research acknowledges the internal diversity of any generation, although word limits preclude systematic consideration of intra-generational differences within this chapter.

'Living in a different time'? Perceptions of prejudice amongst the older generation

Like the European survey on intolerance, prejudice and discrimination (Zick *et al.* 2011), the Living with Difference survey also found that the over-65 generation have less positive attitudes towards a wide range of social groups (homeless; lesbians and gay men; Muslim people; black people; refugees and asylum seekers; and transsexual people). Moreover 62 per cent of the older generation (aged over 65) agreed with the statement that minority groups have too many rights nowadays compared with 43 per cent for the survey population as a whole.

In the course of the interviews, younger generations, drawing on experiences of their intra-familial relationships, frequently characterized the retirement generation as prejudiced, as these quotations demonstrate:

> I think where I live – the older ones ... they're all closet racists to be honest.... My in-laws sometimes are horrendous. It got so bad at one point with my father-in-law, I mean he's 71. So he's obviously completely different generation. He's been brought up to use certain words and certain terms. It got so bad at one point that he was using these words and terms, both me and my husband had to pull him and tell him to stop it, because he were doing it in front of my eight-year-old. I'm like, I don't want her learning words like that. She certainly isn't going to learn them from me.
>
> (Female, 40–44, white British)

> They're quite racist my parents and I think it's because they haven't grown up with what we've grown up with. And anybody with a different skin colour to us ... like my mum will use the 'N' word [a racial slur] quite a lot and I'll tell her to shut up. Because she came here the other day actually and said it, and me laddo upstairs [neighbour in apartment above] is black. I said shut up, my neighbours will hear you.... But I know loads of black people and mixed race people, all my kids have mixed race friends. I think it's just my mum and dad's age, they weren't brought up like that.
>
> (Female, 35–39, white British)

In explaining the perceived differences in generational social attitudes, interviewees argued that the post-retirement generation are in effect living in a different time: a past where population diversity was less common, and where spatial normativities about how people should talk and behave in routine interactions in public space were not informed by understandings of equality and diversity. In this sense, the older generation are represented as lagging behind contemporary society in which processes such as globalization are presumed to have led to a reorganization of social relationships and changed the way other generations experience the world. As such, the older generations' attitudes are represented as ill-informed – lacking in education and understanding of contemporary social norms – rather than as intentionally 'negative' or containing any ill-intent.

> [comparing his own views to those of his parents] But I've just lived a different life, in a different environment *at a different time.*... You get old people who say quite racist things and not because they are necessarily racist, it's just because that's just what you said in 1945 or whatever. You said 'darkie' or this or that, or 'gollywog' or whatever. It wasn't a bad thing to say and there probably wasn't any malice in it. But in 2012 you can't say that and if you do say it, it's considered that it is malicious. Yeah, I think you have to be taught that.
>
> (Male, 30–34, white British)

> I think it's different kind of generations. Because we've been taught about why people seek refuge and all that.... But like my parents don't seem to understand why they're doing that. Because they just give the original – they come here, get the taxes free, NHS all that. It's like, well it's not. Just because you've read maybe an article it just annoys me ... I think personally. I think they haven't been taught about it, maybe as much as we have. I think we're more acceptable [*sic* – accepting], maybe, my age.
>
> (Female, 20–24, white British)

Yet, in representing the older generation in this way, the interviewees implicitly evoke ageist assumptions. In contrast to previous times when older people in traditional and agrarian societies were held in high regard and considered wise by virtue of their age and greater life experiences, society has come to associate older age with negative qualities (McHugh 2003). In particular, older people are often stereotyped by younger generations as having cognitive deficits, being rigid in thought and manner, slow to change, old-fashioned in terms of their morality and skills, and as less socially competent than other generations (Bytheway 2005). These are all characteristics implicitly associated with prejudice. In this way, writing off the older generation as holding redundant social attitudes is one of the ways in which they are treated as 'old' and defined as different or separate from younger people (Minichiello *et al.* 2000). In other words, assumptions that older people are prejudiced becomes a justification for ageism.

Living in a different space? The older generations' account of contact with 'difference'

Interviewees who were over 65 did not identify their own generation as any more or less prejudiced than other generations. However, they did recognize that they have less routine everyday contact with people different from themselves compared with younger generations. Indeed, the Living with Difference survey found that work, school or college – all spaces less likely to be frequented by the over-65 generation – are the places where in day-to-day life the majority of respondents identified they most frequently encounter people of a different ethnicity, different religion and different sexual orientation from themselves. People aged 65 and over who responded to the survey also perceived their residential areas to be less ethnically diverse than younger respondents, and were less likely to see lesbian and gay people in their neighbourhood.

These patterns may, however, reflect several things including: the effective spatial age differentiation and segregation that takes place as a consequence of the different time-space routines people have at different stages of life; the fact that some 'spaces have their own age identities' (Pain 2001: 156); and housing market dynamics which may be producing a growth in age-separated housing arrangements for older people, limiting the contact for some older people with other generations (see Riley and Riley 2000; Pain *et al.* 2000; Hagestad and Uhlenberg 2005; Vanderbeck 2007). In addition, research on ageism also suggests that some older people are themselves marginalized in public space, and subject to exclusionary processes due to the prejudices of other generations, which may further limit their opportunities for wider social interaction.

As a consequence of having limited opportunities to encounter 'difference', older interviewees recognized they were less familiar with ways of being and relating to people from different ethnic backgrounds, cultures and with different lifestyles from themselves than younger generations within their own families. This is significant because research in the discipline of psychology developed a groundbreaking approach, known as the contact hypothesis (Allport 1954), which suggests that bringing people from different social groups together is an effective way of developing mutual respect and thus reducing prejudice. Although this theory has been subject to critique and refinement (Valentine 2008; Valentine and Sadgrove 2014), nonetheless the broad notion of 'encounter' (specifically where the encounter is 'meaningful' rather than banal) remains central to much thinking about prejudice reduction in the social sciences (e.g. Laurier and Philo 2006; Wilson 2011) as well as to strategies to promote age integration (e.g. Riley and Riley 2000; Vanderbeck 2007).

> I've got a granddaughter who's 14 and she's at high school now. But I've never heard her make a comparison and my youngest daughter, there were black people in her school but from her perception never once did I hear her make a comment of any differences … it's like they're taught a little bit about it at school. God forbid you wouldn't have been taught anything like

that at [my] school … [referring back to her granddaughter] the last one through school is probably the most aware – when I looked at her and her group of friends, they're so aware of everything around them, with no pre-conceptions about anybody and what a nice way that is to be…

(Female, 60–64, white British)

The majority of older people live on a fixed income in terms of a pension. Even those who are relatively wealthy described feelings of insecurity or uncertainty in the context of the recent international banking crisis, and subsequent austerity measures. In particular, immigration is perceived to threaten the traditional inter-generational contract that underpins the welfare state. This is based on a prin-ciple of reciprocity (i.e. during its productive years, each generation will support younger and older generations in anticipation that in later life it will receive support from subsequent generations) (e.g. Phillipson 1998). As such, some older interviewees expressed a fear that the welfare benefits which they often erroneously perceive asylum seekers and refugees to receive might threaten the ability of the state to meet their pension entitlements in the future. Immigration was also identified by some interviewees as a potential threat to the future employment prospects and material security of younger generations within their own families (cf. Dench et al. 2006).

Yet, such concerns about immigration were not specific to the older genera-tion but rather were articulated by interviewees from all age cohorts (see Waite et al. 2014), reflecting in part the inaccurate and unbalanced media representa-tions of illegal immigrants living off the welfare state which have come to domi-nate much of the contemporary public debate in the UK (Finney 2003). Indeed, the research also suggests that 'older' citizens only appear to be more prejudiced than younger generations because they are open about their views. Whereas equality legislation – popularly described as a manifestation of 'political correct-ness' – is redefining spatial normativities about how people should talk and behave in routine interactions – particularly in educational contexts and work-places. As a consequence, younger generations may be editing or altering how they appear to relate to others out of an obligation to comply with these norms and because of an expectation that they might be prosecuted and/or morally judged if they fail to do so, rather than because they necessarily believe in, or accept, such normativities (Valentine and Harris forthcoming).

I mean Cameron [British Prime Minister] came in and said we are going to cut immigration down … and it's gone up. I mean we let people in and they go on benefits. I haven't claimed a penny of benefit in my life and I feel I'm being robbed … other people get benefits when they've not done any work and I've worked all my life … I mean – I was always brought up, when I was working, when I paid my national insurance, my tax, I was paying to support pensioners in them days and eventually when I become a pensioner the same would happen then. But now people don't think they should be supporting us.

(Male, 65–69, white British)

I think you get an awful lot of people coming ... who I feel are – scroung-
ers, I suppose is the word. You know, they come here and they accept a lot.
I think I was reading in the paper about somebody in London – Asian I think
– who was living in this really, really very expensive house in London –
you've probably read about it – and getting something like – £8,000 a month
I think it was. And I just thought, this is wrong you know, then of course
you hear about some of the different cultures where they've got two or three
wives and we're paying for that, you know, for them to do this. And I think
that does create a lot of resentment.

(Female, 80–84, white British)

When older people did have opportunities to encounter those different from
themselves they demonstrated the same willingness and capacity to develop
positive and respectful relationships as other generations, as this extract from an
interview and accompanying audio-diary demonstrate.

We've 25 floors on my block and we've 25 floors in the next block – 120
apartments per block. The tallest in the area by a long way, and let me think
of the people that I know, well I often say good morning to each other. It
sounds ridiculous, I can say good morning in Polish, I learnt that one. I
learnt Happy Christmas in Polish as well. Let me see we've got Nigerians,
Kenyans, Ethiopians, Zimbabweans, Bangladeshis, Indians, Pakistanis,
there's Malaya...

(Male, 65–69, white British)

It's later on New Year's Day.... Nobody has rung me. Nobody has been to
see me, but had a nice Christmas card from Ethiopian friends. They stuck
their head through to wish me a Merry Christmas which I thought was
nice.... My brother's not rung me.... The rest of the family just haven't
bothered so I've been on my own most of Christmas and New Year. It's
nice that an Ethiopian family have been nicer to me than my own

(Ibid.)

The space where members of the older generation most frequently identified
they had contact with people different from themselves was within their own
families. Families are increasingly characterized by dissimilarity as a product of
mobility and individualization. Just under half (46.5 per cent) of those who
responded to the Living with Difference survey in Leeds have immediate
(parents, spouse/partner, children, sibling) and/or extended (e.g. grandparents,
uncles, aunts, cousins) family members who they define as 'different' from
themselves in relation to their ethnic background, sexual orientation or (dis)
ability (Valentine *et al.* 2014).

One interviewee, Robert, acknowledged that his social attitudes are conser-
vative and that his adult daughter has labelled him racist. His prejudices were
therefore challenged when she announced that she intended to marry a First

Nation Canadian. Although Robert had reservations about the relationship, he was fearful of hurting his daughter by opposing the marriage. Instead he spent time getting to know his potential son-in-law. Romantic love, according to Johnson and Lawler (2005), emphasizes the importance of *compatibility*, which they argue is made through a process of potential partners morally evaluating one another (i.e. do we share the same values, want the same things?). They further argue that romantic relationships are supposed to be based on authenticity: knowing, caring for and loving another for 'who they are'. In getting to know his future son-in-law, Robert recognized the compatibility and authenticity of this man's relationship with his daughter, acknowledging the shared creative skills and cultural interests which brought them together. At the same time, Robert describes how spending time with his future son-in-law led, in effect, to a similar process of moral evaluation of each other. As a result he came to recognize that despite their ethnic and cultural differences his son-in-law nonetheless had a strong set of moral values, predicated on intergenerational respect. In this way, Robert was able to find a connection across difference with his son-in-law which challenged the fixity of his previous social attitudes. In other words, familial love can be a crucial emotion in bridging 'difference', by creating an emotional connectivity that brings the distant closer.

ROBERT: My daughter is marrying an Aboriginal ... he has a Mohican haircut ... and he does this traditional dancing.

INTERVIEWER: So, what did you think when you first met him?

ROBERT: I had grave reservations. I still do have some – I think, although I don't agree with him, he's been very beneficial to my daughter. It's funny, actually. I can quite understand where he's coming from.... Originally, I was told that I was going to have to hand my daughter across to this fellow, and that really got me going in turmoil. I thought, there's no way I'm going to do that. Yet, if I didn't, it would hurt my daughter ... she used to call me a racist ... [Edit] She's got her MA in photography, model making, film, and so she was involved in photographing all these cultural activities. Of course, he's involved in performing it. So that's where the two got together...

INTERVIEWER: So what did you think when you met this chap she bought into your life?...

ROBERT: It was towards the end of the evening. I know the beer was talking a little bit but it kept coming across that he respected me and I couldn't get over this. I said, what do you mean you respect every word I sort of say? You're making me feel uncomfortable now. He said, you're my elder, I respect everything that my elders tell me ... I says, well don't listen to me [laughs]. But it must be a cultural thing, you've got to respect – which I thought was a nice thing – it would be beneficial here in some ways. It took me back a little bit. Blooming heck, I've never been spoken to like that before [laughs].

(Male, 65–69, white British)

A similar process was also undertaken by several other interviewees who responded positively to a child or grandchild coming out as gay, having a relationship with someone of a different ethnic, religious or cultural heritage or giving birth to a disabled child. Though it is also important to acknowledge that although intra-familial diversity does produce more positive attitudes in public life towards the specific social group that an individual family member is perceived to represent, such positive attitudes do not appear to be translated beyond this specific 'difference' to challenge wider prejudices towards other social groups as well (Valentine *et al.* 2014).

In sum, the evidence of the research is that the reason surveys persistently show that older generations tend to have more negative attitudes towards 'difference' than younger generations is not because they are living in a different time – lagging behind contemporary society; but rather because the spaces within which they live and move offer them fewer potential opportunities to encounter or learn about difference than other generations. Older people's geographies are in turn, at least in part, a product of various processes (age segregation in housing markets, age discrimination and so on) that lead to the marginalization or social exclusion of older people from public space and public life.

'The younger generation today': reflections on generational prejudices

While older people are subject to ageism from other generations, it is equally true that the most common prejudice expressed by older people in this study was not towards minority ethnic groups, religious communities or lesbians and gay men but rather towards 'the younger generation'. In particular, the older generation were critical of teenagers and young people suggesting that they are unruly, undisciplined and disrespectful towards older people in public space (see also Pain *et al.* 2000).

> This really applies to some of the younger generation more than anything else. It doesn't appear to – or it doesn't seem to be affected by their race, creed, colour, it seems to be an age thing. Twenty years ago if you disrespected an adult you'd expect a clip around the ear for it. Today you're lucky if they'll even tell them off.... They were hassling me normally when they'd had a can of beer ... on an evening; certain times of the year when I'd be a little bit wary about going out to various parts of the estate ... these days ... again it's something that's changed completely over the last ten years ... I mean you get a bunch of girls together and they are – teenage girls – and they are just as bad as a bunch of teenage boys, if not sometimes worse.
>
> (Male, 65–69, white British)

Given that there is little public provision of facilities for young people in UK towns and cities – let alone the countryside – public space is therefore an important arena for young people wanting to escape adult surveillance and define their own identities. At the same time, the privatization and commodification of the public realm, as well as broader processes of gentrification, have led to the exclusion of young people, so reinforcing the importance of the street for them as a meeting and leisure space (Vanderbeck and Johnson 2000; Vanderbeck and Dunkley 2004). It offers young people an autonomous space where they can be together (Valentine 1996a; Kjørholt 2003). In this context, underage drinking, petty vandalism and larking about all become ways of intentionally or unintentionally expressing independence from, and resistance to, adultist rules and normativities (Valentine 2004; Maxey 2004; Valentine *et al.* 2010). Although research suggests that young people hanging out on the street do not set out to be intimidating or to cause trouble (Valentine 2004), their non-conformity and disorderly behaviour are often read as threatening not only the peace and order of the street but also the personal safety of older people. Consequently, this is a source of generational conflict (Valentine 1996a; Matthews *et al.* 2000).

> In the front room there I have the television. We used to have gangs of boys used to come down here and shout abuse at you – like you fat old cow and stuff like that – and one day I was in the front room ... and all of a sudden there was five youths at the front there and they came and peered in the window – 'oh come and look at this, it's f'ing great, you know'.... So I went out and I said 'excuse me but you know I'd like you to go please'. I got all this 'F' [profanity] and all the rest of it and what you going to do about it?... and I just said well 'I'm sorry but, I have a dog, I shall have to bring the dog out'.... This dog – it was a Labrador I took the dog out and he'd normally you know, growl at people and strangers, instead of which ... he and went and weed at the nearest lamp post. Well of course they laughed at me, 'Right f'ing dog you've got missus isn't it?' You know and – he said 'I'll get me lighter and I'll burn that f'ing dog's tail'. And he came up with this lighter and the dog went up to him and stood there and he got his lighter, at which time I saw red, and I absolutely – I slashed out [with the dog's lead] – I've never been violent like this in my life – but I got this [dog] chain and I said 'Now get off home. I don't want you here', and I went, I just went mad, and they said, 'that hurt my f'ing face'. I says, 'I'll f'ing kill you', and I've never used that word in my life. And it's the only time in my life that I've lost my rag, and off they went ... I was shaking, you know ... I had a cup of tea and I cried, and I thought oh how awful, me, my age, doing that.
>
> (Female, 80–84, white British)

Indeed, statistics for bullying, joy-riding, teenage crime and youth drinking have often been mobilized by the media to fuel popular anxieties about the unruliness of young people in a series of moral panics about 'hoodies' and 'chavs' (Valentine 1996b; Jones 2012). It is commonly suggested in popular debate that the hierarchal relationship between adults and young people has been eroded, with adults giving up their 'natural' authority over young people, opting instead to develop more equal and emotionally closer relations with their children than previous generations (Wyness 1997). This has been accompanied by a general shift in legal and popular attitudes towards young people away from an 'adults know best' attitude towards an emphasis on children's rights, and that, as a consequence, parents have lost the moral and psychological resources to exercise control over young people. As the quotation below suggests the older generation often hark back to a perceived 'golden age' of their own childhood era when children allegedly had respect for adults (albeit one reinforced by the implicit threat of corporal punishment).

> Just the way they just treat you. They've no respect for you. We used to go on the bus a lot and the school down the bottom of the hill used to get on. They were just idiots, swearing and they were throwing Durex [condoms] up and down, blowing them up and throwing them up and down. I mean a lot of pensioners were on the bus ... I mean if you say 'owt to them you get a mouthful back. If you hit them you're in trouble ... I mean in my day when I were driving the bus, if kids did that they'd have been off the bus.
>
> (Male, 65–69, white British)

The older generations' attitude towards young people, although rarely recognized as a form of prejudice, sometimes shares some of its characteristics. Namely, there is a tendency to make negative judgements about young people in which unfavourable encounters with individuals are generalized across all young people as a social category. In this way, young people are often prejudged in stereotypical and negative terms as a problem, blamed for popular social concerns, and treated as an undifferentiated group regardless of social, cultural, religious or other differences which mirror the way minority ethnic groups, lesbians and gay men, and women can be stigmatized and 'othered' (Sibley 1995). Below, an interviewee from the older generation describes his attitudes towards young people, while an interviewee from a younger generation explains – in both an extract from her interview and her audio-diary – what it is like to be on the receiving end of age-based prejudice.

> Kids are evil these days, terrible ... I sit on a bus and I think shall I stand up for that woman, I'm 68 years old.... You see two young kids in front of me and a [older] woman ... was stood up – why don't they stand up for her? If I'd have been them I'd have stood up [Edit – a bit later he returned to the same thing]. I mean I used to get drunk when I were younger. We'd get drunk but never went round half-naked around the middle of Leeds and things like that. But I mean all this. I mean the police don't do anything. I think it's disgraceful. I know what I'd do with them. The police should arrest them – what they did in the days when I was young. Arrest them, put them in a cell.
>
> (Male, 65–69, white British)

I think the older generation probably just paint us all with the same brush ...
I think younger people are better towards older people than older people are
towards younger people. But if you see all these hoodies on the news, you're
not going to – you're going to feel under fire from them so I can see where
they're coming from, but it is rudeness.

(Female, 20–24, white British)

I really don't like this stigma attached to me and my group of friends. We
aren't horrible people, we aren't chavs, as I say, but it seems that people
will avoid us. I mean, when we see old people who don't have a seat on the
train, we will get up, no problem, but they seem really shocked by it. You
have to change your opinion of something, you can't just read everything
that the *Daily Mail* may say; every person with a hoodie on is going to
attack you or rob you. It's hard to get away from that stigma. Even now
when ... we're 21 – we're not teenagers anymore but it kind of seems like
we're still pushed into that group; that teenagers are going to cause trouble.
I mean ... we're adults. We are just not really seen as adults.

(Ibid.)

Conclusion: reflections on intergenerationality and prejudice

In reflecting on attitudes towards social differences, this chapter has demonstrated
that intergenerationality is an aspect of social identities, with interviewees at least in
part defining themselves on the basis of generational difference. The empirical
research has shown how the older generation is commonly represented by younger
generations as holding old-fashioned, socially backward and prejudiced attitudes.
Correspondingly, the younger generation are characterized by older generations as
disrespectful, ill-disciplined and lacking moral values. In this way, debates about
social attitudes and values are a potential cause of social tensions and conflict
between generations in which disagreements about issues such as sexual orientation,
immigration and racism have the consequence to define older and younger generations as different from each other in ways which, in turn, become a justification for
younger generations' ageism towards older people and, vice versa, older generations'
ageism towards younger people. In other words, individual forms of difference and
associated prejudices always need to be understood in relation to each other.

Given that both social diversity (including anticipated further increases in life
expectancy) and mobility are predicted to continue to grow in future decades as a
product of ongoing demographic change and processes of globalization, it suggests
that such generational tensions – and the consequent ageism they produce – are
likely to become increasingly pertinent and cannot be ignored. Rather, the evidence of this research is that we need to recognize and classify both the labelling
and stereotyping of older people as prejudiced, and young people as disrespectful,
as forms of ageism. In response, we need to develop strategies for addressing
misunderstandings between generations – through, for example, creating
meaningful contact between them (while not forgetting or downplaying the
internal diversity of any generation) – in order to promote a society for all ages.

Acknowledgements

I am grateful to the European Research Council which funded this research through an Advanced Investigator Award (grant agreement no. 249658) entitled *Living with Difference in Europe: Making communities out of strangers in an era of supermobility and superdiversity*.

Note

1 The eight countries were France, Germany, Great Britain, Hungary, Italy, Netherlands, Poland and Portugal.

References

Allport, G.W. (1954) *The Nature of Prejudice*, Reading, MA: Addison-Wesley.

Beck, U. and Beck-Gernsheim, E. (2002) *Individualisation*, London: Sage.

Bytheway, B. (2005) 'Ageism and age categorization', *Journal of Social Issues*, 61(2): 361–374.

Dench, G., Gavron, K. and Young, M. (2006) *The New East End: Kinship, Race and Conflict*, London: Profile Books.

Finney, N. (2003) *The Challenge of Reporting Refugees and Asylum Seekers*, Bristol: Presswise/Information Centre about Asylum and Refugees (ICAR), online, available at: www.icar.org.uk/challengeofreportingreport.pdf.

Ford, R. (2008) 'Is racial prejudice declining in Britain?' *The British Journal of Sociology*, 59(4): 609–636.

Gijsberts, M., Hagendoorn, L. and Scheepers, P. (2004) *Nationalism and the Exclusion of Migrants: Cross-National Comparisons*, Aldershot: Ashgate.

Hagestad, G. and Uhlenberg, P. (2005) 'The social separation of old and young: A root of ageism', *Journal of Social Issues*, 61(2): 343–360.

Johnson, P. and Lawler, S. (2005) 'Coming home to love and class', *Sociological Research Online*, 10(3), online, available at: www.socresonline.org.uk/10/3/ johnson.html (accessed 1 February 2014).

Jones, O. (2012) *Chavs: The Demonization of the Working Class*, London: Verso.

Kjørholt, A.T. (2003) '"Creating a space to belong": Girls' and boys' hut-building as a site for understanding discourses on childhood and generational relations in a Norwegian community' *Children's Geographies*, 1(1): 261–279.

Laurier, E. and Philo, C. (2006) 'Possible geographies: A passing encounter in a café', *Area*, 38(4): 353–363.

Mannheim, K. (1952) *Essays on the Sociology of Knowledge*, London: Routledge and Kegan Paul.

Matthews, H., Limb, M. and Taylor, M. (2000) 'The "street as thirdspace"', in S.L. Holloway and G. Valentine (eds) *Children's Geographies: Playing, Living, Learning*, London: Routledge, pp. 54–68.

Maxey, L.J. (2004) 'The participation of younger people within intentional communities: Evidence from two case studies', *Children's Geographies*, 2(1): 29–48.

McHugh, K.E. (2003) 'Three faces of ageism: Society, image and place', *Ageing and Society*, 23(2): 165–185.

Minichiello, V., Browne, J. and Kendig, H. (2000) 'Perceptions and consequences of ageism: Views of older people', *Ageing and Society*, 20(3): 253–278.

Pain, R. (2001) 'Age, generation and life course' in R. Pain, M. Barke, D. Fuller, J. Gough, R. MacFarlane and G. Mowl (eds) *Introducing Social Geographies*, London: Arnold, pp. 141–163.

Pain, R., Mowl, G. and Talbot, C. (2000) 'Difference and the negotiation of "old age"', *Environment and Planning D: Society and Space*, 18(3): 377–393.

Pettigrew, T. and Meertens, R. (1995) 'Subtle and blatant prejudice in Western Europe', *European Journal of Social Psychology*, 25(1): 57–75.

Phillipson, C. (1998) *Reconstructing Old Age: New Agendas in Social Theory and Practice*, London: Sage.

Riley, M.W. and Riley, J.W. (2000) 'Age-integration: Conceptual and historical background', *The Gerontologist*, 40(3): 266–270.

Rother, N. and Medrano, J.D. (2006) 'Is the West becoming more tolerant?' in P. Ester, M. Braun and P. Mohler (eds) *Globalization, Value Change and Generations*, Leiden: Koninklijke Brill NV, pp. 151–178.

Sibley, D. (1995) *Geographies of Exclusion: Society and Difference in the West*, London: Routledge.

Valentine, G. (1996a) 'Children should be seen and not heard? The role of children in public space', *Urban Geography*, 17(3): 205–220.

Valentine, G. (1996b) 'Angels and devils: Moral landscapes of childhood', *Environment and Planning D: Society and Space*, 14(5): 581–599.

Valentine, G. (2004) *Public Space and the Culture of Childhood*, Burlington, VT: Ashgate.

Valentine, G. (2008) 'Living with difference: Reflections on geographies of encounter', *Progress in Human Geography*, 32(3): 321–335.

Valentine, G. and Harris, C. (forthcoming) 'Encounters and the spatiality of intolerance: Implications for future social relations', available from authors.

Valentine, G. and Sadgrove, J. (2012) 'Lived difference: The transmission of positive and negative attitudes towards others', *Environment and Planning A*, 44(9): 2049–2067.

Valentine, G., Holloway, S.L. and Jayne, M. (2010) 'Generational patterns of alcohol consumption: Continuity and change', *Health & Place*, 16(5): 916–925.

Valentine, G., Piekut, A. and Harris, C. (2014) 'Intimate encounters: The negotiation of difference within the family and its implications for social relations in public space', *The Geographical Journal*, DOI: 10.111/geoj.12095.

Vanderbeck, R.M. (2007) 'Intergenerational geographies: Age relations, segregation and re-engagements', *Geography Compass*, 1(2): 200–221.

Vanderbeck, R.M. and Dunkley, C.M. (2004) 'Geographies of exclusion, inclusion and belonging in young lives', *Children's Geographies*, 2(2): 177–183.

Vanderbeck, R.M. and Johnson, J.H. (2000) '"That's the only place where you can hang out": Urban young people and the space of the mall', *Urban Geography*, 21(1): 5–25.

Waite, L., Valentine, G. and Lewis, H. (2014) Multiply vulnerable populations: Mobilising a politics of compassion from the 'capacity to hurt', *Social & Cultural Geography*, 15(3): 313–331.

Wilson, H.F. (2011) 'Passing propinquities in the multicultural city: The everyday encounters of bus passengering', *Environment and Planning A*, 43(3): 634–649.

Wyness, M. (1997) 'Parental responsibilties, social policy and the maintenance of boundaries', *Sociological Review*, 45(2): 304–324.

Zick, A., Beate, K. and Hövermann, A. (2011) *Intolerance, Prejudice and Discrimination: A European Report*, Berlin: Forum.

12 How do you end racism in a generation?

The Runnymede Trust and Project Generation 3.0 – a multimedia arts project in Birmingham

Adefemi Adekunle

Introduction

This chapter is based on one part of a four-year piece of intergenerational community research instigated by the Runnymede Trust. A team of community workers, researchers, film-makers and arts professionals ran a research project in a number of different areas of Britain (Manchester, London and our focus here, Birmingham). Each part of the project was based around local answers to the question 'How do you end racism in a generation?' After providing some background on the aims of the project and giving some local context from Birmingham, we will show the subtlety of current constructions of race and racism. By accentuating the intergenerational aspect, we will also show how an intergenerational slant to community work can give a flavour of 'lived experience' to seemingly abstract notions of race and ethnicity as well as manifest an innovative and locally engaged set of outputs. Accordingly, there are two streams to this chapter: the main focus here is the community research, but it is also important to reflect on our dissemination strategy, a multimedia arts project with aims of both community and educational engagement. A research outcome was to see how the stories that people told amongst their cohort created or hindered a shared sense of understanding and how this affected 'community' – however our participants conceived of it.

The project background: Generation 3.0 in context

The Generation 3.0 project was conceived by the Runnymede Trust as a road map to eliminate racism. Using the SS *Empire Windrush*'s docking in 1948 as a historically symbolic (albeit arbitrary) starting point, there have been three generations of migration since this time. In the years that have elapsed since *Windrush* docked at Tilbury, and the 'first' generation of immigration there has been considerable progress in eliminating racism and discrimination of all forms (Philips and Philips 1998). Nonetheless, racism persists. Runnymede's aim is to ensure that the fourth generation will see dramatically less racial discrimination than their forebears.

Generation 3.0 was conceived as a project that intertwined arts and multi-media; as a piece of applied and engaged community research; and as a space for intergenerational debate and discussion to catalyse and advance community action. We aimed to challenge uncontested local ideas on the salience of race, its meaning and its operation in society, documenting how, if and when it has changed over time. Our thinking was based around constructing a better under-standing of how race is socially constructed, dynamic and changeable. Accordingly, Generation 3.0 aimed to engage younger and older community leaders and activists and 'ordinary' people to discover how ideas about race and racism differ and what tactics are most successful across generations (see www.genera-tion3-0.org for more about the project).

The intergenerational challenge

Within the intersection of age and race it is interesting to see how responses to novel situations often take the form of reference to old situations, or which establish their own legitimacy by reference to tradition: the manner in which the past has its own momentum (Hobsbawm and Ranger 1983). Rather than as an academic or purely theoretical issue, we took our cue from the Carnegie's Young People Initiative (Carnegie UK Trust 2008: 30) which concluded:

> There is also a new emerging phenomenon – that of adults and young people being in competition ... for power and influence over public decision-making. The nature of challenges in society – community cohesion, environ-mental concerns, and responsive public services, amongst others – requires co-operative relations between generations and communities, made up of dialogue, tolerance, awareness and understanding including a realization of the assets, talent and potential of young people within and across different groups in communities including religious and ethnic dialogue.

We would suggest that this is an area ripe for further academic attention. Polling has shown a consistent increase in tolerance among younger generations (Steeh and Schuman 1992). Young black and minority ethnic people are often charac-terized as victims of racism without the tools to challenge these seemingly embedded patterns of disadvantage – in spite of the legacy of the struggle against racism which has been a feature of the post-war years (Sharp and Atherton 2007). The literature has also suggested discrimination on the basis of race also persists with some white young people drawn to far right rhetoric and action on the basis of racism (McDowell *et al.* 2014). If racism is to be eliminated within a generation, a central task is to activate young people to move from voicing an experience to engaging in social action to challenge injustice. There is a serious challenge to be met in opening a dialogue between different generations, so that they can exchange ideas and offer advice based on their various experiences.

The importance of such an issue remains primal since, for organizations like Runnymede or any other that is trying to enact social change, there is a need to know that we are not still fighting yesterday's battle. It has now been three generations since landmark moments like the docking of SS *Empire Windrush*. What has changed? What has stayed the same? Creating an intergenerational cohort of research participants thus creates an easy longitudinal 'snapshot' of representations by the careful comparison of social constructions like place and space and their relation to racism (Gill and Sveinsson 2011). In aligning our work within the community research and scholarly work that has seemingly only begun to recently exploit the potential of intergenerational relations (see in particular the Beth Johnson Foundation) we use place as a constant (Mitchell and Ellwood 2013; Vanderbeck 2007).

The local context: Birmingham

In fulfilling our research objectives, location was paramount and there were a number of very good reasons why our project was situated in Birmingham not least since Birmingham has long been considered to be Britain's second city and has a concomitant history of race based protest. Within this history lies Runnymede's first report – Gus John's (1972) influential *Race in the Inner City*. Revisiting the area meant not only engaging with Runnymede's own history but also engaging with a mature canon of community research in the area (Back and Solomos 1992). It also meant returning to the site of racial tension that could and had resulted in overt violent conflict between different communities (John 2005).

Furthermore, it meant situating the project at the cutting edge of debate not least because the area has one of the highest populations of ethnic minorities in the country (ONS 2011). Indeed, many parts of Birmingham can now be described as 'super-diverse' or an environment where a multitude of different groups live side-by-side within the city limits (Vertovec 2007). Birmingham is one of the most ethnically diverse cities in the country and by 2011, Barrow Cadbury predicts it will become a 'majority minority' community – wherein no one ethnic group has an absolute local majority (in Benjamin 2006). The city has a long history of migration, a trend that has continued to this day (Myers and Grosvenor 2011). New groups – including Somalis, Kurds, Eastern Europeans and others – have recently arrived adding to more established African-Caribbean and Asian communities. In many areas – such as Handsworth, Lozells and Small Heath – it is no longer appropriate to think of ethnic diversity in terms of black, Asian and white. The reality is more complicated, a fact that local young people were acutely aware of. It has been noted that ethnic identities are necessarily contextual and change over time (Okamura 1981). To again stress the intergenerational aspect, it is more important to investigate the context within which individuals and groups identify themselves and others, rather than focusing on the 'cultural stuff' of each ethnic group.

At the time of our initial research, Birmingham had a population of 1,016,800 rising to 3,683,000 if surrounding commuter towns are taken into account (ESPON 2007). Until the 1990s, Birmingham was a major industrial centre.

With the decline of manufacturing and a shrinking public sector (Dale 2010), unemployment is a chronic problem in Birmingham. The city contains five of the most deprived wards in the country, including the three most affected by unemployment: Ladywood, Hodge Hill and Hall Green. By 2010, unemployment in Birmingham reached over 12 per cent, more than double the national average; its distribution has been far from equal (NOMIS 2010). The youth unemployment rate was double that of the city average at 24 per cent (NOMIS 2010). During the time of our primary research, in the lead up to the 2010 general election the *Birmingham Post* reported that Muslim men were on average three times more likely to be unemployed than their white counterparts (Dale 2010). African-Caribbean, Asian and Irish men and women are all more likely to be unemployed than the white British population. It should be noted that this part of the project was conducted in Birmingham between June and August of 2010, well before the August riots of 2011 (Runnymede Trust 2011).

Methodology

To gain an accurate intergenerational snapshot, we chose a number of overlapping methods to give us different local traction with the different generations. These included walking interviews which underlined our focus on place (Evans and Jones 2011); focus groups and participant observation in a number of venues; as well as visiting forums favoured by the number of different participating groups (youth clubs, schools, OAP (old-age pensioner) walking groups and activist organizations, etc.). The younger set included people aged 16–25 from the Bangladeshi Youth Forum, the Small Heath Community Centre and the Birmingham Youth Parliament. As well as being ethnically and religiously diverse these groups included young activists as well as members of local community centres. The older participants – defined here as those involved in activism or who grew up in the latter half of the twentieth century – were similarly diverse. We talked to members of the Indian Workers' Association (IWA) the Afro-Caribbean Millennium Centre (ACMC).

We recruited the groups through a variety of methods; desk research enabled us to identify community centres and organizations that would be suitable for our research. In addition, contacts that we made with the council, or with organizations involved with an associated project who helped us to identify and contact participants including schools, youth groups and even OAP walking groups. The majority of participants were engaged in some way in community activism or anti-racist politics and a significant tranche of those who participated at their local community centre, since it was revealing to hear their views as well.

Groups like the Bangladeshi Youth Forum (BYF) are situated in a youth centre primarily for Bangladeshi and Pakistani teenagers in the Lozells area of the city. Lozells was the location for the notorious 2005 riots and the BYF is situated minutes away from Lozells Road, the centre-point of the violence. The area thus stands as a silent character within our intergeneration debate. Other groups we contacted included the Birmingham Youth Parliament (BYP) – a group of young activists who had been elected from amongst their peers to take part in the project run by Birmingham Council's participation unit. As far as possible we asked these groups the same questions about race, racism and activism in Birmingham. These groups included participants from the African-Caribbean, Bangladeshi and Pakistani, Indian, Somali and white British communities and covered ages from 18 to 90+. In addition we conducted participant observation in a comprehensive school and a number of youth clubs in the Handsworth area; this allowed us to contextualize the responses we received from young people within the focus groups. It also gave us a better perspective on how young people from different communities interacted in school and informal environments, broadening our understanding of how they interacted both with each other and older generations. The historical insights gained were also extremely useful as the 'backstory' to community tensions that remain vital and pressing, particularly to older participants.

Finally, it is to the use of one particular methodology that we would like to draw particular attention. We used a professional film and production company to interview a representative tranche of 70 of our participants in a series of interviews. It was as an arts/media platform that allowed each interviewee to tell their own story in the manner in which they felt the most comfortable. A semi-structured interview proved a methodology loose enough to be applied to the 35 school children and 35 OAPs that we recruited for this part of the study to consider inter- and intra-generational interactions. The same format also made comparisons easier and highlighted the importance of place. (Interviews are available at: www.generation3-0.org, with the website's comments forum creating a further space for comment and debate.)

Findings and outputs

Race and racism: old and young

A primary finding was how different conceptions of race and racisms across generations acted as a potential barrier to community engagement. The first thing that should be highlighted is the rich and nuanced understanding young people have of race, and how racism operates in society. For the young people we interviewed, the idea that race is a social construct came naturally. They voiced strong opinions that there is nothing biological or natural about racial inequalities, no inherent genetic qualities that make one group superior to another. At the same time, however, they demonstrated a deep awareness that 'race' is real in society and shapes how we see ourselves and others, and that the ways in which racial boundaries are drawn can impact greatly on the lives of individuals and groups. When asked the question 'What is race?', all groups discussed different forms of race and racism, including discrimination based on religion, culture, identity and ethnicity.

As a social construct, race and racism can take many different forms, be mobilized in various ways and change rapidly, a fact that many interviewees commented on. In spite of this fluid and complicated nature of race, the young people nevertheless had a strong sense of racism and how it operates. When asked whether race is important today, many responded along these lines:

> I think it is but it shouldn't be. Because, I go back to when you're a kid, when do you actually find out you're black? When do you find out you're white? Who actually tells you? That person is trying to define who you are, why can't you define who you are yourself? Why does race have to define who you are? Why can't how smart you are define who you are? How good at sport you are? Why is it race? I think that when you say someone's Bengali Indian, Caucasian, anything like that, it's like a label.
>
> (Black male, BYP)

Our young participants described living in a world of flux – where their focus was on a social reality that changed very quickly and which they had to adapt instantly. Similarly, and related to the development of a super-diverse inner Birmingham, young people need to navigate several social spaces, which requires them to develop multiple identities and different codes of conduct for their peer group, family and school for instance. These realities mean young people tend to be more fluid in their definitions. Terms like 'black', 'English', 'Asian' and 'British' shifted meaning depending on the context, including or excluding groups like the Asian/Muslim/BME[1] community. In this sense, it could be useful to talk of super-diverse identities, where young people engage in code-switching according to the social environment: young people's concept of identity was shaped by the super-diverse environment in which they grew up. In particular they expressed concern over the potential for racism to affect their job prospects.

Nonetheless, the younger participants showed a deep appreciation of the anti-racist struggles of previous generations. There were other differences with our older contributors. Our older participants were often keen to stress the difference between structural racism and individual prejudice though they were highly critical of both. They, also, had a flexible conception of identity but this, rather, was rooted more in a static understanding. To illustrate: discussions of local authority funding to the 'Asian' as opposed to the 'African-Caribbean' or 'black' communities often gave the impression that these were more solid and tangible definitions than our younger interviewees. Their perspectives were, unsurprisingly, dominated by contemporary politics on the one hand and the political struggles of the 1970s and 1980s – when participants became activists – on the other. The growth and establishment of multicultural discourses and its implications for the administration and funding of community groups by the local authority was an especially important cause for concern. Arguments within our older group often operated at two levels. On a wider structural level, many blamed the council for mismanagement or discrimination towards one or another community. However, on a more personal day-to-day level, significant antagonism was directed towards those communities they felt received preferential treatment from the council. Themes of daily interaction and structural relations are therefore not easily divisible. Much discussion focused on domination of the council by one ethnic group or another.

All interviewees, however, described their identities in complex terms, in many ways reflecting the complexity of the social worlds they inhabit. While participants frequently expressed strident views, these were often accompanied by more nuanced interpretations of the issue at hand providing a foundation for future discussion. However, this should not obscure the serious tensions between groups and discomfort about issues like distribution of resources and relations with new migrants. Even when all participants expressed such views they were accompanied by discussion of the historical context and awareness that this discomfort was inconsistent with opposition to racism. Participants of all generations were at pains to stress that such reactions themselves caused them discomfort and often emphasized a lack of antipathy towards the new communities, objecting instead to the manner of their arrival.

Community relations amongst and between two generations

The site of greater difference between generations was rather community relations. Young people's perception of their community was again defined by their experience of super-diversity. Central to their narratives was a strong sense of competing expectations. The interviewees described how they had to balance demands from different parts of their social lives – like family, peer group, school, church or mosque, and agencies such as the police and youth workers – who all had a certain set of expectations, not only how they should behave, but also how they should define themselves in relation to wider society. These demands could often stand in direct conflict with each other and were not easily reconcilable. When asked about whether the older generation understood how young people now mixed with each other, one set of participants answered:

FEMALE 5: [Mixing is] kind of more encouraged...

FEMALE 4: But then again it's not because, for example my mum; because she and her parents have experienced so much serious racial discrimination and if I was to, when I first went to college, because I went to an all-girls Asian school. When I went to college it was hard for me as well, because I've, you know, found some amazing friends who were not even Asian but ah, to tell my mum that I have a white friend, when I first said to her 'My friend, so and so, blah blah'. She didn't attack it, she just kind of, you know 'But why are you going out with white people? They'll just hate you and...'. I think that 'cos of what they'd experienced 'cos of, kind of to protect us they're telling us not to do it. But also, as [she] just said there are some who understand that our generation has mixed and mingled and that we find it easier to get on with people who are not from your same culture and whatnot. So there is still that barrier, but as I say for my mum to say that, it's 'cos of lack of knowledge.

While opinions differed, most offered an image of a rough acceptance of the various communities amongst their peers. Some even argued that communal tensions tended to be more problematic amongst older generations. Nonetheless, it is difficult to determine the sincerity of the sentiments these young people expressed. Our young participants' accounts were at times contradictory or at least paradoxical. They described race relations in their areas as tense, while simultaneously stating that race no longer matters. Given the complex social world the young people live in, and how they receive different signals and messages from parents, school, peer group and so on, this is perhaps not a surprise. Importantly, however, the fact that some groups still have a strong and stable sense of community was not necessarily seen as a problem or a threat to cohesion. Indeed, some participants argued that a strong sense of community can be an important anchor in an otherwise unstable social world. A Small Heath participant said:

> I come from an area in Birmingham where most of the people in that area are Asian. So there's not much, and because we go to a local school, about 80 per cent of that school is Asian. I would say I don't think I experience racism on a daily basis because I am so used to the people around me that there's no point in having that racism there.
>
> (Asian male, BYP)

The participants of the Small Heath discussion deliberated that although community relations are often tense, young people were increasingly mixing with people of different backgrounds and linked this with the process of becoming an adult.

> RESEARCHER: But why do you think there's more mixing going on? Do you know why that's changing?
>
> MALE 4: It's time, innit?
>
> FEMALE 3: I think it's 'cos you meet new people when you move out of a certain place. Like you leave school, you go to college, that's out of the area and multicultural places, you start to learn what different people are about. You learn their values even if there are differences you learn to accept them, you've got to live with them 'cos you can't change them. So I think that's what kind of opens your eyes to it all.
>
> MALE 3: You learn the similarities as well that different cultures have. They do similar...
>
> MALE 1: When you're at school you're sort of forced to mix. So people are just like 'I thought that culture was bad before, but now that I've met someone...'.
>
> MALE 3: It's like work, everywhere now you've got to work with people from different backgrounds than you. It's equality and everything you've got to work with people from different backgrounds than you...

It should also be noted that the young people also articulated the limits of inclusive super-diversity. They acknowledged that new immigrant groups can suffer racist abuse and social exclusion from more established BME groups. It is clear from discussions with the young people that race still matters to their lives. It is equally clear that they are very aware of this and have a nuanced understanding of how racism and discrimination has an impact on their life chances. Interestingly, however, they seemed less concerned about overt everyday and/or violent expressions of racism, and generally had more to say about the structural discrimination they face. In many ways, this may be a fair reflection on how many areas of inner Birmingham have changed since their parents and grandparents first arrived there. The young people we interviewed have grown up in areas where ethnic minorities are a majority, and many suggested that this provided them with safety in numbers.

By comparison our older participants also depicted an image of increasingly tense community relations. Members of the Asian and African-Caribbean communities referred to specific causes of disagreement between them, though discussion of the white population was notably absent. Among older participants there were signs of considerable tension between African-Caribbean and Asian – often Muslim – communities. This is perhaps not surprising given that at the time of research less than seven years had passed since the civil disturbance erupted on Lozells Road, which ran in close proximity to several of the locations of our study. The causes and effects of these tensions still had an imprint as can be seen in the opinion expressed by one of our older participants at the ACMC:

> From the black perspective, we are diverse. We have different factions and everybody's got different views on what you believe or what you don't believe. From an Asian perspective it's either you are a Sikh, a Hindu or a Muslim.... And they help each other. But our community: we are in factions.

By contrast, one of the IWA participants expressed almost the opposite view. One crystallized the point by asking the simple question: 'How many Sikhs employ Muslims into their shops?' Members of the IWA considered relations within the Asian community to be increasingly undermined by the same divisions that as those believed to separate them from the African/Caribbean community. However, even those who were clear and vehement in their complaints expressed sympathy, solidarity and understanding with other communities and the challenges they were facing.

Despite this, all groups expressed sympathy for the struggles facing others in Birmingham, and the desire to see greater community harmony. Some interviewees were of the opinion that the civil disturbances could happen again – that there are still tensions between communities that could erupt, and that community relations are fragile. This view, however, was contradicted by the opinion expressed by many interviewees that young people mix fairly well together, and that the picture painted of segregated communities are an exaggeration. There are problems with community cohesion, they argued, but there is still mixing going on, especially amongst younger people.

Race and identity amongst the younger generation

One noticeable difference between our cohorts, largely due to super-diversity and the degree of mixing that went on in their peer groups, many younger interviewees voiced the opinion that race does not matter. As one of our focus group participants expressed:

> FEMALE 2: I think I'm actually the only non-Asian one here. My mum's white and my dad's black, my boyfriend's Asian and I've got a half-Asian brother, so I'm all mixed up. But I don't really like, just chill with mixed race people, just black people or white people, a lot of my friends are Asian. Not because I wanna be with the Asians, it's where I grew up, the people you grow up around, the friends that you make.

Paradoxically, it quickly became clear that race has a strong impact on how the young people see their place in the world. In a sense, the parts of their identities that they chose to highlight in the focus groups were those on the basis of which they feel discriminated against.

> FEMALE 4: It's kind of sad really because erm, you wouldn't believe it but I'm mixed race also. And when I get white people or black people coming to me making racist comments I'm thinking 'Well, my nan's white so you're just making a fool out of yourself, because you're white' but obviously they don't know that. But like [she] mentioned earlier on it's hard to establish one's race, culture and identity on the basis of their colour because evidently they didn't realize that I had white in me whereas I do, and it's just not made any difference because they've just continued to make the same racist comments.

Despite the relative harmony amongst the younger generation, as one of the youth workers we spoke to pointed out, there is a danger that these young people would not want to move out of the areas where they feel secure, comfortable and in control. This was consistent with what the young people told us, as most of the participants talked about racist abuse happening elsewhere, in mostly white areas, and therefore not necessarily of any great consequence to them. This becomes a problem in that upward social mobility might demand that they venture out of the comfort zone. Staying in areas with a high proportion of ethnic minorities may very well confine people to a certain type of education and a limited range of jobs. In this, the area was ripe for some intergenerational dialogue. In this sense, their apathy about the racism that happens elsewhere is harmful, as they may not develop the necessary knowledge and courage to challenge racist behaviour. Our challenge was in linking this with older generations who continued to hold valuable experience and knowledge. Nonetheless there was often a disconnect between them and the younger generations of residents. What is required is a set of neutral forums and spaces where all people can meet and exchange ideas, knowledge and experience.

Outputs: how, when and where

In attempting to move the project beyond being a simple research exercise certain questions had to be directly tackled. Since our aim was to catalyse community action, the practical issue of where and how to do this were key. In attempting to construct a space purpose built for an intergenerational dialogue the importance of place was a recurring motif. The churches/mosques/temples that housed some of our initial research encounters stood as part of the cultural infrastructure and so had certain connotations for different parts of the community. In solving this equation, we managed to 'borrow' a shop that had not been used for an extended period and presented all our findings back to the community as a 'pop-up' exhibition area. Transient, empty space was for the days we exhibited transformed into a dynamic seat of intercultural and intergenerational debate and discussion.

At the centre of our exhibition were the 70 interviews we had undertaken previously (see methodology). It was our procedure for catalysing community participation by presenting our findings back to the community in the form of an arts project – a format that could be equally enjoyed by all age cohorts. By invoking the curiosity of each interviewer and their friends and families – How did they look on screen? How did they answer the questions we asked? – we managed to generate a significant footfall of participants throughout the days when the project was exhibited. Our careful recruitment of a representative tranche of locally prominent residents ensured that we were provoked a great deal of neighbourhood interest. By carefully scheduling a series of events throughout the period we had school groups mixing with OAPs talking over cake and tea.

At the same time, we also came up with a learning resource – a set of games, debates and topics for discussion that could be used in schools, youth clubs and OAP groups designed to underline how, based around an honest debate of issues, hard topics can be tackled. It is designed for teachers, youth workers, and those who work specifically with older people's groups. It can, of course be extended to any other professionals who work with young people growing up especially young people, aged 11–14 attending youth clubs, youth groups or projects run by voluntary sector organizations and older people aged 55+ attending older people's groups at community centres, luncheon clubs etc. (A copy of the learning resource can be also be found on the project website: www.generation3-0.org.)

In traversing the gap between community engagement and social research, we suggest that the learning resource is the foundation for something locally nuanced and specific. One example of this was the work we did with one local school. By using mapping we succeeded in informing different generations of their views of race and racism, inviting pupils to ask questions or raise any concerns they had. By encouraging pupils and members of the public to contribute ideas or expertise to the planning of this project the space was cleared for something that had community tangibility. One method of doing this was through mapping, since a map is someone's way of getting you to look at the world his or her way, by capturing each participant's views, assumptions and 'facts' about their area. This approach seemed more appropriate than a purely quantitative intervention and this meant that there were some interesting outputs once the project had ended. Our objective was to unpick the idea of a map as a static representation of fact, and use it as a basis for further action. As a research tool, there something simple yet effective about just asking people what their story is and building upon this.

Conclusions

In research terms, there were important similarities in generational attitudes to racism. However, there are equally important, though often more subtle, differences. If our findings are in any way representative, the social realities of many young people growing up in Britain's cities today are in many ways more complicated than they used to be. Whether they described segregation in an integrated world or integration in a segregated world, it was clearly a situation where a modicum of intergenerational guidance and/or understanding would have made their situation easier to navigate. Within a call for this action, both the more established groups focused far more on the necessity for institutional or structural interventions to address these issues.

Of course, Generation 3.0 is not a solution to racism per se; however, it is a blueprint for the generation of such solutions. A number of general observations can be made from the research – the exceptions to these conclusions are also illuminating. Founded upon these and the community engagement aspect there was a widespread feeling that racial divisions are artificial rather than inherent. This did not mean that they were considered unimportant. Tensions were discussed between Asian and African-Caribbean communities, and even among the more socially distant Sutton Coldfield community, concern was expressed about the detrimental impact of communities that never interact. While members of the Sutton Coldfield group expressed greater reservations about the impact of large-scale ethnic diversity on community cohesion, they also expressed greater faith in the ability of everyday interaction to resolve tensions.

As a community engagement exercise, the many discussions, impassioned debates and the various modes of public discussion we used should act as testimony to the various ways one can approach talking about something as difficult as race (Hitchings 2012). It shows how intergenerational community work has a capacity to tackle hard issues like class, inequity and racism. It should also show how a community consultation with an intergenerational edge can and will challenge the idea that young people are unmotivated or politically apathetic. Indeed our experience of the young people was that they displayed frustration that they do not see many avenues that they could follow in order to effect changes they imagined. It must be acknowledged that within what 'should' happen and what can be done next, the opportunity for young and older people to do the talking and policy makers to listen to them is limited. In future, we would suggest that this must change.

Note

1 BME is 'black and minority ethnic'.

References

Back, L. and Solomos, J. (1992) 'Black politics and social change in Birmingham, UK: An analysis of recent trends', *Ethnic and Racial Studies*, 15(3): 327–351.

Benjamin, A. (2006) 'Mixed responses', *Guardian*, 13 December 2006, online, available at: www.guardian.co.uk/society/2006/dec/13/communities.guardiansocietysupplement2 (accessed 10 February 2014).

Carnegie UK Trust (2008) *Empowering Young People: Final Report of the Carnegie Young People's Initiative*, London: Carnegie UK Trust.

Connolly, P. (2002) *Racism, Gender Identities and Young Children: Social Relations in a Multi-Ethnic, Inner City Primary School*, London: Routledge.

Dale, Paul (2010) 'Unemployment is the main election issue for Birmingham', *Birmingham Post*, 2 May 2010, online, available at: www.birminghampost.co.uk/ news/local-news/unemployment-main-election-issue-birmingham-3930873 (accessed 10 June 2014).

ESPON (2007) *ESPON Project 1.4.3 Study on Urban Functions*, Final Report. Luxembourg: The ESPON Monitoring Committee, online, available at: www.espon.eu/export/ sites/default/Documents/Projects/ESPON2006Projects/StudiesScientificSupport-Projects/UrbanFunctions/fr-1.4.3_April2007-final.pdf (accessed 10 February 2014).

Evans, J. and Jones, P. (2011) 'The walking interview: Methodology, mobility and place', *Applied Geography*, 31(2): 849–858.

Gill, K. and Sveinsson, K. (2011) *Fair's Fair: Equality and Justice in the Big Society*, London: Runnymede Trust, online, available at: www.runnymedetrust.org/ uploads/ publications/pdfs/FairsFair-2011.pdf (accessed 10 June 2014).

Hitchings, R. (2012) 'People can talk about their practices', *Area*, 44(1): 61–67.

Hobsbawm, E. and Ranger, T. (1983) *The Invention of Tradition*, Cambridge: Cambridge University Press.

John, A. (1972) *Race in the Inner City*, London: Runnymede Trust.

John, A. (2005) 'This conflict has been 30 years in the making', *Guardian*, 26 October 2005, online, available at: www.theguardian.com/uk/2005/oct/26/ ukcrime.race (accessed 10 June 2014).

McDowell, L., Rootham, E. and Hardgrove, A. (2014) 'Politics, anti-politics, quiescence and radical unpolitics: Young men's political participation in an "ordinary" English town', *Journal of Youth Studies*, 17(1): 42–62.

Mitchell, K. and Elwood, S. (2013) 'Intergenerational mapping and the cultural politics of memory', *Space and Polity*, 17(1): 33–52.

Myers, K. and Grosvenor, I. (2011) 'Birmingham stories: Local histories of migration and settlement and the practice of history', *Midland History*, 36(2): 149–162.

NOMIS (2010) Labour Market Profile West Midlans, online, available at: www. nomisweb.co.uk/reports/lmp/gor/2013265925/report.aspx (accessed 10 June 2014).

Okamura, J.Y. (1981) 'Situational ethnicity', *Ethnic and Racial Studies* 4(4): 452–465.

ONS (Office of National Statistics) (2011) 'Local area labour markets: Statistical indicators April 2011', online, available at: www.ons.gov.uk/ons/rel/subnational-labour/ local-area-labour-market/april-2011/local-area-labour-market.pdf (accessed 13 June 2014).

Parekh, B.C. (2000) *The Future of Multi-Ethnic Britain: Report of the Commission on the Future of Multi-Ethnic Britain*, London: Profile Books.

Phillips, M. and Phillips, T. (eds) (1998) *Windrush: The Irresistible Rise of Multi-Racial Britain*, London: HarperCollins.

Runnymede Trust (2011) *The Riot Roundtables: Race and the Riots of August 2011*, online, available at: www.runnymedetrust.org/publications/174/32.html (accessed 10 June 2014).

Sharp, D. and Atherton, S. (2007) 'To serve and protect? The experiences of policing in the community of young people from black and other ethnic minority groups', *British Journal of Criminology*, 47(5): 746–763.

Steeh, C. and Schuman, H. (1992) 'Young white adults: Did racial attitudes change in the 1980s?' *American Journal of Sociology*, 98(2): 340–367.

Vanderbeck, R.M. (2007) 'Intergenerational geographies: Age relations, segregation and re-engagements', *Geography Compass*, 1(2): 200–221.

Vertovec, S. (2007) 'Super-diversity and its implications', *Ethnic and Racial Studies*, 30(6): 1024–1054.

13 One roof, different dreams

Lives of Shanghai teenage girls and their fathers

Qiong Xu

Introduction

Intergenerationality is a spatial concept (Luzia 2010) and space matters to family life, especially in young people's daily lives. Morgan's (1999) concept of 'family practice', which emphasizes individuals 'doing family', provides a useful lens to understand the changing family. The home place is central to understanding the practices of the family, given that it is a primary site of familial activity (Luzia 2010). Geographical theories of place are useful for conceptualizing how the home place is both constituted by and constitutive of family practices. As McDowell (1999: 4) has argued, 'places are contested, fluid and uncertain. It is socio-spatial practices that define places and these practices result in overlapping and intersecting places with multiple and changing boundaries, constituted and maintained by social relations of power and exclusion'.

Previous research has shown that children's outdoor play has decreased considerably over time in a number of contexts (Pinkster and Fortuijn 2009; Tomanović and Petrović 2010; Karsten 2005). Children's freedom of movement and their spatial autonomy have become increasingly limited due to parental concern over the safety of public space (O'Brien *et al.* 2000). Moreover, parents fear the exposure to other children may be potentially harmful for their own (Valentine 1996).

The tension between parental control and children's autonomy therefore has become one of the important features of modern childhood (Brannen 1996; Scott *et al.* 1998). Both parents and children have different and sometimes opposing goals so there is an ongoing process of negotiation and struggle between them (Finch 1989). In response to parental effort to limit children's spatial autonomy, children often negotiate with their parents and find their own way to get access to public spaces (Punch 2004). This does not always result in full agreement between all parties but nevertheless it is through negotiation that young people may gain more independence from their parents.

There are also significant gender (Valentine 1999) autonomy seeking differences in term of children's autonomy seeking. Research evidence from Western countries shows that boys tend to be granted more freedom than girls. Girls' movements are limited not only by adult strangers but also male peers (Matthews

2003; Tucker and Matthews 2001). There are also differences between girls' and boys' perceptions and uses of space (Skelton 2000; Tucker and Matthews 2001). More specifically, girls tend to be more fearful and constrained, while boys feel more comfortable and have a wider spatial range (Pain 2006). Parents' fear about sexual assault often results in the limiting of girls' use of public space (Scott *et al.* 1998). Furthermore, Furlong and Cartmel (1997) argue that the gender distinctions are greater than those associated with class or race, in terms of young people's leisure and lifestyles.

While most studies focus on exploring one major aspect of intergenerational relationships (most commonly, the mother–child relationship), there is very little work exploring how fathers negotiate issues of spatial authority and autonomy with their children (Luzia 2010). Furthermore, the literature has been dominated by research in Western contexts, and there is a need to recognize culture differences in terms of children's autonomy seeking (Punch 2001). For example, a previous study of young people's access to public space in Singapore indicates that there is a degree of negotiation between young people and parents, although 'the notion of protection is very heavily supported in Singaporean discourses about young people' (Skelton and Hamed 2011: 208).

This chapter examines the nature of negotiations between fathers and daughters over girls' spatial autonomy during their adolescent years within the context of rapid social change in China. The research focuses on two particular aspects of girls' spatial use of the home: going out and use of the internet. The chapter will start with introducing the context of Chinese intergenerational relationships in relation to the home. Following this, a brief introduction of the research methods will be provided. The chapter then examines girls' and fathers' respective understandings of parental authority in contemporary Chinese society and girls' daily practices. It concludes with a brief discussion of the main findings of the research and their significance.

The context of Chinese family in relation to home

The Chinese word *guan* (管) is a culture-specific idiom which means taking care of but also controlling, governing, monitoring and interfering (Wu 1996; Chao 1994). The word *guan* (管) vividly suggests the importance of parental authority as well as parental responsibility in Chinese family life. One of the key aspects of parent–child relationships in Chinese families is the expectation of reciprocity from children. As Chinese parents sacrifice whatever they can to care for their children, children are expected to obey and respect them in return (Xu *et al.* 2005). However, contemporary Chinese families are experiencing tremendous changes: the growth of the economy, the speed of social change and the consequences of the one-child policy. These transformations are inevitably affecting family relationships in modern Chinese families.

In addition, most young people have been brought up with strong Western cultural influences: McDonald's, chips (fries) and pizza are as familiar to many as rice and noodles. Many young people now spend considerable time playing on PCs and using the internet as a means of communication, entertainment and study. The internet has become an essential component of many young people's lives, especially for only children who do not have siblings with whom to play.

As a result of this dramatic increase in contact with Western technologies and ideologies, urban Chinese families are facing rising conflicts as they move from a traditional lifestyle to new ways of living. For example, some parents still impose strong authority and do not wish to grant more freedom to their children. A study of only children in China showed that more than half of the parents tended to like their children to make friends with high-achieving students and avoid making friends of the opposite sex (Sun and Zhao 2005). Although Chinese parents impose stronger authority and grant less freedom than is typical in many Western contexts, this does not mean that Chinese adolescents do not negotiate with their parents to seek more autonomy. Indeed there are studies suggesting that Chinese adolescents' practices of autonomy seeking are similar to those found in Western countries. Yau and Smetana's study (1996) of Hong Kong adolescents found that they conflicted with their parents over everyday matters such as doing housework, making friends and doing homework.

The fieldwork: fathers and their daughters in Shanghai

This study confines itself to fieldwork in Shanghai, one of the biggest cities in China. In an effort to gain a comprehensive understanding of father–daughter relationships in the context of a changing Chinese society, the research design included both fathers and daughters. A multi-method research design was employed: four focus groups conducted in schools, a questionnaire survey with girls ($N = 773$) and their fathers ($N = 598$), and 17 in-depth interviews carried out separately with daughters and their fathers (more details see Xu and Yeung 2013). The interview samples are a sub-sample of the questionnaire sample (see Table 13.1) and it includes eight father–daughter pairs and one girl whose father did not take part in the study. In this chapter, the data are mostly from the interviews.

'I start to lock my door': home as a private space for teenagers?

Perceptions of home differed considerably between fathers and daughters. Home is a place that is perceived to belong to adults, and it is also where they relax, particularly for men. Among the 401 fathers who responded to the open-ended question in the questionnaire survey 'What do you like to do at home?', 34 per cent answered 'relaxation and entertainment' and 28 per cent liked to 'do housework'. Reading and using the internet were also popular, with 13 per cent and 8 per cent respectively. Other things like chatting, pursuing particular interests, doing exercise and helping their daughters with studying were mentioned less often.

Table 13.1 Details of interview participants

Surname	Daughter's age	Father year born	Father occupation	Father education	Home size	Properties owned	Notes
Shen	16–17	1965	University student support officer	Diploma (3-year diploma)	3 bed	2	Father Shen has two children, (twins, one girl, one boy)
Li	16–17	1964	Senior manager in trading company	Master's degree	2 bed	2	
Chen	16–17	1959	Officer in foreign trading company	Senior high	2 bed	1	
Cai	16–17	1963	Civil servant	BA	3 bed	2	
Xiao	13–14	1973	Labourer	Senior high	1 bed	Rented	Xiao's parents migrated to Shanghai to work
Zhan	13–14	1967	Engineer	Master's degree	2 bed	3	
Wang	13–14	1961	Administrative staff (in a restaurant)	Junior high	2 bed	1	Father Wang is divorced and lives with his daughter and his mother
Mei	13–14	1968	Manufacturing worker	Junior high	2 bed	1	
Liu	13–14	1965	Owner of private enterprises	BA	3 bed	2	Liu's father did not take part in the interview

While fathers see home mainly as a place to relax, they often perceive themselves as the ultimate power in the family. In contrast, their children are often limited in what they can and cannot do. A few girls in the interview complained that their parents, especially their mothers, controlled too much of their lives.

> I think my parents *guan* too much. For example, my mum is very sensitive. Every time my classmates call me, she will ask whether they are boys or girls. Whenever I chat online, she only allows me to talk to girls. Whenever I talk about things happening at school, or when I talk about teachers, she will think I do not respect teachers and she just gets very angry. I think she is worried that I will do extreme things. But I won't. I think most of the time, it is not necessary to worry.
>
> (Daughter Zhan, father has a master's degree and
> worked as an engineer)

> I think everyone has his/her own things. I wish that my mum wouldn't *guan* me all the time. She is very strict with me. For example, in the morning, she will check that I have read English before breakfast. And my dad won't ask such tiny and detailed things. I think everyone has her/his own life. Don't *guan* me too much. I once read an article saying that Chinese parents are born for their children. I totally agree with that. My mum is a full-time mum and she just *guan*s me all the time. She starts to talk and ask the moment I come back from school.
>
> (Daughter Liu, father has a university degree and
> owns his own company)

Most girls in my sample have their own separate rooms. As in my questionnaire sample, most families lived in two- or three-bedroom flats; only 12.8 per cent lived in a one-bedroom flat, 44.6 per cent lived in a two-bedroom flat, 22.5 per cent lived in three-bedroom flats and 20.1 per cent lived in houses ($N = 751$). So girls often have their own bedroom. This particular private space belongs to the girls themselves. However, some parents still feel entitled to interfere in their children's living space as they are the authority in this territory. Daughters Li and Zhan talked about their parents invading their privacy.

> DAUGHTER LI: Now I start to lock my door. Not like before I have to bear that. My dad will come in and interfere with whatever I was doing. And he just comes in to my room and starts to talk about studying. Recently, he came into my room and took my exam paper out while I was asleep. I was very upset about this. No one would allow his/her own room to be entered without permission, not least her stuff being looked at. I was very angry and feel uncomfortable. But I am just so used to it now.
>
> QIONG: So how about your father, does he know that?

DAUGHTER LI: He is very traditional and stubborn. So he has not thought that people have their own rights. He is not treating me as an individual. He always thinks of people as a group. He won't think I have my own needs as well.

(Daughter Li, father has a master's degree and worked as manager in a trading company)

QIONG: Now you are more grown up, do you think you deserve more freedom?
DAUGHTER ZHAN: I have thought about that. I used to write diaries. A long time ago I had a diary. But my mum secretly looked at it while I was asleep. I was very angry and told her that I had my own privacy. My mum argued that you were just a child how could you have your own privacy. And she also looked at my bags, my cupboards. She recently found my new diary, and we had an argument.

(Daughter Zhan, father has a master's degree and worked as an engineer)

Parental discourses of family inseparability are influenced not only by their perception of themselves as the authority in the family, but also a result of the only-child policy. Therefore, they feel it necessary to strengthen the bond between parent and child. For example, Father Li believed that his daughter will always belong to the same family even when she is married. As he said,

FATHER LI: As long as we are alive, this is one family, it will be like this forever. I can't make clear line with her. It is impossible for me.
QIONG: So you wish you are always together?
FATHER LI: It is not what I wish, it is the fact. In parents' minds, children will always be children, no matter how old their children are. Although their children are 40, or 50, parents will always treat their children as children as long as they are alive.

(Father Li, master's degree, manager in a trading company)

The girls in my sample stress the importance of their bedroom as a place to 'keep secrets from parents', 'being left alone', which is similar to Lincoln's (2004) research. In a sense, it offers the opportunity for young people to develop their reflexive project of self (Giddens 1991) and it acts as a boundary between different family members. Having one's own bedroom means more privacy and individual control of what happens in that space. However, parents sometimes still regard young people's rooms as part of their space and have the power to restrict and control young people's use of the room. It shows that the notion of individual privacy at home is a relative concept. Home provides elements of privacy from outsiders; however, the boundary between family members is vague, especially when the notion of family togetherness and parental authority is strong. I would argue that space for young people is often a fluid concept and they often have to negotiate with their parents to gain a degree of power for themselves.

'The society is very dangerous': visible public space for girls

Public space plays an important role for establishing young people's peer relationships and it is often a main concern for parents. Some girls reported that their fathers placed many restrictions on where and with whom they could go out. None of the girl interviewees mentioned they were totally free to go out or socialize with friends.

> When I want to go out, my parents will ask many questions and try ever so hard to find reasons to stop me going out. I just feel that you can go out anytime, why can't I? I just feel they *guan* too much. My mum is okay but my father *guan*s too much. He is like hoping that I would stay at home all the time. For example, if someone phones me, he will ask my mum to listen without me knowing. Sometimes, he pulls the landline out. If I go to the supermarket to buy something, he will think perhaps I go out to make phone calls secretly. So he follows me. If I am late to come home from school for one minute, he will phone my teacher to check. Sometimes, I think he doing this for my own sake. But sometimes, I think he just doesn't trust me. Really, I feel, I am your daughter, why don't you trust me? Who could you trust then? Sometimes, I am trying to comfort myself by saying to myself that he is my father. If I don't listen and obey him, who else would I listen to? He is doing this for me.
>
> (Daughter Xiao, father went to high school, worked as a labourer)

Concern about girls choosing the wrong friends and therefore affecting their studies are the main reasons for parents restricting their daughters' friends. Most parents did not want their daughters to be distracted from their studies by leisure activities. Parents normally preferred their daughters making friends with high achievers, as Daughters Liu and Zhan talked about their own experiences when asking for permission to go out. Father Xiao confirmed this

> It is normal to go out and play with your classmates. But the point is who you are playing with, good students or bad students. After all, you should go out with good students. We say 'one who mixes with vermilion will turn red, one who touches pitch shall be defiled therewith' (a Chinese old saying).

As Father Shen pointed out, nothing should affect children's study and children eventually have to give up in the face of authority. As he admitted, 'Not allowed is just not allowed. In the end, she has to give up. After all, we are the adults; we are the powerful ones'.

Safety is another major concern. Parents often feel that the outside world is not safe for their teenage children. As Daughter Liu described,

> I think my parents are very strict in every aspect. For example, during the school holidays, they don't let me go out with my classmates. They think

the society is very dangerous. My mum is so funny that she will threaten me saying that if I go out I will meet some bad people. She keeps talking about that and it does scare me. I know these things are probably not totally true. But sometimes, it happens. So I normally won't go out at night and I will definitely be back home by 1800.

(Daughter Liu)

As Valentine (2004: 28) suggests, 'rather than deal with the complexities of children's vulnerability it is easier for schools and parents to paint a broad picture of the "stranger" in the street who kidnaps children'. As the children were left alone to judge or make sense of the knowledge, it is often like 'a jigsaw puzzle of knowledge from which many pieces are missing' (Scott *et al.* 1998: 700). It is clear from Daughter Liu's response that she accepted the message and felt threatened by her parents' portrait of the outside world. The hostile image portrayed by the parents is often effective in stopping girls from spending more time in public space.

As mentioned before, there are significant gender differences in young people's use of public space in different contexts (Beazley 2002; Evans 2006; Tomanović and Petrović 2010). It is also evident in my data as some fathers thought their daughters are more vulnerable simply because they are girls. Father Wang thought being a girl meant more danger; therefore, his daughter should return home early. Father Xiao thought more restrictions should be enforced to protect his daughter,

I would let her do whatever she wants to do if she is a boy. It doesn't matter. I would not worry that much even if he is fighting with others. What worse can happen to a boy, apart from being caught in a fight? Maybe, stealing or something. But girls are different, if she makes one mistake, then it will be a disaster.

(Father Xiao, senior high school, worked as a labourer)

When the girls talked about going out and making friends, they did not tend to distinguish between fathers and mothers. Often the girls used 'they' to refer to both mothers and fathers which is consistent with Allatt's study of young people (1996). Also, none of the girls in my study reported lying to parents, which was seen as morally wrong by Chinese adolescents. However, four girls reported concealing information from parents so that they could go out, all four from families with a high social-economic background. By not volunteering information, it avoided parental disappointment and intervention. It seems that girls negotiate their spatial autonomy by making choices of when and how to conceal the information since they still wanted their own privacy and freedom. They would tell only when asked, and chose the details they wished to tell their parents.

Daughters Zhan and Shen talked about how they managed to do whatever they wanted to do and then reported it afterwards, or did whatever they planned to do by ignoring their parents. In addition, this daring approach of 'just doing it' without permission may be due to daughters' assumption that there would not be serious punishment or consequences. Zhan described how she got away with it:

> Normally they will not be against it. They will ask who I am going out with. Especially my mum, she will be thinking it over, of course she was thinking whether the person is good at study, whether the person is morally good, can I go out together with her? After she had finished thinking, I just went back to my room. I just waited until that day, I just said 'bye bye' and left.
>
> (Daughter Zhan)

In relation to young people's practices of using outdoor space, parents tend to portray a risky picture so that they can protect them from any potential risk from outside. It also throws light on some of the tensions that emerge from the idea of being a good parent. Parents' intentions to protect their children are often in contrast to children's desires for freedom. On the other hand, parental perceptions of risk may influence young people's autonomous use of space, including peer interactions and outdoor activities. The findings from this study have resonance with the concept of 'collective anxiety' in which fear is a collective and social experience (Pain and Smith 2008). It highlights: 'a phenomenon can lead to powerful social change even among those who are not personally anxious, thereby affecting wider social groups and orderings through shared experience with sites of social anxiety converging with discourses of risk' (Foy-Phillips and Lloyd-Evans 2011: 382).

The world is one click away: internet use at home

> The living environment that the 90s generation live in is closed. I feel that the relationships between people are closed up, and their virtual world can't be possibly compared with ours. Ours was just the opposite. We didn't have the internet world. The living conditions of our children are now much better and more independent. They get home and they can close their doors if they are not happy. For example, my son's study room, apart from sometimes entering to get books, we cannot get in. They all have their own independent space. And what he thinks and feels is mostly reflected in the internet world. He will chat online. I only have my work email address, that's all. And I don't understand their internet world.
>
> (Father Shen, college education, university staff)

The internet has become an essential part of young people's lives. It has also caused huge concern for the parents while there is the increasing presence of modern technologies within the home. Among the nine girls interviewed, three daughters, Xiao, Wang and Mei had no internet access at home (Mei had internet access before she became addicted to the internet and her father subsequently stopped the access). All three girls were from junior high school and all were from families with relatively low family resources. The girls who had computers at home reported that they used the internet for many daily activities, including looking at the news, chatting, reading novels and looking at cartoons. Sometimes, they also used the internet as a tool to study, to check things and to look at the notes that teachers posted online. Only one family (Zhan) had more than one computer and the girl has her own computer to use.

Fathers seemed to be experiencing a media-fuelled panic about their daughters' use of the internet and all the fathers had negative views of the internet. Moreover, they all preferred to have control over their daughters' use of the internet. Their reasons were: the internet could not be trusted, children could easily learn 'bad things' and, more importantly, would spend too much time on it, affecting their studies. Father Wang explained why he did not agree to buy his daughter a computer:

> Why won't I buy her one? When she is using the internet, I don't know what kind of website she is looking at. I won't come and check her computer. I don't know things about the computer. How could I check her? I'll say the internet has everything inside, whatever you want. There is so much in it, how could I know what you are playing inside?... When you grow up, you know what you really need and you know things, I can buy you one.
> (Father Wang, junior high school, administrative staff in a restaurant)

The unfamiliarity of some fathers with the internet provides another reason that they want to limit their daughters' access. As a result, some fathers used all possible ways to stop girls using the internet. Some fathers made rules about how much time and when the girls could use it. In Daughter Chen's case, she could use the internet for around one hour at the weekend as a way to relax. In Daughter Zhan's case, her father set a password on the computer so that she had to ask for permission every time she used the computer. However, while not all fathers were so restrictive, they still monitored their daughters' activities occasionally to make sure they did not do anything they perceived as bad. Father Chen checked the websites his daughter had visited in the past two years, and he sometimes glanced at the website when his daughter was using the internet. He wanted to reassure himself that everything was fine. Although he did not set any obvious limitations on internet use, he was not totally at ease.

Fathers' firm attitudes and control of the internet did not seem to stop the girls using every possible opportunity to use the internet. Two daughters from the older group (Daughters Li and Cai) talked about how they secretly used the internet while their parents were not at home. Liu used her mother's mobile to get the access as her father did not wish her spend too much time using the internet. Her mother did not seem to be as strict as her father in terms of internet use. She lent her mobile phone for her daughter's internet chatting without letting the father know. Xiao once had internet access at home, but soon she got addicted to it. She spent most of her time on the internet chatting and wanted to quit school. Her father therefore stopped internet access at home. However, this did not stop her looking for other options. She started to go to internet cafes and even stayed there overnight. She hung out with friends and even gangsters she met online and ran away from home a few times. Her parents had to search for her in many internet cafes in order to find her. Although she had other family and school issues, it shows that parental controls are sometimes ineffective once a child is determined to fight for her own way, especially in the case of the internet.

The presence of the internet at home means that the children are able to expand their personal territory and space without physically going out. Connections to the outside world thus become a click away. Therefore, fathers were more worried and tried to take control of it. However, fathers were less confident in taking control of their daughter's use of the internet because they sometimes lacked technological skills themselves. The data also reveals that a lack of family access to ICT promotes social exclusion from activities online (Valentine *et al.* 2006).

Similar to daughters' use of public space, young people were again seen as vulnerable in cyberspace and protection/restrictions are often provided by the parents. However, the girls in my study still tried different ways to manage their access to the internet as they did with their outdoor activities (Valentine 2004). In this way, the girls developed their social competence both in public space and online space, alongside their parents' interference. However, for the families without any computer at home, this process could not happen and children were simply excluded from using the internet.

Conclusion

Traditionally, children were perceived as less developed, more vulnerable and less competent than adults (Waksler 1991). However, children often challenge this traditional way of constructing parent–child power relations when they adopt different strategies in their daily lives, as much previous research has shown (Valentine *et al.* 2000). The home therefore is an important site for reconstructing and negotiating parent–child relationships.

In this study, fathers often act in a protective role over young children and therefore limit the spatial autonomy of their daughters. They worry about their daughters' safety, the potential negative influences from peers, and educational achievement. However, I argue that fathers seek to maintain a spatial supremacy as they still see themselves as authorities in the family and try to control their daughters' daily lives, including practices of going out and using the internet. In Chinese society, there is a cultural expectation that children have an obligation to their parents. However, many teenage girls challenge the traditional idea, and they fulfil their own independence by negotiating the boundaries imposed by parents and use their own initiative to gain a degree of spatial autonomy.

The introduction of the internet to the family space, to a certain extent, has increased this power struggle between parents and children. As many have argued, the new technologies such as media and the internet have transformed traditional family relations and power structures between children and adults (Holloway and Valentine 2003, 2001). Children are often more knowledgeable

and literate than their fathers in terms of the use of the internet and new technology. Fathers regulate their children's practices of going out and set times for
them to return home. However, with internet access at home, they are less able
to limit their children's connections with the wider world. The home, therefore,
is more connected with society and fathers feel less empowered to have control.
Many fathers see the internet as a threat, which brings potential risk into the
family space. In a way, fathers' attitudes towards the internet represents their
own anxiety and confidence of parenting and the relationship. This was particularly true with low-income and less educated families.

References

Allatt, P. (1996) 'Conceptualizing parenting from the standpoint of children: Relationship
and transition in the life course', in J. Brannen and M. O'Brien (eds) *Children in Families: Research and Policy*, London: Falmer, pp. 130–144.

Beazley, H. (2002) '"Vagrants wearing makeup": Negotiating space on the streets of
Yogyakarta, Indonesia', *Urban Studies*, 39(9): 1665–1687.

Brannen, J. (1996) 'Discourses of adolescence: Young people's independence and autonomy within families', in J. Brannen and M. O'Brien (eds) *Children in Families:
Research and Policy*, London: Falmer Press, pp. 114–129.

Chao, R.K. (1994) 'Beyond parental control and authoritarian parenting style: Understanding Chinese parenting through the cultural notion of training', *Child Development*,
65(4): 1111–1119.

Evans, R. (2006) 'Negotiating social identities: The influence of gender, age and ethnicity
on young people's "street careers" in Tanzania', *Children's Geographies*, 4(1): 109–128.

Finch, J. (1989) *Family Obligations and Social Change*, Cambridge: Polity Press.

Foy-Phillips, P. and Lloyd-Evans, S. (2011) 'Shaping children's mobilities: Expectations
of gendered parenting in the English rural idyll', *Children's Geographies*, 9(3/4):
379–394.

Furlong, A. and Cartmel, F. (1997) *Young People and Social Change: Individualization
and Risk in Late Modernity*, Buckingham: Open University Press.

Giddens, A. (1991) *Modernity and Self-Identity: Self and Society in the Late Modern Age*,
Cambridge: Polity.

Holloway, S.L. and Valentine, G. (2001) '"It's only as stupid as you are": Children's and
adults' negotiation of ICT competence at home and at school', *Social & Cultural Geography*, 2(1): 25–42.

Holloway, S.L. and Valentine, G. (2003) *Cyberkids: Children in the Information Age*,
London: Routledge.

Karsten, L. (2005) 'It all used to be better? Different generations on continuity and
change in urban children's use of space', *Children's Geographies*, 3(3): 275–290.

Lincoln, S. (2004) 'Teenage girls bedroom culture: Codes versus zones', in A. Bennett
and K. Kahn-Harris (eds) *Beyond Subculture: Critical Commentaries in Contemporary
Youth Culture*, Basingstoke: Palgrave, pp. 94–106.

Luzia, K. (2010) 'Travelling in your backyard: The unfamiliar places of parenting', *Social
& Cultural Geography*, 11(4): 359–375.

Matthews, H. (2003) 'The street as a liminal space: The barbed spaces of childhood', in
P. Christiansen and M. O'Brien (eds) *Children in the City: Home, Neighbourhood and
Community*, London: RoutledgeFalmer, pp. 101–117.

McDowell, L. (1999) *Gender, Identity, and Place: Understanding Feminist Geographies*, Minneapolis, MN: University of Minnesota Press.

Morgan, D. (1999) 'Risk and family practices: Accounting for change and fluidity in family life', in E.B. Silva and C. Smart (eds) *The New Family?* London: Sage.

O'Brien, M., Jones, D., Sloan, D. and Rustin, M. (2000) 'Children's independent spatial mobility in the urban public realm', *Childhood*, 7(3): 257–277.

Pain, R. (2006) 'Paranoid parenting? Rematerializing risk and fear for children', *Social & Cultural Geography*, 7(2): 221–243.

Pain, R. and Smith, S. (2008) *Fear: Critical Geopolitics and Everyday Life*, Aldershot: Ashgate.

Pinkster, F. and Fortuijn, J. (2009) 'Watch out for the neighborhood trap! A case study on parental perceptions of and strategies to counter risks for children in a disadvantaged neighborhood', *Children's Geographies*, 7(3): 323–337.

Punch, S. (2001) Negotiating autonomy: Childhoods in rural Bolivia', in L. Alanen and B. Mayall (eds) *Conceptualizing Child–Adult Relations*, London: Routledge.

Punch, S. (2004) 'Negotiating autonomy: Children's use of time and space in rural Bolivia', in V. Lewis, M. Kellett, C. Robinson, S. Fraser and S. Ding (eds) *The Reality of Research with Children and Young People*, London: Sage.

Scott, S., Jackson, S. and Backett-Millburn, K. (1998) 'Swings and roundabouts: Risk anxiety and the everyday worlds of children', *Sociology*, 32(4): 689–705.

Skelton, T. (2000) 'Nothing to do, nowhere to go? Teenage girls and public space in the Rhonda Valleys', in S.L. Holloway and G. Valentine (eds) *Children's Geographies: Playing, Living and Learning*, London: Routledge.

Skelton, T. and Hamed, N. (2011) 'Adult anxieties versus young people's resistance: Negotiating access to public space in Singapore', in L. Holt (ed.) *Geographies of Children, Youth and Families: An International Perspective*, Abingdon: Routledge.

Sun, Y. and Zhao, X. (2005) 'Only children in China', in J. Xi, Y. Sun and J.J. Xiao (eds) *Chinese Youth in Transition*, Aldershot: Ashgate.

Tomanović, S. and Petrović, M. (2010) 'Children's and parents' perspectives on risks and safety in three Belgrade neighbourhoods', *Children's Geographies*, 8(2): 141–156.

Tucker, F. and Matthews, H. (2001) ' "They don't like girls hanging around there": Conflicts over recreational space in rural Northamptonshire', *Area*, 33(2): 161–168.

Valentine, G. (1996) 'Angels and devils: The moral landscapes of childhood', *Environment and Planning D: Society and Space*, 14(5): 581–599.

Valentine, G. (1999) ' "Oh please mum, oh please dad": Negotiating children's spatial boundaries', in L. MacKie, S. Bowlby and S. Gregory (eds) *Gender, Power and the Household*, Basingstoke: Palgrave.

Valentine, G. (2004) *Public Space and the Culture of Childhood*, Aldershot: Ashgate.

Valentine, G., Holloway, S.L. and Bingham, N. (2000) 'Transforming cyberspace: Children's interventions in the new public sphere', in S.L. Holloway and G. Valentine (eds) *Children's Geographies: Playing, Living, Learning*, London: Routledge.

Valentine, G., Holloway, S. and Bingham, N. (2006) 'The digital generation? Children, ICT and the everyday nature of social exclusion', *Antipode*, 34(2): 296–315.

Waksler, F. (1991) 'Studying children: Phenomenological insights', in F. Waksler (ed.) *Studying the Social Worlds of Children: Sociological Readings*, London: The Falmer Press.

Wu, D.Y.H. (1996) 'Parental control: Psychocultural interpretations of Chinese patterns of socialization', in S. Lau (ed.) *Growing Up the Chinese Way: Chinese Child and Adolescent Development*, Hong Kong: Hong Kong Chinese University Press.

Xu, Q. and Yeung, W.J. (2013) 'Hoping for a phoenix: Shanghai fathers and their daughters', *Journal of Family Issues*, 34(2): 184–209.

Xu, Y., Farver, J., Zhang, Z., Zeng, Q., Yu, L. and Cai, B. (2005) 'Mainland Chinese parenting styles and parent–child interaction', *International Journal of Behavioural Development*, 29(6): 524–531.

Yau, J. and Smetana, J. (1996) 'Adolescent-parent conflict among Chinese adolescents in Hong Kong', *Child Development*, 67(3): 1262–1275.

Part IV
Education, work and care

14 Negotiating intergenerational relations and care in diverse African contexts

Ruth Evans

Introduction: intergenerational relations, ageing and development

This chapter discusses how intergenerational relations and care may be changing in a range of African contexts.[1] Following an overview of the connections between intergenerational relations, global population ageing and development, I discuss the theoretical concepts of 'generation' and the 'generational bargain'. Drawing on empirical research with families affected by HIV in Tanzania and Uganda and among those who have experienced bereavement in Senegal,[2] I explore everyday caring practices and intergenerational dynamics. The spaces of 'home' and the 'community' are analysed as key sites where gendered and age-based identities and generational power relations are reproduced, negotiated and contested. I also analyse how inter- and intra-generational caring relations shape young people's transitions to 'adulthood'. To conclude, I reflect on the contribution of 'intergenerationality' to geographies of care, family relations and life transitions.

Concerns about spaces of age segregation and generational separation in people's lives (Vanderbeck 2007) appear to reflect the interests of research and policy in the global North. Yet, issues of population ageing and intergenerational relations are particularly pertinent to the global South. The older population of the world is increasing rapidly as a result of sustained gains in longevity and declines in fertility over the last century. While the highest proportions of older people (aged 60 and over) as a percentage of the population currently live in the global North, almost two-thirds of the world's older population live in the global South (54 per cent live in Asia) (UN 2009). Furthermore, the increase in the older population in the twenty-first century will be greatest and most rapid in the global South (Zelenev 2008), where countries are already facing enormous challenges associated with economic restructuring, industrialization, urbanization, changing household and family structures, environmental degradation and climate change. Older people in the global South, especially older women, are often perceived as particularly vulnerable to poverty, ill health, disability and abuse. Formal safety nets and social protection mechanisms for older and young people are very limited and concern has been expressed that some countries 'will grow old before they grow rich' (Powell 2010: 3). Older people's active contributions to society, strong familial responsibilities towards older relatives and the respect older people often command within age hierarchies in many cultures and social institutions, however, reveal the dangers of constructing older people as a passive, homogeneous vulnerable group.

Despite this global context, development agencies and researchers have only recently started to investigate the connections between age, intergenerational relations and development in the global South. Reflecting the tendency noted in the global North to 'fetishize' the margins (the young and the old) and to overlook the 'middle generation' (Vanderbeck 2007; Hopkins and Pain 2007), research has tended to focus either on children and youth, or, to a lesser extent, older people in the global South.

'Generations' in Africa

In diverse contexts in Africa, the everyday lives of different generations are often integrated within families and communities, based on an implicit 'intergenerational contract' (Kabeer 2000; Collard 2000) and generational hierarchies (Alber *et al.* 2008). In many African countries, there is a long history of child fosterage practices and household structures are complex and fluid (Young and Ansell 2003; Notermans 2008). From Senegal to South Africa, many families live in large, multigenerational households, especially in rural areas (Bass and Sow 2006; Schatz and Ogunmefun 2007). These may be headed by an older or middle-aged male 'patriarch' who is generally expected to provide shelter and financial support, while women are usually responsible for basic household needs, including domestic chores, food production and preparation and caring for children and sick or elderly relatives (Oheneba-Sakyi and Takyi 2006). Men and women often have separate budgets, investments and different responsibilities for household expenditure, and wives, daughters and other female relatives may have considerable autonomy to earn their own income and to provide for their children. Children and young people in Africa have long been expected to contribute to the household from an early age, according to gendered and age-based norms of productive and social reproductive work (Evans 2010). The family home is thus a crucial site where gendered and age-based identities and generational power relations are reproduced and contested (Evans 2012a, 2014).

Processes of urbanization and economic crisis have led to a substantial increase in marital dissolution in many African countries in recent years, and growing numbers of women head or principally maintain households, particularly in urban areas (Oheneba-Sakyi and Takyi 2006). There is also considerable mobility between rural and urban households and intra-household exchanges of resources. For example, family members of different generations move to live with relatives for temporary or longer periods to gain access to care, material resources, education, training and employment opportunities (Van Blerk and Ansell 2006; Evans 2005; Skovdal 2011). Evidence from 21 African countries, including those with high and low HIV prevalence, suggests that there has been a shift towards grandparents taking an increased child-care responsibility in recent years, especially where orphan rates are growing rapidly (Beegle *et al.* 2010).

As Oheneba-Sakyi and Takyi (2006: 14) note, historically, the relationship between the generations in African societies has been based on 'the value of the knowledge, respect and experience that the younger generation receives from the older generation'. Indeed, these values and recognition of children's place within generational hierarchies are enshrined in the African Charter on the Rights and Welfare of the Child (1990). Article 31 – 'The Responsibility of the Child' – states that, among other duties, 'the child, subject to his [*sic*] age and ability, shall have the duty to work for the cohesion of the family, to respect his parents, superiors and elders at all times and to assist them in case of need'; and 'to preserve and strengthen African cultural values in his relations with other members of the society, in the spirit of tolerance, dialogue and consultation and to contribute to the moral well-being of society'.

Anthropological literature on generational relations and transitions between different life phases is useful in conceptualizing notions of 'generation' and relatedness. Whyte *et al.* (2008: 4–6) identify three main understandings of 'generation'. The first, most common conceptualization within African anthropology is as a *genealogical relation of kinship*, which refers to an individual's generational position within families, focusing on descent, filiation and succession. In classic studies of genealogical generations, conflict between generations over the allocation of resources and within a generation over rights to succeed to office and property were prominent themes. The second, overlapping understanding is of 'generation' as a *principle for structuring society*, beyond the specific links of kinship. This understanding is similar to age, but rather than referring to chronological age through the lifecourse which usually corresponds to biological ageing, it refers to the ways that 'maturational age, life stages or age grades are socially marked and recognized as milestones or phases that do not always correspond to chronological age' (Whyte *et al.* 2008: 4). It is thus a relational term that may denote social categories of seniors and juniors in society and which is often used alongside the terms 'age and 'life stage'. The third concept of generation is the idea of *historical generation* or *generation as cohort* which is often traced back to Karl Mannheim's [1923] seminal essay 'The problem of generations'. Whyte *et al.* (2008: 6) argue that analysis of historical generations shows how change occurs unevenly, since differences and conflicts are found both within generations and between them. Recent literature on youth in contemporary African societies has drawn on this understanding of generation to explore the historical circumstances of a particular cohort affected by economic crisis, urbanization, migration, conflict, the HIV epidemic and other development challenges (Cruise O'Brien 1996; Christiansen *et al.* 2006; Diouf 2003).

Building on Whyte *et al.*'s (2006) first and third understandings of 'generation', generational transfers and intergenerational caring relations in the global South have been theorized using the concept of a 'generational bargain' (Collard 2000) and 'intergenerational contract' (Kabeer 2000). The bargain is that the most economically active 'middle generation' makes transfers to the young with the expectation that resources will be reciprocated to them in old age when they require care and support, while also fulfilling their obligations to support their elderly parents. Collard (2000: 456) suggests that the generational bargain relies on each generation making 'such transfers as are consistent with each cohort having a good life-prospect', which can break down if the middle generation is unable to make the necessary transfer of resources to the young and old.

Research suggests that the 'generational bargain' is coming under increasing pressure in much of the global South due to societal transformations associated with a range of global processes. These include globalization, neoliberal economic restructuring, the HIV epidemic and the consequences of reduced public health spending, user fees and policies that emphasize home-based care (Ogden *et al.* 2006), rapid urbanization, high levels of transnational and rural–urban migration, greater emphasis on education, changing family structures and the individualization of kinship responsibilities (Kabeer 2000; Van Blerk and Ansell 2007). Such changes affect the ability of the middle generation to provide care and support for younger and older generations in a range of familial contexts. The ways that intergenerational relations are changing in contemporary African societies are explored in more depth through a focus on the everyday socio-spatial and temporal contexts of caring relations in Tanzania, Uganda and Senegal.

Changing inter- and intra-generational relations in homespace

In eastern and southern Africa, three decades of the HIV epidemic have taken a considerable toll on the ability of families and communities to care for large numbers of people living with a chronic life-limiting illness, orphaned children and youth, and older people whose adult children have died (Evans 2010). The loss of the parental 'middle generation' has led to the emergence of new household forms, such as 'skipped-generation households' (where the older generation live with orphaned children without any co-resident middle-aged adults) and 'child- and youth-headed households' (where siblings live independently without a co-resident adult relative) (Samuels and Wells 2009; Evans 2011). In the context of stigma and changing kinship responsibilities, grandmothers play a crucial role in caring for widows living with HIV and children who have been orphaned by AIDS (Van Blerk and Ansell 2007; Schatz and Ogunmefun 2007; Nyambedha *et al.* 2003).

A growing literature has demonstrated the significance of children's roles in caring for chronically ill parents, younger siblings and other community members in HIV-affected communities in eastern and southern Africa (Bauman *et al.* 2006; Robson *et al.* 2006; Skovdal 2011). Children's responsibilities to provide care for adults call into question conventional norms of generation and parenting. Adopting an ethic of care (Tronto 1993) perspective, my research has shown, however, that caregiving is not a one-way process, but rather children, parents and other family members provide reciprocal care within the household (Evans and Becker 2009; Evans and Thomas 2009). Indeed, caring relations and responsibilities may strengthen emotional connections between members of different generations, which may help to protect 'caregivers' from negative outcomes. The reciprocity and mutual emotional support that characterized parent–child caring relations refutes negative assumptions about the competence of parents living with HIV to provide good care for their children and notions of 'parentification' (that caregiving children become their 'parent's parent') that have been discussed by some authors (Bauman *et al.* 2006; Oduaran and Oduaran 2010).

Furthermore, mothers with HIV and grandparents maintained their position of moral authority in the household and were responsible for decision-making, the allocation of resources and providing emotional support, discipline and informal education for children, despite children's responsibilities for income-earning and domestic, health-related and personal care tasks (Evans 2010). Unsurprisingly, adults continued to occupy positions of power and have higher social status than children, despite their illness and/or infirmity in old age. This highlights the methodological importance of researching the perspectives of both 'caregivers' and 'care receivers' involved in intergenerational caring relations.

While care is shaped by wider gendered and generational power relations, dominant norms may also become more fluid within the home. For example, in Tanzania, almost as many boys provided care for their mothers/female relatives as the number of girls. Some boys provided personal care for their mothers, despite the fact that this subverted dominant gender norms and preferences for 'gender matching' between 'caregiver' and 'care-receiver' (Evans and Becker 2009). Boys who took on caring roles that were considered to transgress culturally accepted gender boundaries, however, sometimes faced significant social consequences, such as bullying and ostracism within the community.

Anthropological literature has highlighted the cultural importance of older siblings' roles in caring for younger siblings and in socialization and informal training in many African societies (LeVine *et al.* 1996). Very few geographical studies have focused on sibling relations to date (see Punch 2001; Klett-Davies 2008 for notable exceptions). In the context of the HIV epidemic in Tanzania and Uganda, sibling relationships may be crucial in ensuring the survival of the sibling group and access to inherited assets following parental death (Evans 2013, 2012b). Orphaned siblings heading child- and youth-headed households usually developed interdependent caring relations and shared domestic duties within the home (Evans 2011). The household was reconfigured as a more autonomous space free of the usual intergenerational hierarchy. In contrast to previous negative experiences in foster relatives' households, young people were able to manage the household independently of adult control. In participatory workshops in Tanzania, siblings identified close relationships and freedom and autonomy to make decisions as positive aspects of living together, in addition to highlighting the poverty, lack of basic needs and loneliness they experienced.

It is important not to romanticize such intra-generational relations, however, since they are shaped by wider social norms and often reproduce hierarchies of gender, age and sibling birth order. Older siblings often sought to maintain such hierarchies in order to manage younger siblings' misbehaviour, which some-times led to tensions and conflict. Young people also reproduced conventional gender norms in the allocation and negotiation of household chores and care work between siblings. Young women heading households reported spending more time each week doing domestic duties and managing the household, while young men heading households tended to share the care work and allocate household chores to younger siblings (particularly girls). Young men spent many more hours than young women each week engaging in paid work and other income-generation activities to support the family financially (Evans 2012a). This was linked to gendered constructions of domestic duties and care, in addi-tion to gendered inequalities in access to employment opportunities.

Many eldest siblings in Tanzania and Uganda perceived caring for their sib-lings in terms of taking on an adult role as the breadwinner and took pride in their role as head of the family. Hamisa (aged 19, Tanzania) said:

> Me? I see myself as an adult because I look after my young sister and I look after myself without depending on my mother, without saying that 'my mother will buy me clothes, will cook for me, I will ask her for a pen'. I know how to manage my finances as a father or mother.

Young people also identified themselves as a parent/guardian for their younger siblings because of their role in providing moral guidance and teaching to their siblings, attempting to fulfil the parental role their mother or father would have performed if they were still alive, as research with child-headed households in Zambia also found (Bell and Payne 2009). Some young men appeared to enjoy their 'adult' responsibilities in managing the household and commanding respect from their younger siblings. Rickson emphasized his authoritative role in the family: 'I see myself as a grown up person because I am the one leading the family. I have authority over my siblings.' However, young people's accounts also highlighted the ambiguous position that they occupied as an elder sibling taking on a full-time parental caring role. Several young people expressed contradictory feelings about assuming 'adult' responsibilities whilst they still considered themselves a 'child'.

As a West African country with a much lower prevalence of HIV, Senegal provides an interesting contrast to the East African context. My research with Serer family members of different generations who had experienced an adult rel-ative's death in recent years in rural and urban Senegal reveals how a death may represent a 'vital conjuncture' (Johnson-Hanks 2002) that reconfigures intergen-erational caring relations and family practices within the home (Evans 2014). Young people saw their roles predominantly in terms of 'caring about' and 'taking care of' (Tronto 1993) bereaved family members through assuming responsibility and providing financial support to meet their needs. However, many were also involved in practical 'caregiving' activities within the home, such as child care, informal teaching and disciplining of younger siblings, undertaking domestic chores (young women only) and providing emotional support for a grieving parent and other family members. Young women also engaged in paid domestic work to earn an income to support their families or to pay for their school expenses during the school vacations. For 'double orphans', the death of both parents could result in a reconfiguration of kinship relations and a move to live with maternal relatives if paternal relatives were unwilling to accept their conventional responsibilities. This enabled family members to redefine kinship relations and to de-emphasize some familial ties, revealing the fluid, contingent nature of kinship that is lived and remade through shared family practices and intergenerational care (Morgan 2011; Carsten 2004).

Young people's involvement in caring for bereaved family members in Senegal was associated with agency (Tronto 1993) and reaffirmed their position within the household, their sense of responsibility towards their younger siblings and their role in the intergenerational contract. In the context of very limited social protection to support children and older people in many African countries, older youth and adult offspring were expected to support their bereaved mother and siblings following the death of a male head of household. Several young people articulated a heightened awareness of the mortality of close, surviving family members and were committed to fulfilling their role in the intergenerational contract. Anna (aged 27), for example, had completed teacher training and was about to start her first teaching post at the time of the interview. She commented:

> even if I have brothers and sisters who do so already, I must also support her. Beside my father, I was a child, but now, I have set myself a goal: to help my mother because she's the only one I have left, so I must do all I can for her.

Following parental death, eldest sons living in rural areas in Senegal gained from intergenerational transfers of wealth, such as inherited land and livestock, and were able to re-assert their role within the household. This sometimes led to shifts in power relations and the fulfilment of the intergenerational contract, as the eldest son became the *Kilifa*, or head of the household, and was regarded as the moral authority, responsible for decision-making over household expenses and providing for younger siblings, step-mothers and their children and other relatives. Young people's roles in the intergenerational contract also influenced their ties to the family home and their future aspirations and life transitions. Samba (aged 27), whose mother had died, for example, wanted to do an apprenticeship in construction work with a relative in Dakar. He said that he would nevertheless return to the village during the rainy season to help his father with the agricultural work and then leave again for the city when the crops had been harvested. He anticipated marrying and bringing his future wife to live in the family home, since he would be the eventual heir, as he explained: 'If I marry,

I will bring her here to live together with the family because the house belongs to me, I am there with my father and the children, so, this house belongs to me.' A wife and children would enhance Samba's status within the household and eventually he would be expected to assume greater responsibility as head of the household, as his father aged and became less able to sustain the household through physically demanding farm work. This demonstrates how a parent's death and subsequent life transitions may reaffirm young people's roles in the intergenerational contract and the family practices (Morgan 2011) in which they engage.

Changing intergenerational relations in 'community' spaces

As Hopkins and Pain (2007: 289, italics in original) comment, an intergenerational perspective foregrounds the understanding that 'identities of children and others are produced *through* interactions with other age/generational groups and are in a constant state of flux'. Although some young people in Tanzania and Uganda enjoyed their position of power within the household and were proud of their 'adult' responsibilities as the head of the family, young people often found that they had low social status within the community and were not treated as 'adults' in decision-making processes. Rickson (aged 19, Tanzania) felt that it was better to adopt the position of a child within the community:

> In the village, when it comes to participating in things, I don't get involved, I position myself as a child, a student who goes to school, I become young. I don't get involved with parents. I become like a child, although in our house, I become like parents, but when with others, I become young, like a young child.

As Bushin *et al.* (2007: 76) note, 'young people construct their identities in part through the discourses that are available to them', including amongst others, 'ontological narratives (used by individual social actors to make sense of their lives)' and 'public narratives (used by institutions varying from family, workplace, church, government, nation)'. Rickson's self-identification as a 'young child' in the village can be interpreted in the context of wider international, national and local discourses of orphanhood in Tanzania (and other sub-Saharan African countries) that seek to identify the 'most vulnerable' children in need of support according to strict age criteria, levels of poverty and relations with adult caregivers (MHSW 2008). In order to continue to access support and resources from NGOs and members of the community, young people, particularly those who exceed the usual age-boundaries of qualifying for support, may seek to emphasize their vulnerability as children lacking adult supervision rather than demonstrating their competencies as youth who were 'managing their own lives'. This reveals the ways that young people perform their 'social age' (Solberg 1997) and demonstrate or disguise their competencies according to the potential benefits associated with appearing 'older' or 'younger'.

Global discourses of orphanhood co-existed alongside local constructions of young people's 'deviancy' based on the ways that sibling-headed households transgressed African socio-cultural norms of 'the family', kinship responsibilities and household formation. While many young people developed supportive relationships with neighbours, relatives, community leaders and others who they often turned to for advice, company and material support, several young people had direct experiences of stigmatization linked to poverty, orphanhood and AIDS. Young people found themselves in a weak bargaining position to negotiate a fair payment for casual work, with people sometimes refusing to pay them for work they had completed. Young people in Uganda who had inherited property experienced difficulties securing rental income from tenants or found that relatives often took this money, leaving them with just a small amount of food instead. Several young people said that they were regularly insulted and intimidated by neighbours who, according to the young people, were trying to force them out of their home to seize their land.

Young people's narratives also reveal the strategies they adopted to resist marginalization and harassment, such as seeking protection from community leaders and neighbours, developing supportive friendships with their peers and interdependence and solidarity between siblings, or suppressing their emotions and becoming highly self-reliant. Hamisa (aged 19, Tanzania) saw community leaders as a key source of protection:

> Some bad neighbours come at night and throw stones at my roof, so I'm too scared to sleep. But I went to our local leader and informed him about this, they started to guard us. So I don't have this problem anymore.

This reveals the ambiguous nature of young people's agency (Robson *et al.* 2007) and 'place' within the community, as they are positioned and position themselves simultaneously as vulnerable dependents in need of protection and support from adults and as competent social actors able to manage their lives and to call on the resources of others.

Changing transitions to 'adulthood'

In many African countries (Ansell 2004; Langevang 2008), marriage continues to be considered the key marker of 'adulthood' for young men and women, particularly in rural areas. The research from Tanzania, Uganda and Senegal suggests, however, that marriage was not regarded as an immediate priority for most young people interviewed, despite its importance as a marker of adult status. Young people's caregiving roles for siblings and other family members represented a long-term commitment which was sustained as young people grew older, in spite of such socially expected lifecourse transitions. Both young men and women heading households in Tanzania and Uganda regarded completing their educations and achieving a level of financial security as much higher priorities for themselves and their siblings than marriage. Young women in all three countries expressed concerns about being economically dependent on a husband. They instead articulated a strong desire to obtain a good job that enabled them to support themselves and their family before they considered marriage and motherhood.

However, the steps towards achieving 'successful adulthood' (Langevang 2008) did not necessarily always happen in the order young people wished. Khady (aged 20, Senegal), for example, had a three-year-old daughter and was unmarried, but was able to return to school to complete her Baccalaureate. She hoped to marry the father of her child and obtain a job in nursing or teaching in future. The notion that young people posed a threat to the generational order (Diouf 2003) was evident in assumptions and unease about the unregulated bodies and sexuality of young people, particularly young women, who had children outside of marriage and/or lived independently and headed households without 'adult' supervision. In Tanzania and Uganda, young women heading households and living alone with their siblings were considered to be particularly vulnerable to sexual abuse, coercion, pregnancy and/or early marriage. The stigma and sexual harassment they faced reveal public disquiet about their 'in-between' place between adolescence and adulthood as 'not-quite adults' who were taking on 'adult' roles.

Some young people in Senegal articulated a desire to travel and migrate to Europe or the US if employment opportunities were available. As Hertrich and Lesclingand (2013) argue in relation to Mali, labour migration has increasingly become regarded as a key event in youth transitions to adulthood for both genders in the West African context. Mobility between urban centres within countries may also represent a key strategy for young people to develop the skills, social networks and access to employment opportunities needed to achieve future financial security (Langevang 2008; Punch 2002; Evans 2012a). Young people's experiences in Tanzania and Uganda suggest, however, that sibling caregiving may prevent young people from taking up employment opportunities and pursuing migration strategies that would help to reduce their vulnerability to chronic poverty over the lifecourse (Evans 2012a).

Although marriage and having the financial means to provide for a wife and children was strongly associated with 'successful adulthood' for young men in rural areas in Senegal, older youth who had only studied for a few years of primary school expressed a strong commitment to continue to support their younger siblings and other family members in future. Issa (aged 25) for example, had paid the bridewealth for his fiancée, but she had not yet joined the marital home. He explained that the marriage celebration would not take place until he had enough money the following year or longer if necessary. He emphasized that his responsibilities towards his widowed step-mother and younger siblings came first:

> Today, where I would like to be, I don't want to abandon them. I will always fulfil my duties towards them [...]. If you do things which exceed your means, you are going to really rack your brains. I have to look after the

daily needs of the home and other things and when you have responsibilities, you have to go step by step, go very gently.

Hertrich and Lesclingand (2013) among others have argued that young men often delay marriage for financial reasons, leading to young people occupying the liminal category of 'youth' for increasingly prolonged periods (Cruise O'Brien 1996; Christensen *et al.* 2006). My research suggests that both young men and women may also delay marriage and migration and may not take up educational or employment opportunities due to their inter- and intra-generational caring responsibilities.

Conclusion

Rather than becoming 'fractured', as some have argued (Collard 2000; Van Blerk and Ansell 2007), my research suggests that the 'generational bargain' continues to underpin intergenerational relations and familial caring responsibilities in East and West Africa. Nevertheless, the generational bargain is coming under increasing pressure in the context of rapid processes of economic and social change in many African countries, including economic restructuring, high levels of transnational and rural–urban migration and urbanization (Hertrich and Lesclingand 2013; United Nations Human Settlements Programme 2013), the global HIV epidemic and the policy emphasis on 'home-based care' (Ogden *et al.* 2006), limited formal safety nets and social protection for children, older people and persons living with disability and chronic illness (UNICEF 2009) and an increasing individualization of familial responsibilities (Kabeer 2000; Evans and Becker 2009). The loss of the 'middle generation' due to AIDS-related and other deaths, in addition to these wider structural processes, mean that young people in contemporary African countries are called on to take responsibility, and provide care, for older generations at a much younger age than usual. Young people may have limited access to informal and formal support, in addition to less human capital (education, training and life and livelihood skills) to draw on than middle-generational cohorts who supported the young and old in previous historical periods.

The examples from Tanzania, Uganda and Senegal reveal the dynamic, constantly shifting nature of the generational categories of 'childhood', 'youth' and 'adulthood', which are negotiated within specific temporal and spatial contexts. Young people affected by parental death may be regarded as a moral threat to the generational order (Diouf 2003) when they engage in activities conventionally associated with adulthood, such as heading households without a co-resident adult relative or young parenting outside of marriage. This may result in marginalization and intergenerational tensions that impact on young people's well-being and future life transitions (Evans 2011, 2012a). It is important to note, however, that the death of a close adult relative is not necessarily always associated with biographical disruption and a reconfiguration of intergenerational caring relations; it may be associated with continuities in family practices, reaffirming young people's roles in the intergenerational contract and leading to few, if any, changes in material circumstances or living arrangements (Evans 2014). Thus, while changes may be occurring in response to different pressures in many African countries, generational hierarchies, intergenerational caring relations and respect for elders are also enduring features of African societies.

This chapter has shown that 'intergenerationality' provides useful insights into the spaces and temporalities of care, family practices and lifecourse transitions. Geographers have much to contribute to analyses of how inter-generational caring relations may be changing across time and space. However, a focus on *inter*generational geographies should not lead to the neglect of analyses of *intra*-generational relations, such as those between siblings and peers, which may be significant throughout the lifecourse (Punch 2002; Evans 2012a, 2014; Skovdal and Onyango Ogutu 2012). This reveals the importance of drawing on related concepts, such as the lifecourse and intersectionality, in developing future relational geographies of age (Hopkins and Pain 2007).

Notes

1 This chapter synthesizes some of the arguments made in Evans (2011, 2012a, 2014) and Chapter 4.5 'Ageing and Development' in Potter *et al.* (2012).
2 For research methods used in these studies, see Evans (2011, 2012a, 2014). I am grateful to the Royal Geographical Society (with the IBG) and the University of Reading for funding these studies.

References

Alber, E., Van de Geest, S. and Whyte, S.R. (eds) (2008) *Generations in Africa: Connections and Conflicts*, Berlin: Lit Verlag.

Ansell, N. (2004) 'Secondary schooling and rural youth transitions in Lesotho and Zimbabwe', *Youth and Society*, 36(2): 183–202.

Bass, L. and Sow, F. (2006) 'Senegalese families: The confluence of ethnicity, history and social change', in Y. Oheneba-Sakyi and B. Takyi (eds) *African Families at the Turn of the 21st Century*, Westport, CT: Praeger, pp. 83–102.

Bauman, L., Foster, G., Johnson Silver, E., Gamble, I. and Muchaneta, L. (2006) 'Children caring for their ill parents with HIV/AIDS', *Vulnerable Children and Youth Studies*, 1(1): 1–14.

Beegle, K., Filmer, D., Stokes, A. and Teirerova, L. (2010) 'Orphanhood and the living arrangements of children in sub-Saharan Africa', *World Development*, 38(2): 1727–1746.

Bell, S. and Payne, R. (2009) 'Young people as agents in development processes: Reconsidering perspectives for development geography', *Third World Quarterly*, 30(5): 1027–1044.

Bushin, N., Ansell, N., Adriansen, H., Lähteenmaa, J. and Panelli, R. (2007) 'Reflecting on contexts and identities for young rural lives', in R. Panelli, S. Punch and E. Robson (eds) *Global Perspectives on Rural Childhood and Youth: Young Rural Lives*, Abingdon: Routledge, pp. 69–80.

Carsten, J. (2004) *After Kinship*, Cambridge: Cambridge University Press.

Christiansen, C., Utas, M. and Vigh, H. (2006) 'Youth(e)scapes', in C. Christiansen, M. Utas and H. Vigh (eds) *Navigating Youth, Generating Adulthood: Social Becoming in an African Context*, Stockholm: Nordiska Afrikainstitutet.

Collard, D. (2000) 'Generational transfers and the generational bargain', *Journal of International Development*, 12(4): 453–462.

Cruise O'Brien, D. (1996) 'A lost generation? Youth identity and state decay in West Africa', in R. Werbner and T. Ranger (eds) *Postcolonial Identities in Africa*, London: Zed Books.

Diouf, M. (2003) 'Engaging postcolonial cultures: African youth and public space', *African Studies Review*, 46(2): 1–12.

Evans, R. (2005) 'Social networks, migration and care in Tanzania: Caregivers' and children's resilience in coping with HIV/AIDS', *Journal of Children and Poverty*, 11(2): 111–129.

Evans, R. (2010) 'Children's caring roles and responsibilities within the family in Africa', *Geography Compass*, 4(10): 1477–1496.

Evans, R. (2011) '"We are managing our own lives…": Life transitions and care in sibling-headed households affected by AIDS in Tanzania and Uganda', *Area*, 43(4): 384–396.

Evans, R. (2012a) 'Sibling caringscapes: Time-space practices of caring within youth-headed households in Tanzania and Uganda', *Geoforum*, 43(4): 824–935.

Evans, R. (2012b) 'Safeguarding inherited assets and enhancing the resilience of young people living in child- and youth-headed households in Tanzania and Uganda', *African Journal of AIDS Research*, 11(3): 177–189.

Evans, R. (2013) 'Young people's caring relations and transitions within families affected by HIV', in J. Ribbens McCarthy, C. Hooper and V. Gillies (eds) *Family Troubles? Exploring Changes and Challenges in the Family Lives of Children and Young People*, Bristol: The Policy Press, pp. 233–243.

Evans, R. (2014) 'Parental death as a vital conjuncture? Intergenerational care and responsibility following bereavement in Senegal', *Social and Cultural Geography*, 15(5): 547–570.

Evans, R. and Becker, S. (2009) *Children Caring for Parents with HIV and AIDS: Global Issues and Policy Responses*, Bristol: The Policy Press.

Evans, R. and Thomas, F. (2009) 'Emotional interactions and an ethic of care: Caring relations in families affected by HIV and AIDS', *Emotions, Space and Society*, 2(2): 111–119.

Hertrich, V. and Lesclingand, M. (2013) 'Adolescent migration in rural Africa as a challenge to gender and intergenerational relationships: Evidence from Mali', *Annals of the American Academy*, 648(1): 175–188.

Hopkins, P. and Pain, R. (2007) 'Geographies of age: Thinking relationally', *Area*, 39(3): 287–294.

Johnson-Hanks, J. (2002) 'On the limits of life stages in ethnography: Toward a theory of vital conjunctures', *American Anthropologist*, 104(3): 865–880.

Kabeer, N. (2000) 'Inter-generational contracts, demographic transitions and the "quantity-quality" trade-off: Parents, children and investing in the future', *Journal of International Development*, 12(4): 463–482.

Klett-Davies, M. (ed.) (2008) *Putting Sibling Relationships on the Map: A Multidisciplinary Perspective*, London: Family and Parenting Institute.

Langevang, T. (2008) '"We are managing!" Uncertain paths to respectable adulthoods in Accra Ghana', *Geoforum*, 39(6): 2039–2047.

LeVine, R., Dixon, S., LeVine, S., Richman, A., Leiderman, P., Keefer, C. and Brazelton, T. (1996) *Child Care and Culture: Lessons From Africa*, New York, NY: Cambridge University Press.

MHSW (Ministry of Health and Social Welfare) (2008) *The National Costed Plan of Action for Most Vulnerable Children 2007–2010*, Department of Social Welfare, MHSW, United Republic of Tanzania.

Morgan, D. (2011) *Rethinking Family Practices*, Basingstoke: Palgrave Macmillan.

Notermans, C. (2008) 'The emotional world of kinship: Children's experiences of fosterage in East Cameroon', *Childhood*, 15(3): 355–377.

Nyambedha, E., Wandibba, S. and Aagaard-Hansen, J. (2003) 'Changing patterns of orphan care due to the HIV epidemic in Western Kenya', *Social Science & Medicine*, 57(2): 301–311.

Oduaran, A and Oduaran, C. (2010) 'Grandparents and HIV and AIDS in sub-Saharan Africa', in M. Izuhara (ed.) *Ageing and Intergenerational Relations: Family Reciprocity From a Global Perspective*, Bristol: The Policy Press, pp. 95–110.

Ogden, J., Esim, S. and Grown, C. (2006) 'Expanding the care continuum for HIV/AIDS: Bringing carers into focus', *Health Policy and Planning*, 21(5): 333–342.

Oheneba-Sakyi, Y. and Takyi, B. (2006) 'Introduction to the study of African families: A framework for analysis', in Y. Oheneba-Sakyi and B. Takyi (eds), *African Families at the Turn of the 21st Century*, Westport, CT: Praeger, pp. 1–23.

Potter, R., Conway, D., Evans, R. and Lloyd-Evans, S. (2012) *Key Concepts in Development Geography*, London: Sage.

Powell, J. (2010) 'The power of global aging', *Ageing International*, 35(1): 1–14.

Punch, S. (2001) 'Household division of labour: Generation, gender, age, birth order and sibling composition', *Work, Employment and Society*, 15(4): 803–823.

Punch, S. (2002) 'Youth transitions and interdependent adult–child relations in rural Bolivia', *Journal of Rural Studies*, 18(2): 123–133.

Robson, E., Ansell, N., Huber, U.S., Gould, W.T.S., Van Blerk, L. (2006) 'Young caregivers in the context of the HIV/AIDS pandemic in sub-Saharan Africa', *Population, Space and Place*, 12(2): 93–111.

Robson, E., Bell, S. and Klocker, N. (2007) 'Conceptualising agency in the lives and actions of rural young people', in R. Panelli, S. Punch and E. Robson (eds) *Global Perspectives on Rural Childhood and Youth: Young Rural Lives*, Abingdon: Routledge, pp. 135–148.

Samuels, F. and Wells, J. (2009) 'The loss of the middle ground: The impact of crises and HIV and AIDS on "skipped-generation" households', *ODI Project Briefing No. 33*, London: Overseas Development Institute.

Schatz, E. and Ogunmefun, C. (2007) 'Caring and contributing: The role of older women in rural South African multi-generational households in the HIV/AIDS era', *World Development*, 35(8): 1390–1403.

Skovdal, M. (2011) 'Examining the trajectories of children providing care for adults in rural Kenya: Implications for service delivery', *Children and Youth Services Review*, 33(7): 1262–1269.

Skovdal, M. and Onyango Ogutu, V. (2012) 'Coping with hardship through friendship: The importance of peer social capital among children affected by HIV in Kenya', *African Journal of AIDS Research*, 11(3): 241–250.

Solberg, A. (1997) 'Negotiating childhood: Changing constructions of age for Norwegian children', in A. James and A. Prout (eds) *Constructing and Reconstructing Childhood: Contemporary Issues in the Sociological Study of Childhood*, London: Falmer Press, pp. 123–140.

Tronto, J. (1993) *Moral Boundaries: A Political Argument for an Ethic of Care*, New York, NY/London: Routledge.

UN (2009) *Population Ageing and Development 2009*, Department of Economic and Social Affairs, United Nations Population Division, online, available at: www.un.org/esa/population/publications/ageing/ageing2009.htm (accessed 10 June 2014).

UNICEF (2009) *La Protection Sociale des Enfants en Afrique de l'Ouest et du Centre. Etude de cas du Sénégal*, Dakar: UNICEF.

United Nations Human Settlements Programme (2013) *State of the World's Cities 2012/2013*, Nairobi: UN-Habitat.

Van Blerk, L. and Ansell, N. (2006) 'Children's experiences of migration: Moving in the wake of AIDS in southern Africa', *Environment and Planning D: Society and Space*, 24: 449–471.

Van Blerk, L. and Ansell, N. (2007) 'Alternative care giving in the context of AIDS in southern Africa: Complex strategies for care', *Journal of International Development*, 19(7): 865–884.

Vanderbeck, R.M. (2007) 'Intergenerational geographies: Age relations, segregation and re-engagements', *Geography Compass*, 1(2): 200–221.

Whyte, S.R., Alber, E. and van de Geest, S. (2008) 'Generational connections and conflicts in Africa: An introduction', in E. Alber, S. Van de Geest and S.R. Whyte (eds) (2008) *Generations in Africa: Connections and Conflicts*, Berlin: Lit Verlag, pp. 1–23.

Young, L. and Ansell, N. (2003) 'Fluid households, complex families: The impacts of children's migration as a response to HIV/AIDS in southern Africa', *The Professional Geographer*, 55(4): 464–476.

Zelenev, S. (2008) 'The Madrid Plan: A comprehensive agenda for an ageing world', in *Review Report 2008 of The Madrid International Plan of Action on Ageing*, New York, NY: United Nations, pp. 1–17.

15 Splintered generations

Difference, the outdoors, and the making of 'family' at an American wilderness therapy camp

Cheryl Morse

Introduction

Scholars of intergenerational relations have called for more research on the socio-cultural and spatial interactions of differently aged people, and have urged researchers to pay close attention to the operation of difference amongst and within age groups (Hopkins and Pain 2007; Tarrant 2010; Valentine 2008). In response, this case study considers how age difference and generational divides are produced through the workings of multiple modes of difference. The research for this chapter took place at a residential wilderness therapy programme for troubled youth where the age difference between the participants and the camp counsellors was small. However, I argue that the programme exploited class and lifecourse differences to enable staff members – who were only slightly older than their clients – to exert authority over, and promote behavioural change in, the troubled youth they served. Despite the narrow age gap, the camp produced a generational difference between clients and counsellors, not least through the mimicking of a kind of fictive familial relationship that positioned one group as 'child' and the other as 'adult'. Much of the difference operationalized in this therapeutic setting became visible in the staff members' and campers' disparate perceptions of nature. Wilderness narratives emerged as a mode of difference, alongside several others, that served to fracture this group of young Americans, making two generations from one.

Vanderbeck's (2007) survey of age-focused research in geography called for more work on the role of space in intergenerational interactions. He also noted a dearth of research on extrafamilial relationships and pointed to the need for additional investigation into cultural difference by asking 'How do the geographies of intergenerational relationships vary between social groups and contexts?' (2007: 202). This chapter addresses these three elements – space, extrafamilial relationships, and varying social contexts – however, it approaches each from an unexpected direction. The wilderness therapy programme I researched (Camp E-Wen-Akee) was located in a remote rural location and it drew on a range of boundary-keeping tactics to closely surveille the participants. Rather than looking at the ways that spatial configurations promote or foreclose intergenerational contact, I consider how the co-location of a single generation within an intimate and bounded space produced generational difference. Camp E-Wen-Akee aimed to mimic a family environment by substituting staff members who are trained in behavioural therapy for a participant's 'true' family members. Through practicing at familial relationships, participants were meant to adopt new, socially appropriate ways of relating with others. This case study afforded the opportunity to examine extrafamilial relationships occurring within something that looked like a family space but was in fact an institutional space, amongst two groups who (by some definitions) could be assumed to share a generational position but who rather showed evidence of something clearly akin to intergenerational difference in power, authority and roles. Finally, this case study provided insights into how social difference manifests within this form of institutional setting, bringing into view the varying impacts such difference has on the lived experiences of individuals occupying privileged and less-privileged subject positions.

Intergenerational geographies and difference

Lay definitions of 'generation' assert that the term refers to a group of people born and living in the same time range, and suggest that 30 years is the duration of a generation (Oxford Dictionary On-Line 2013). Academic treatment of the term has been more nuanced; for example scholars have drawn on Mannheim's (1952 [1923]) multiple definitions of generation (for example: Mayall 2001; Vanderbeck 2007). Tucker (2003) summarized Mannheim's three-staged analysis of formation as: generational location, where it is recognized that people living in the same time and place will be exposed to similar events; 'actual generation' that takes into account the differentiated experience of generation as related to class position; and the 'generational unit' wherein people produce goals as an outcome of shared experience. Tucker critiques these definitions for imagining class as a fixed position, and for the exclusion of other modes of difference in generation formation.

The analysis of the wilderness therapy programme that I offer here was influenced both by Tucker's observation of diversity amongst a generation of teenaged girls, and by Hopkins and Pain's (2007) article on intergenerationality, intersectionality and lifecourse. Hopkins and Pain point out that much of the work on age conducted by geographers has focused on the very young or the old, and that little attention has been paid to the interactions between generations. They propose a 'relational geographies of age' in which age and identity are understood as produced from interactions between different people (p. 288). Intersectionality is concerned with how modes of difference, such as sexuality, gender, class and (dis)ability interact, and acknowledges that no single aspect of identity works in isolation. This echoes previous assertions made by feminist geographers that assigning identity based on one category of difference can mask the influence of other differences, as well as diversity within a social group (Bondi and Davidson 2003; Gibson-Graham *et al.* 2000), and that an individual can occupy or move between more than one class position, for example (Pratt 1998).

Geographies of wilderness therapy

Wilderness therapy (WT) attempts to make deliberate interventions into the lives of individuals who exhibit anti-social or self-destructive behaviours. Typically, a wilderness therapy programme involves bringing a small group of clients (usually youth) into a remote area and conducting recreational experiences such as hiking, rafting and rock-climbing. Participants move from place to place, carrying food, water and shelter with them; programmes last from several weeks to months. Material geographies of the outdoors are put to work in the therapeutic process; nature is meant to both physically distance the individual from the individual's original social world, and to provide conditions and experiences which focus the individual on solving at-the-moment problems such as keeping warm and dry, averting accidents and meeting one's basic needs of food and shelter (Beringer 2004). Cultural constructions of wilderness are also actively enlisted in WT programmes. In the United States, the wilderness has been imagined as an unspoiled space (Cronon 1995), Eden (Nash 1982), a location to experience the sublime (DeLuca and Demo 2001), and an environment to test one's personal fortitude (and oftentimes masculinity) (Haraway 1989). The American wilderness is understood as an ideal place to relocate people away from polluted social worlds, prompting them to encounter their 'true' nature within, in effect resocializing them to live healthier lives.

Americans have long regarded the outdoors as a healthy place for children. Beginning in the late nineteenth century, American children were sent to summer camps in rural areas to address a range of societal anxieties from the fear that city life was making elite children 'soft', to the concern that unsupervised immigrant children were overtaking the crowded streets of industrial cities (Smith 2002; Van Slyke 2006). Summer-stay programmes and camps which remove the child from an urban environment for 'healthy' experiences in rural spaces continue to thrive in the US today. Paris (2000, 2001) has argued that class position and gender have always influenced which children go to which camps and for what outcomes.

Therapeutic camps, programmes intended to heal children of mental illness and behavioural problems, emerged following the Second World War as a result of the merging of developments in psychology, progressive educational theory, and the tradition of summer camp (McNeil 1957). Different from summer camps which operate only during the warm months, and provide children and young teens with instruction in specific skills or a set of outdoor recreational experiences, therapeutic camping programmes are year-round experiences designed to change youth behaviour. Similar to summer camp, therapeutic programmes often hire counsellors who are close in age to the campers; in both cases, young adults are charged with the care and discipline of slightly younger campers.

Camp E-Wen-Akee

Camp E-Wen-Akee was a therapeutic camping programme that also drew on practices of wilderness therapy in their treatment of troubled boys and girls, ages

13–17. The camp was owned by the private non-profit Eckerd Youth Altern-
atives which operated a string of camps along the US east coast from 1969–2012;
this camp was located in rural Vermont in the north-eastern US. Camp E-Wen-
Akee mainly served adjudicated youth who were sent by the state of Vermont's
social services division for therapeutic treatment. A total of 30 campers lived in
the camp and were divided into three separate groups, one for girls, one for boys,
and one for male sex-offenders. Campers entered and exited the camp on a continuous basis. Each group had three or four primary counsellors, called 'chiefs',
who were responsible for providing everyday therapy, leading the group through
daily tasks such as cleaning, preparing firewood and organizing recreational
activities like group games and swimming. They were also responsible for disciplining campers primarily through instant group meetings called 'huddles' but
occasionally through physical restraints. Additional staff persons at camp
included 'master counsellors' (MCs) who provided supervision and a host of
other people who ran the camp's administration, individual therapy, education
and facilities. Individuals working in each of these positions directly engaged
with campers for varying lengths of time, but only the chiefs lived full-time with
the group members, literally moving with them through day and night.

The campers and chiefs lived in campsites comprised of several winterized
tents, which the groups built themselves, each containing a woodstove for heat.
They had no electricity or running water and relied on primitive outhouses.
Meals were taken in a group cafeteria at the centre of the camp, or were cooked
by the groups in their respective campsites. Several times a year groups left the
camp to go on extended canoe or hiking trips. The imagined geographies of rural
nature and wilderness were both drawn upon in this programme, with wilderness
trips providing campers the sense they were travelling to even more remote locations than their own rustic outdoor sites back at camp.

Methods

While the broader project drew on several ethnographic and qualitative research
methods, the most salient information came from participant observation. I lived
with the groups for two to three days at a time, participating in activities, watching the relationship dynamics as they unfolded, and speaking with campers and
staff as we moved through events. The fieldwork took place over seven months
in 2003. I received consent from each staff member and camper for their inclusion in the research. All identifying information has been changed to protect the
identities of the participants.

Campers and chiefs

The campers and their counsellors all identified as white and had grown up in
the United States. The campers ranged in age from 13 to 17 years. The youngest
chiefs I interviewed were 23 and the eldest was 32 years old. The majority of
chiefs were 23–26 years of age. Both chiefs and campers would classify as

'youth' in the United Nations definition of the term (UNESCO 2014). As Arnett (2010) has pointed out, recent research suggests that people in Western countries increasingly identify the start of 'adulthood' as occurring in the mid-20s. In Mannheim's terms, the campers and chiefs would occupy a common generational location, meaning they together had lived through common events such as September 11, 2001, and were familiar with the same popular culture symbols such as 'Mutant Ninja Turtles' yet they did not cohere as an 'actual generation'. Beyond the obvious fact that chiefs held authority over the campers, personal histories and class differences, along with the perceptions of the very outdoor setting in which they lived varied too greatly to constitute a single generational category. A closer look at the series of events that brought campers and counsellors to Camp E-Wen-Akee demonstrates the ways in which social difference works to separate people, even those of the same age.

Campers

With one exception, all of the campers I met at Camp E-Wen-Akee were legal wards of their home state. Some were already in the foster-care system before they were accused of a legal offence, and others entered the system only once charged with an offence. All were adjudicated, which means that their cases were put before a judge who determined their educational, health and restitution needs ('adjudicated' does not connote guilt of a crime, however). Offences and behavioural problems amongst campers included theft, truancy, drug use and assault. Some of the campers in the sex offender group had raped or attempted rape, and others had been caught stalking or exhibiting other threatening sexual behaviours. Nearly all of the campers had experienced some form of neglect and/or abuse in their lifetimes. The mother of one girl had 'traded' her as a baby to a woman in exchange for drugs and an old car. Another boy, according to his chief, 'was raped every day of his life for five years'. Many knew what it was like to have insufficient food or warm clothing. Nearly all of the students struggled academically and all had difficulty interacting with others. Each camper was on an individual therapy plan that identified specific behavioural skills they were working to master. Every camper was sent to camp by adults who did not consult them in the decision-making process.

Chiefs

Each of the chiefs held a bachelor's degree and all expressed a keen interest in outdoor recreation. Most had grown up in what they referred to as 'middle- or upper-class' families and had not been exposed to wilderness recreation until they attended college. Most enjoyed hiking, rock-climbing, mountain biking, camping, skiing and similar activities. Many constructed their identities in relation to outdoor activity. One woman told me she wanted to be a 'wildernessy sort of gal'. Several said that when they graduated from college they knew they did not want to work indoors and the draw to work at this wilderness therapy programme was primarily its outdoor setting. Only two to three staff people told me they were drawn to Camp E-Wen-Akee to work with troubled youth and none was trained in therapy or social work prior to their arrival. Typically, they were fresh out of a liberal arts college and were eager to do good and be outside.

Class, lifecourse and what nature is good for

The job of a chief was a combination of counsellor, disciplinarian and care provider. It was physically and emotionally demanding work, especially given that chiefs sometimes worked five 24-hour days or more without time off from the campers. Under any circumstances, it can be trying to live with ten behaviourally challenged youth, especially considering the chiefs were youth themselves. Adding to these challenges, the campers often did not hold the same enthusiasm for living outdoors as did staff members. Therefore, the role of cheerleader-for-the-outdoors got thrown into the chiefs' job too. In fact, chiefs' and campers' strikingly different perceptions of nature were one of the first signs of the class and lifecourse differences between them. The following excerpt from my field notes illustrates this difference.

> Chiefs Stanleigh, Stella and Alice and I are quietly talking in the chiefs' tent while the campers have 'siesta', their mandatory quiet time in the afternoon. The girls' group is preparing to go on a multi-day hiking trip. I ask Stanleigh, 'How are you feeling about the trip?' She replies, 'I have anxieties about how bitchy they [the campers] are going to be and how many times we'll hear "I can't"'. Stella, a new chief says, 'I feel it's easy to counteract. We're just not going to listen. These are people who have never worn a pack.' I ask Stella if she has done a lot of hiking, she says she has. She adds, 'I am excited. It will be a nice break in the routine of the group and the organization. It'll be an opportunity for coming into myself as a new leader. Hiking, I already feel comfortable with that – it's the other things that I'm not [meaning working with the group]'. Chief Alice who has worked her way up to a master counselor and has been at camp for several years muses, 'I ask myself, "Will taking them on an extended hiking trip ruin it for me?"' The women talk about how arduous it is to motivate the girls on a trip. 'Yeah, it's really important to me [being outdoors] and I don't want to make it so I dread it,' adds Chief Stanleigh.

The young women explained that nature provides them 'a backdrop for building relationships with people I'm traveling with', 'serenity' and 'beauty'. Chief Stanleigh added, 'I love to go out at night and look at the stars. It's my one on one time.' They experienced wilderness as aesthetically pleasing, a place to sense interconnectedness, and as a site for recreation. Implicit in the camp's protocols and practices was the notion that outdoor pursuits were good for the campers, but it took work to convince the campers of this. In the following field notes, notice how Stanleigh attempts to steer the campers toward a positive view of living outside:

The campers are preparing to eat dinner. As the group waits for their plates to be handed to them, I describe a research project I'd like the girls to work on with me. Because there are new girls in the group, I begin by describing my research question: 'Why use nature as the site for therapy programmes for teenagers'? All at once a conversation about the deficiencies of the nature setting erupts. Sydney says, 'The ones [programmes] not in the woods are better.' She says that the programmes that take place in houses and towns are better 'cause they're in civilization.' Another girl says they are 'separated' and 'segregated' in the Eckerd programme. Other girls join in, listing the problems with living outside: 'nasty bugs', 'getting eaten alive', 'you don't have to be afraid to go to the bathroom [because there are wolf spiders in the outhouses]'. Chief Stanleigh looks surprised and says to the group, 'Remember the chapel [a values-based time for reflection] we had on this? You guys said you could concentrate better here [in nature].' The girls deny this now and say that they could run this programme inside and it would still work.

The dinner conversation provoked me to ask in my field notes, 'if living outside is so wonderful, why is it something that the counsellor needs to teach?'

Each of the chiefs I spoke with discovered wilderness recreation while in college, often as part of a university-led activity wherein participants are taught the skills necessary to travel through remote locations and are issued high-performance outdoor gear. When they returned home from a trip they had hot showers, food and a warm sleeping space waiting for them. Contrast these outdoor experiences to those of two campers: Emma and Curtis. I asked Emma about her experiences in the outdoors prior to coming to camp. She said that she spent some time living in the woods in a tent and a bus with her 21 family members when they had no house to live in. Curtis lived alone for three months in winter when his father abandoned him at age 13. Living in a heated tent at camp was an improvement over his unheated trailer at home (winter temperatures can dip to $-20°F$ where he lived). One chief recounted that even when it was chilly and the firewood they had to burn was green, Curtis would say, 'Oh, at least there's some heat.'

Emma and Curtis had never been exposed to the experience of outdoor living as a recreational pursuit; for them, it was a survival strategy. Many campers had moved frequently during their childhoods, from home to home and between family members and foster families. Many had spent a good deal of time unsupervised by adults, and in this way their spatial experience of home differed significantly from the chiefs, who described more stable home lives. Jamieson and Milne (2012) found that the spatial mobility of British young people growing up in disadvantaged neighbourhoods was more similar to that of children living in Majority world contexts than their Minority world privileged peers who experienced more adult supervision and spatial segregation. The campers, like the children in Jamieson and Milne's research, may have spent more time outside than their middle-class counsellors, but it was not time spent engaged in formalized recreational activities such as skiing, hiking and the like.

Environmental scholars have asserted that nature narratives have classed dimensions; the notion that wilderness is beautiful, rejuvenating and peaceful has historically been held by an elite, literate, white cultural group which did not have to make its living in the wild (Bullard 1993; DeLuca 1999; Nash 1982). Richard White (1995), for example, contended that wilderness debates are fuelled by differing notions of what nature is good for, and that these opinions could be categorized as urban/middle-class notions that the environment is a site for leisure, and rural/working-class ideas of nature as a site for work, not play. I was reminded of White's essay when Chief Lulu said to me, 'When I see a deer I think, "oh cool" and the campers think "I want to shoot it".' Implicit in her comment was the idea that for the rural working-class, deer are venison (food) and for the middle- and upper-classes, deer are a delight to behold (aesthetic experience). Many of the campers were enthusiastic about hunting but none of the chiefs mentioned hunting as a recreational activity. This reinforced my initial impression that a classed process was at work at camp, dividing the young campers from the slightly older chiefs.

Some staff members were sensitive to the fact that they had greater access to material resources and other privileges than the campers, while others were remarkably insensitive. Chief Chuck, a camp administrator, explained to me that when he worked with the campers he made it a point to wear the same military-issue jacket that they gave to all campers, while many of the younger staff wore their high-performance gear, seemingly oblivious to the fact that their gear made them more physically comfortable than the campers. Chief Dexter spent an educational session showing campers the new mountain bike he purchased, explaining the extensive research he did to be sure he was making the right purchase. The mountain bike cost thousands of dollars. While Dexter's intention may have been to model sensible consumer practices and introduce campers to a healthy activity, he seemed unaware of the fact that such a purchase was far beyond the means of any of the campers. I thought of Emma who said that her life aspirations were to have a home, a computer and telephone, as her family lived without a phone even when they had a house.

Each of these moments seemed instances when class difference fractured the 13–30-year-old camp residents into two disparate social groups: the urban, educated 'haves' who valued the outdoors for recreation, and the rural, uneducated 'have nots' who saw the outdoors as both a dangerous place and a site from which to extract resources like venison and firewood. An 'intersectional' perspective on this dynamic shows multiple modes of difference at work: identities based on residence, education and class. As tempting as it is to identify class as the main process in this story, there are problems with the class perspective, the first of which is that cultural practices and values do not always align with income or livelihood categorizations. In Vermont, for example, people of all income levels hunt deer for venison and recreation (Boglioli 2009). Further, one of the hazards of using a hierarchical class-based categorization is that it assumes the existence of an 'underclass', a group that is defined as producing surplus labour, deviant behaviour and chronic poverty (Gans 1999; MacDonald 1997).

Deviant behaviour is pervasive throughout society and cannot be accurately linked to a class position; to do so can elicit discussions about the deservedness or unworthiness of the poor (Gans 1999). It could be argued that most of the campers operated outside of class definitions because without a family and a home, there was no place to affix their class identity, and 'underclass' would be the only conceivable option in this model.

Gibson-Graham (1996) have argued for a different definition of class in which class is neither an exclusive nor a fixed category of difference. Instead, they proposed that class be conceptualized as a set of processes which operate through various contexts. Class-as-process is useful in this case study because it sheds light on how wilderness narratives are crafted as outcomes of personal life experiences, which in turn are influenced by a range of other factors and conditions including cultural practices, access to material resources, and events that are unique to the individual. As well, the notion that class and other social positions are fluid and offer ways to build new identities and forge new alliances provides the opportunity to see how class- and age-based divides can be mended. 'Class' remains a helpful category in so far as it roughly collects and drafts a set of cultural values which tend to coalesce around privilege, income and position, yet it should always be applied with an awareness of the ways in which it interacts with other modes of difference. In the following paragraphs, I suggest that lifecourse differences served to divide the youthful residents of Camp E-Wen-Akee but also elicited instances of caring and support in which age and class were trumped by genuine relationship-building in an extrafamilial setting.

Authority, ordering and life experience

In the dinner discussion cited above, I noted that the campers used the words 'segregated' and 'separated' to describe their experience of camp. At other times I heard campers use the terms 'kicked out' and 'sent away' to refer to earlier experiences of home life, school and foster care. As Sibley (1995) has pointed out, marginalized populations are often sent to the geographic peripheries of human societies; wilderness therapy draws on this spatialization to separate youth from specific social influences and instil new social behaviours and norms. The campers understood they were not welcome in the centre of their communities and had been relocated to camp in order to become 'good citizens' (as one social worker put it). Camp's therapeutic process was made possible by what can be considered an artificial or manufactured ordering (for a Foucauldian analysis of this disciplinary order at work, see Dunkley 2009). An essential piece of this ordering was to empower staff with disciplinary control and to render campers compliant to the therapy strategies of the programme. In order for this to work, the 'youth' residing at camp were classified into two. The classification of an individual as a chief or camper resulted from their lifecourse history. Chiefs had the opportunity to attend university, the skills and networks to find work, and the wherewithal to move through their lives without resorting to anti-social behaviours (or perhaps, were simply not caught engaging in 'bad' behaviours). Campers, generally speaking, had shallow personal networks with few material, familial or other supports to navigate their lives. Traumatic experiences and anti-social behaviours punctuated their lifecourses. Much of the time, the rigid social division of chiefs and camper was maintained, yet there were moments when the border between the two dissolved, showing how modes of difference operate temporally and unevenly.

Camp's basic therapeutic procedure was to engage students in everyday activities and when behavioural problems manifested, the activities were ceased and identification and resolution of the problem was performed by the entire group. If a person's behaviour threatened others physically, or if a camper began to run away, the chiefs were authorized to physically restrain a camper. Chiefs, therefore, held a great deal of power and drew on the network of master counsellors, security guards and therapists to exercise this power. In such moments, it was clear that chiefs, regardless of their age, were authority figures. Yet, in some situations, the campers held authority. Campers lived from one to three years at E-Wen-Akee. Chiefs, on the other hand, cycled in and out of the position frequently, often leaving their job before the two-year contract was complete. As long-term residents of camp, the campers often knew more about camp practices and procedures than their chiefs. One day I watched as Stephen cut perfect one-inch pieces of wood from a small log with an axe. The wood pieces were used to light the small campfire for a 'pow wow' ceremony that opened and closed each day, one of the most important therapeutic practices in camp life. Chief McBragg said, 'He's better than I am', referring to Stephen's expertise with the axe. One day I heard a chief ask a long-time camper how much cream cheese they should order for the week's supply. In moments such as these, longevity, local knowledge and skill overrode lifecourse, age and authority privilege, and provided campers the opportunity to know they were valued.

Mimicking family: transgressing difference through relationship

In his analysis of an American programme that sends inner-city youth (mainly children of colour) for summer stays in the home of rural and suburban (mainly white) families, Vanderbeck (2008) identified notions of racial difference, belief in the healing power of nature, and narratives about the 'moral order of the family' as the dominant discourses embedded in media coverage of the programme. The 'family' emerged as essential framework at Camp E-Wen-Akee as well, and as in Vanderbeck's study it was assumed that this new 'family' would be somehow better than the camper's original family. Here 'family' was not an existing unit of parents and children, but a set of practices. Role playing at family relations was intended to instil new moral behaviours amongst campers, and to demonstrate authentic care for others.

When I asked her what made camp work, Chief June, a highly respected, long-time staff member replied, 'relationships'. Despite the fact that some chiefs came and went faster than some of the campers, chiefs were often the first people to consistently demonstrate care for campers. They sat with campers in the rain as someone relived a painful memory. They put baby powder (talc) in wet hiking boots, and played endless games of capture the flag. Chief Chuck explained, 'Just having someone wake them up in a kind way is therapy for these kids.'

The campers notice these kindnesses. When the girls' group returned from their hiking trip, I asked them to describe how it went. They told me that it was difficult but one night the chiefs relieved the campers of chores and cooked dinner for them in the rain. Another day a chief hiked miles off the trail and into town to buy them candy bars. I asked how those actions made them feel, and Julie said 'It made me feel that they do care for us.' I also noticed in that discussion subtle efforts that chiefs made to promote positive self-esteem amongst campers by reframing the narration of events. For example, Austasia said she was so miserable that she complained for the first half of the week and then stopped talking the second half of the week just so she would keep herself from complaining. Chief Stella would not let Austasia's account end on that note and pointed out, 'I thought the last night you were really positive. You made extensive journal entries like how funny the day was.' Like a parent might, chiefs noticed and affirmed the progress campers made, even in the smallest ways.

Chiefs sought to provide campers with a stable routine and a set of therapeutic practices that were consistently applied as difficulties arose (in essence engaging in many of the kinds of behaviours that a 'good parent' is meant to provide for their children). Camp used such ordering to both separate disruptive youth from society, but also to put them in close proximity to other young people who were charged with their care. As Valentine (2008) has shown, proximity or co-location of people from different social groups does not inevitably lead to mutual respect or an end of prejudice. In the case of Camp E-Wen-Akee, co-location was the first step in reproducing something close to an intimate family setting. The second step was the engagement of staff and campers in a range of activities which encouraged meaningful contact. It was important that campers saw a chief gracefully lose a game of cards or acknowledge that a camper was more skilled than themselves in chopping wood or drawing. Such interactions opened an opportunity for campers to see how others dealt with adversity and difference, and to catch a glimpse of their own value, potential to change and ability to overcome difficulty. The director of the camp shared with me a story that dramatically illustrated this process. While doing a high element of a ropes course, one of the campers 'froze' on the platform. He said he was too afraid to walk out onto the rope suspended high above the ground, and too afraid to climb down the tree he had just climbed up. The director, who was then a chief, climbed up to the platform and stayed with the camper while the group waited below. They stayed for hours, even as dark descended. Then, without warning, the boy stepped out on the rope and was belayed down to the ground where he burst into tears. 'Now I know how I made my victims feel', he told his chief.

Conclusions

Camp provided a physically and emotionally safe environment, an institutional site comprised of non-family members which became home and family for many young people who never had experienced a stable home location or predictable family relations. When I visited the boys' group which was spending time at an Eckerd facility in New Hampshire, I asked Josh how he was doing on the trip. He said, '[I'm]homesick. Because I've been at camp so long that it's my home. And I'm happy.'

The story of Camp E-Wen-Akee illustrates the fact that like the pieces of wood Stephen split apart from a single log, a generation can be splintered by any number of social factors. Camp hired privileged, educated and outdoor-oriented youth to engage underprivileged, educationally challenged and anti-social youth in activities that would lead the troubled youth into socially acceptable lives. A critique of this process as simply the imposition of dominant middle-class norms onto the 'underclass' is easily done, but unfair. What I hoped to have shown in this chapter are the complex ways in which small age differences, class position, personal histories and place perception strongly influence the shape of young people's lives, even those who share a generation. As well, it has been my aim to highlight instances when 'splintered generations' can overcome difference through acts of genuine connection, upending normalized and static notions of generation, family and home.

References

Arnett, J. (2010) 'Oh, grow up! Generational grumbling and the new life stage of emerging adulthood-commentary on Trzesniewski & Donnellan (2010)', *Perspectives on Psychological Science*, 5(1): 89–92.

Beringer, A. (2004) 'Toward an ecological paradigm in adventure programming', *Journal of Experiential Education*, 27(1): 51–66.

Boglioli, M. (2009) *A Matter of Life and Death: Hunting in Contemporary Vermont*, Amherst, MA/Boston, MA: University of Massachusetts Press.

Bondi, L. and Davidson, J. (2003) 'Troubling the place of gender', in K. Anderson, M. Domosh, S. Pile and N. Thrift (eds) *The Handbook of Cultural Geography*, London/ Thousand Oaks, CA/New Delhi: Sage Publications.

Bullard, R.D. (1993) 'Anatomy of environmental racism and the environmental justice movement', in R.D. Bullard (ed.) *Confronting Environmental Racism: Voices from the Grassroots*, Boston, MA: South End Press.

Cronon, W. (1995) '"The trouble with wilderness" or, getting back to the wrong nature', in W. Cronon (ed.) *Uncommon Ground: Rethinking the Human Place in Nature*, New York, NY: W.W. Norton & Company, Inc.

DeLuca, K. (1999) 'In the shadow of whiteness: The consequences of constructions of nature in environmental politics', in T.K. Nakayama and J.N. Martin (eds) *Whiteness: The Communication of Social Identity*, Thousand Oaks, CA/London/New Delhi: Sage Publications.

DeLuca, K. and Demo, A. (2001) 'Imagining nature and erasing class and race: Carleton Watkins, John Muir, and the construction of wilderness', *Environmental History*, 6(4): 541–560.

Dunkley, C.M. (2009) 'A therapeutic taskscape: Theorizing place-making, discipline, and care at a camp for troubled youth', *Health & Place*, 15(1): 88–96.

Gans, H. (1999) 'Deconstructing the underclass', in P. Banjeree, M.D. Gupta and C. Rosati (eds) *Readings in Social Problems*, third edition, Needham Heights, MA: Pearson Custom Publishing.

Gibson-Graham, J.K. (1996) *The End of Capitalism (As We Knew It): A Feminist Critique of Political Economy*, Cambridge, MA: Blackwell Publishers.

Gibson-Graham, J.K., Resnick, S.A. and Wolff, R.D. (2000) 'Introduction: Class in a poststructuralist frame', in J.K. Gibson-Graham, S.A. Resnick and R.D. Wolff (eds) *Class and Its Others*, Minneapolis, MN: London: University of Minnesota Press.

Haraway, D. (1989) *Primate Visions: Gender, Race, and Nature in the World of Modern Science*, New York, NY/London: Routledge.

Hopkins, P. and Pain, R. (2007) 'Geographies of age: Thinking relationally', *Area*, 39(3): 287–294.

Jamieson, L. and Milne, S. (2012) 'Children and young people's relationships, relational processes and social change: Reading across worlds', *Children's Geographies*, 10(3): 265–278.

MacDonald, R. (1997) 'Dangerous youth and the dangerous class', in R. MacDonald (ed.) *Youth, the 'Underclass' and Social Exclusion*, London/New York, NY: Routledge.

Mannheim, K. (1952 [1923]) *Essays on the Sociology of Knowledge*, London: Routledge and Kegan Paul.

Mayall, B. (2001) 'Introduction', in L. Alanen and B. Mayall (eds) *Conceptualizing Child–Adult Relations*, London: RoutledgeFalmer.

McNeil, E.B. (1957) 'The background of therapeutic camping', *The Journal of Social Issues*, 13(1): 3–14.

Nash, R. (1982) *Wilderness and the American Mind*, third edition, New Haven, CT: Yale University Press.

Oxford Dictionary On-Line (2013) 'Generation', online, available at: www.oxforddictionaries.com (accessed 30 October 2013).

Paris, L. (2000) 'Children's nature: Summer camps in New York State, 1919–1941', PhD dissertation, University of Michigan.

Paris, L. (2001) 'The adventures of Peanut and Bo: Summer camps and early-twentieth-century American girlhood', *Journal of Women's History*, 12(4): 47–76.

Pratt, G. (1998) 'Grids of difference: Place and identity formation', in R. Fincher and J.M. Jacobs (eds) *Cities of Difference*, New York, NY/London: The Guilford Press.

Sibley, D. (1995) *Geographies of Exclusion: Society and Difference in the West*, London/New York, NY: Routledge.

Smith, M.B. (2002) '"And they say we'll have some fun when in stops raining": A history of summer camp in America', PhD dissertation, University of Indiana, Bloomington, IN.

Tarrant, A. (2010) 'Constructing a social geography of grandparenthood: A new focus for intergenerationality', *Area*, 42(2): 190–197.

Tucker, F. (2003) 'Sameness or difference? Exploring girls' use of recreational spaces', *Children's Geographies*, 1(1): 111–124.

UNESCO (United Nations Educational, Scientific and Cultural Organization) (2014) 'What do we mean by "youth"?' Online, available at: www.unesco.org/new/en/social-and-human-sciences/themes/youth/youth-definition/ (accessed 23 February 2014).

Valentine, G. (2008) 'Living with difference: Reflections on geographies of encounter', *Progress in Human Geography*, 32(3): 323–337.

Van Slyke, A. (2006) *A Manufactured Wilderness: Summer Camps and the Shaping of American Youth, 1890–1960*, Minneapolis, MN/London: University of Minnesota Press.

Vanderbeck, R.M. (2007) 'Intergenerational geographies: Age relations, segregation and re-engagements', *Geography Compass*, 1(2): 200–221.

Vanderbeck, R.M. (2008) 'Inner-city children, country summers: Narrating American childhood and geographies of whiteness', *Environment and Planning A*, 40(5): 1132–1150.

White, R. (1995) '"Are you an environmentalist or do you work for a living?": Work and nature', in W. Cronon (ed.) *Uncommon Ground: Rethinking the Human Place in Nature*, New York, NY: W.W. Norton & Company, Inc.

16 Place-responsive intergenerational education

Greg Mannion and Joyce Gilbert

Introduction

In this chapter we outline two premises that provide a viable theoretical basis for intergenerationally focused, place-responsive education. On their basis, *intergenerational learning* is said to have occurred when people of more than one generation respond to generational differences within a place. The contribution we seek to make comes from bringing together theories of the socio-material (Barad 2003, 2007; Grosz 2005; Edwards 2010; Hultman and Taguchi 2010), place and learning (Casey 1993; Nespor 2008; Somerville 2010; Somerville *et al.* 2011; Gough and Price 2004), embodied experience (Grosz 1994; Ellsworth 2005; Semetsky 2013), non-representation (Thrift 2003, 2008; Lorimer 2005; MacPherson 2010), and intergenerationality (Mayall 2000; Vanderbeck 2007; Hopkins and Pain 2007; Mannion 2012; Field 2013).

The chapter is structured as follows. First, we provide some context by outlining some selected policy issues around intergenerational programming and education. After providing a definition of intergenerational education, we set out two premises that inform our view of intergenerational education:

1 people and places are reciprocally enmeshed and co-emergent, and
2 people learn through making embodied responses to differences.

We will provide sources for our rationale for these two premises. The two premises support a few key ideas emerging in the theory and practice of intergenerational practice: the importance of reciprocity of relations between generations, and the importance of 'place' in everyday embodied encounters between different generational members, and the idea that intergenerational learning is made possible through responding to differences which are found in the relations between generations and the relations between people and the places they inhabit. Whilst the premises will have wider application, we apply them here to our consideration of intergenerational education and derive some consequences for practice. One key consequence is that communication and meaning-making in (intergenerational) education can be viewed as a non-representational practice or reconfiguring ourselves within the material world.

Space here does not allow for an in-depth analysis of any empirical data from our ongoing research which are described elsewhere (see Mannion and Adey 2011; and 'Stories in the Land' blog: www.storiesintheland.blogspot.co.uk).

Policy contexts

International policy regimes are shifting radically along generational lines. Policy terms such as 'active ageing', 'troubled youth' and 'family learning' flag the desire of governments to take greater account of generational difference and the desire to harness intergenerational encounters and programmes in pursuit of social goods. In a non-exhaustive manner, it is perhaps sufficient here to notice three ways in which policy appears to validate intergenerational approaches to education.

First, we notice the many policies regarding the changing demographics of developed world populations. Policies here centre on the desire for the development of a 'society for all ages' as the age structure of the population is changing considerably with lifespan increases. Currently, in the UK, there are ten million people over 65 years old. This number is projected to nearly double to around 19 million by 2050. A policy agenda has arisen within the UK, EU and OECD countries around the needs of this increasing population for active ageing, meaningful work if they desire it, and their participation in activities that enhance wider social cohesion and inclusion. The focus on intergenerational relations in policy arises in the US in the 1960s (Sanchez 2007) amid concerns about a widening generation gap, a rise in services for the young and old and, as a result, threatened social harmony and cohesion. Since then, an interest in intergenerational *practice* within and outside of formal education has grown since it is seen as ameliorating some of these challenges and social problems.

The second policy area pertains to the ongoing integration of the UN Convention on the Rights of the Child into services, including educational services, leading to wider acceptance of children and young people as participants in society in their own right. In education, schools are seen as being remarkably resistant to fully taking on board these children's rights agendas. In theoretical debates, commentators now advance the idea that young people's participation cannot be understood outside of a consideration of place and generation (Mannion 2009).

Lastly, in international educational policy, in the face of globalizing economies, there are concerns to improve the quality of school-based provisions. In policy terms in Scotland, where the authors presently work, the local expression of this is found in the new Curriculum for Excellence initiative (Scottish Executive 2004). This curriculum emphasizes 'skills for learning, life and work' alongside the development of 'capacities' of successful learners, confident individuals, responsible citizens and effective contributors. New approaches to teaching and learning are advocated such as outdoor methodologies and are encouraged to harness a wider array of partners in the delivery of education (including, for example, voluntary organizations, NGOs, youth workers, families, employers, health, social work, police and community workers). A new kind of curriculum-making is evoked that harnesses new places, new partners and new processes linking schooling to new contexts beyond the classroom.

In policy terms, schools seem ripe for new and innovative experiments in intergenerational approaches, but the authors have found making intergenerational place-based education less easy to realize on the ground in part because professionals and locals alike are unsure about what constitutes place-related pedagogies for linking schools to local places and local people. In redress, this chapter seeks to delineate two viable premises for place-responsive intergenerational education and the consequences for practice thereof.

Defining intergenerational education

Internationally, within education, there have been efforts to define intergenerational learning and education. For example, the European Network for Intergenerational Learning (www.enilnet.eu), takes intergenerational *learning* to occur when people of different generations work *reciprocally* in gaining skills, values, and knowledge. Coull (2010) defines formal intergenerational learning as any planned activities between generations that results in achieving set objectives *for each generation involved.*

While some viable definitions of intergenerational education are emerging, theoretically informed discussions of these have been lacking in the literature until recently. Understandably, much of the earlier commentary and research on intergenerational practice has set out to *describe* practice and look at *outcomes* in health, leisure, educational, public service and personal development (Ames and Youatt 1994; Brown and Ohsako 2003). Kaplan (2002) notes, however, that even though one generation may be nominally the provider and another the recipient of some service, the outcomes may be reciprocally experienced.

In empirical intergenerational research, Hammad (2011) has inquired into the lived experiences of people in places of conflict, showing how differences and connections across the generations are sustained. In schools, there is some empirical evidence that participants from all generations can benefit through learning via intergenerational encounter and mutual engagement especially in environmental learning (for example, Mannion and Adey 2011; Duvall and Zint 2007; Peterat and Mayer-Smith 2006). Out of earlier empirical work on place-based intergenerational practice, Mannion (2012) offered the following definition of intergenerational education. It reads:

> Intergenerational education (a) involves people from two or more generations participating in a common practice that happens in some place; (b) involves different interests across the generations and can be employed to address the betterment of individual, community, and ecological well-being through tackling some problem or challenge; (c) requires a willingness to reciprocally communicate across generational divides (through activities involving consensus, conflict, or cooperation) with the hope of generating

and sharing new intergenerational meanings, practices, and places that are to some degree held in common, and (d) requires a willingness to be responsive to places and one another in an ongoing manner.

(Mannion 2012: 397)

Mannion (2012) goes on to suggest intergenerational education would aim to promote *greater understanding and respect between generations*, since, it is argued, without this outcome, almost any form of education that involves different age groups could claim the label. Hence, *improved intergenerational relations is a key distinctive outcome for intergenerational learning.*

But Mannion (2012) argued improved intergenerational relations are *not sufficient* for an educational programme since improved relations need themselves to be purposeful in some additional manner. Intergenerational programmes are always located some 'where', so they will generate new meanings, practices and places through their work. Granville and Ellis (1999: 236) argue that a truly intergenerational programme must show a benefit and value for both generations *and* 'demonstrate an improvement in the quality of life for both, and from that, an improvement in the quality of life for all'. Similarly, Mannion (2012) argues that intergenerational education will be a planned and progressive programme requiring the ongoing production of new and improved *relations* between adults and children *within and through place-change* processes. Behind the above definition lie many theoretical ideas and some key debates about agency, person–place relations, and meaning-making which this chapter seeks to explore in more depth. We commence this exploration with our two premises.

Two premises and their consequences

In this section, we explore two connected premises:

1 people and places are reciprocally enmeshed and co-emergent, and
2 people learn through making embodied responses to differences.

We consider that these premises will have wider application to other studies of person–place interaction, but we apply them here to intergenerational education. Taken together, we will argue that they provide support for the above definition of intergenerational education, for onward programming and practice and for researching forms of place- and generation-responsive education in formal, non-formal and informal settings.

Premise 1: People and place are reciprocally enmeshed and co-emergent

A range of theorists from across the arts and human sciences have been arguing quite cogently that people and places are best viewed not as separate objects, or separate objects in a distal relationship but, rather, are deeply reciprocally enmeshed and co-emergent. Educational theorists, anthropologists, geographers and social scientists alike often draw upon some key theoreticians such as Deleuze and Guattari, Latour, Haraway, Ingold, and Barad to support a broadly socio-materialist or posthumanist view where the person and their context are seen as both in process and co-emergent in different ways. Feminist materialist positions, too, provide an account of this relational position promoting the idea that people and places are linked in a reciprocally emergent and processual manner (Hultman and Taguchi 2010; Grosz 2005). Of the many sources we could explore, Tim Ingold and Karen Barad provide the main rationales for our first premise.

Ingold (2003), drawing upon Deleuze and Heidegger, asks us to understand people-and-place as a contingent unfolding interacting process. Ingold reminds us that people and places are relationally emergent through the activities of both people and many other entities and processes that allow life to unfold (including the weather, the activities of animals as much as humans). Ingold suggests that all living beings act within a unified field of relations (similar to Deleuze's singular plane of immanence). Thus, Ingold brings the non-human/ecological together with the human social world within a relational field of action where nature and culture interpenetrate. Within this view,

> organisms 'issue forth' along the lines of their relationships, then each organism must be coextensive with the relationship issuing from a particular source. It is not possible therefore for any relationship to cross a boundary separating the organism from the environment.
>
> (Ingold 2003: 305)

We characterize Ingold's position as ontologically reciprocal.

Material feminists and 'new science' scholars alike have made strong claims about the links between social and material cultures. Barad (2003) has argued that discursive practices and material phenomena need to be understood in a linked manner. By discursive practice, Barad means 'specific material (re)configurings of the world' (2003: 819). All languaging of the world happens in the world and is part of its ongoing reconfiguration. Rocks, soil, and people too, are seen as dynamic and undergoing change. For Barad, all 'things' including texts, people and all kinds of materials are going through similar dynamic processes of reconfiguration. Barad's position can be characterized as a unification of epistemology and ontology – an onto-epistemology, as she puts it.

Among the many issues that arise in theorizations of person–place, the question of where agency lies gives rise to much debate, particularly in socio-material and actor network (ANT) literature. Ingold (2008) draws on Deleuze and Guattari's notion of 'lines of becoming' to argue for a focus on the power of relations in meshworks (rather than nodes or entities in networks – as in the case of some ANT theorizations). By this view, 'action is not so much the result of an agency that is distributed around the network, but emerges from the interplay of forces that are conducted *along the lines* of the meshwork' (Ingold 2008: 212, italics added). In refocusing his definition of agency in this flowing, relational way – it is the actions of the spider that enlivens the agency of the web – Ingold reminds us that a form of agency may be the effect of technologies and objects but is likely to be revealing of a flow of others' agencies 'down the line': others such as animals and humans who may have set technologies into motion. Agency no longer 'resides' in one location or in the human; it proceeds dynamically along some connective tissue which allows for the enactment of a relationship; within these interpenetrating socio-material relations, people and place co-emerge.

Premise 2: People learn through making embodied responses to differences

Next, we wish to show how a reciprocal onto-epistemology of becoming can be aligned with the idea of making embodied responses to difference. The second premise is therefore concerned with learning in ways that take account of place and difference in reciprocal relation. Put simply, the premise is that learning is understood as responding in an embodied way to differences that arise in a place through reciprocal processes of people–place.

The onto-epistemological stance (Barad 2003) sees change in people and place in a coupled way: knowing (or learning) is ongoing creative, active and made possible through activity *within places*. Within an reciprocal ontology of becoming, learning is always *situated* and is an ongoing happening and, therefore, could be said to always be locally 'performed' (Thrift and Dewsbury 2000), in all cases, as a result of the responses people make within a particular person–place assemblage or enmeshment (Ingold 2011). Learning is influenced by what people do in and to a place and how places act back on them over time. Our enmeshment, as Mulcahy (2012) puts it, brings a sense of 'collective responsibility' for what ensues. This responsibility is distributed across technologies, bodies, teachers, learners, desires and places intersubjectively linking the social with the material in teaching and learning.

Barad's 'onto-epistemology' (Barad 2003) suggests learning is brought about by change in relation to other entities which are themselves in process of becoming *different*. Using Barad's (2003) account we can begin to build a posthumanist account of *difference as in-built into human–environment relations*. In part, like Ingold, this is the idea that humans are always in some place *intra*-acting with it, and reciprocally being affected by *changes* in the environment. Because humans are part of nature, intra-acting with it, the human's ways of knowing are inherently part of the unfolding of *differences* in nature. Hence, responding to these differences is central to learning and learning is part and parcel of becoming as an organism in a place.

Drawing on Ingold (2011), because processes are always unfolding and we exist within the ongoing events in educational processes, we suggest learning can only be understood to have happened in hindsight. It is only through looking back at events that we can create a story about how our knowing emerged in a connected way within the unified field of relations of our experience as a living breathing organism – the basis of the ontology of becoming outlined above. Following Ingold (2011), learning is a form of place-responsive enskillment that happens within and from our experiences of place-enmeshment. There are resonances here with pragmatism. As Dewey put it: 'There is no such thing as sheer self-activity possible: because all activity takes place in a medium, in a situation, and with reference to its conditions' (Dewey 1902, cited in Quay and Seaman 2013: 84). Quay and Seaman (2013) show us, through a useful re-reading of Dewey, how knowing, doing and being in a place are intimately connected indivisible evolutionary processes of organism-in-environment. Further, our embodied immediate, direct, emotional, aesthetic experiences are connected to reflection and to ethical action. Even reflection itself is a kind of experience in relation in a place.

The second aspect of this premise is that learning happens within our *embodied* state as organisms located within places. We suggest, alongside other authors (Perry and Medina 2011; Ellsworth 2005; Grosz 1994; Semetsky 2013) that embodiment is not an optional choice for learners but is an essential and pre-given corporeal, biological, sensual, social, cultural feature of our experience. After Deleuze and Guattari (1988), Semetsky likens all learning to a form of groping, embodied experimentation akin to learning to swim for the first time. We learn through experience which brings body and mind, person and context together in some new practice: 'a body actualizes in practice the multiplicity of its virtual potentialities' (2013: 82). It is through our bodies that we learn in relation to differences found in various structures discourse, times, places and other people. Following Perry and Medina (2011), we note that in diverse cultures learners will be able to draw upon various 'conventions of embodiment' in order to participate in cultural meanings and generate critiques. They argue this is made possible through pedagogies of embodied performance. Somerville (2010), too, argues for the body being the centre of our experience of places. Building on a similar relational ontology, Somerville suggests that places are the zone of cultural contact and that people and place are dialogically interpenetrative through stories. Through stories, the self is in process, becoming 'other' within a socio-material dynamic interaction within a place. As Somerville (2010: 342) puts it, 'This becoming-other is a relational ontology that includes the non-human and inanimate "flesh of the world" as well as human others'.

Consequences for intergenerational practice and education

From constructivism to perceiving and apprehending the world

Our premises suggest some challenging consequences for how people come to know and for the elements that are commonly assembled to make any curriculum possible (such as texts, images, accounts, narratives, histories and other kinds of 'content'). Mostly, these consequences stem from premise one, the idea that humans are not separate from places (or nature). We do not sit separately as part of culture mentally 'constructing' nature as a detached spectator view of the world 'objectively seeing' it. As Hekman (2008: 109, italics added) puts it 'concepts and theories have *material* consequences' but 'there is a world out there that shapes and constrains the consequences of the concepts we employ'. A first consequence, therefore, of premise one suggests educational programmes of intergenerational and place-based education might do well to consider teaching and learning in a non-constructivist manner. This is in fact demanded by accepting a reciprocal ontology of becoming where agency is best seen as an effect of relations in a socio-cultural meshwork. Replacing the ideas of mentalist human-derived constructivism is entirely possible however. Ingold (2011) draws inter alia on Heidegger and Deleuze to offer the view that people are engaging in continuous acts of *perception*, at once *apprehending* the world, *transforming* it, and *being transformed by it* (as a plant drawn to the light).

From representations to performative reconfigurations

Gould and Ingold's views challenge us to think again about the communication processes we commonly think of as *representing* the world. Gould (2012) contrasts language used as representative with language used as 'performative enactment'. In the former view 'language based on representation attempts to represent aspects of reality as accurately and completely as possible by tethering new information to old, limiting teaching and learning to what teachers and students already know (Deleuze 1994)' (Gould 2012: 197). In contrast, when language is considered as a performative act, 'teaching and learning consists of responses to readings' (Gould 2012: 199). 'Reconfiguration' is another useful term here. In educational processes, harnessing Barad's ontology of becoming, there are no fixed things. Representations, therefore, cannot be possible (in a correspondence view). The world needs to be understood as undergoing continual re-enactment, reconfiguration or performance where individual agency is only made visible in dynamic interactions or 'reconfigurings'.

Taken together, and pushed to their limit, the two premises suggest there is the need to give up on traditional mimetic views of how representations work. One way forward is to accept that all representations are to some degree ongoing presentations, performances (see MacPherson 2010) or reconfigurings (Barad 2003). Ross and Mannion (2012) suggest the world becomes meaningful for its inhabitants through *active inhabitation* and *not through cognitive representation*. Within our active inhabitation (alongside other organisms) we come to know in an entangled or interlaced manner in relation to other species and the environment wherein organisms are 'points of growth'. In place of representing the world 'as is', we must continually bring it into being, 'performing' (Thrift and Dewsbury 2000) it through perception and interaction within environments.

With our premises in mind, our performative reconfigurings clearly form part of the ongoing flow of life. But this means continually challenging the boundaries around which entities (such as a curriculum) can be circumscribed. As Somerville says, 'individual representation is conceived as a pause in an iterative process of representation and reflection, and as contributing to assemblages of such artefacts whose meanings are intertextual' (Somerville 2010: 342). Gould (2012) similarly argues that if an educational activity does not catalyse a response via some *new and multiple readings* of some aspect of the world, it is likely to be an impoverished form of educational activity merely requiring some form of mirroring and repetition.

A new set of verbs for the workings of educational discourses and texts, and images is required; performing is one. Within a non-representational approach, knowledge (in texts, or stories, plays or websites and so on) is there not to be represented, but to be performed, or 'witnessed' or 'narrated' (Jones 2008) and it is through these communications experiences can be 'gained' or responses 'evoked' or 'solicited'. Gough and Price (2004) (after Deleuze) suggest other verbs as replacements for 'representation', including 'implicate', 'propagate', 'displace', 'join', 'circle back', 'fold', 'de-/re-territorialize'. Whichever performative verbs are used, within a non-representational view, there are no non-neutral educational probes to simply mirror a world as it is. The non-representational view of teaching and learning asks that we know by doing and through ongoing journeying *with/in* places, places that are themselves changing.

Open-ended experimentation via assembling

Ross and Mannion (2012) note that within Ingold's process 'dwelling' ontology, the business of teaching and learning must be seen as an *open-ended* relational activity. As Somerville puts it, '[a]ny pedagogy of place must remain open and dynamic, responsive to the interaction between specific people and their local places' (Somerville 2010: 342). The need to experiment is another consequence. Edwards (2010: 13) suggests 'that a post-human condition could position learning as a gathering of the human and non-human in responsible experimentation to establish matters of concern'. The educator's job might, therefore, be to help learners with raising intergenerational matters of concern in an experimental way. Drawing on Hultman and Taguchi (2010) and Edwards (2010), we might suggest the educator's role is not so easily limited to skill acquisition or the imparting of knowledge, but rather to *evoke responsibility* in their own work, and in learners through *assembling* the human and non-human in new ways and demanding some form of response via active engagement within places through working with issues that span generational boundaries. Unfortunately, these consequences mean we cannot easily legislate for either the assured outcomes of education nor for the approaches we should take to get there. Our position is not to say that learners *should* encounter difference through, say, going outdoors and meeting local people from other generations, but rather, that we *could* or *might* benefit from such encounters and that these encounters are inexorable since we live in a unified field of relations as humans in places. In our view of education, *relations* (rather than individually held skills, knowledge or attitudes) become more significant. Clearly, at a time when educational outcomes are prespecified in often decontextualized generic skills-based terms, this presents a set of challenges for the posthumanist educator wishing to work in a place- and generation-responsive manner.

From controlling to responding to place

Another consequence relates to the importance of place in teaching and learning. Mannion *et al.* (2013) sign some of the consequences of place-responsive pedagogy which can be extended to intergenerational contexts. They suggest responding to place involves explicitly teaching by-means-of-an-environment with the aim of understanding and improving human–environment relations. After Mannion *et al.* (2013), we suggest place- and generation-responsive curriculum-making will involve intra-actions among (1) educators' own experiences and dispositions to place and generation, (2) learners' dispositions and experiences of place and generation, and (3) the ongoing contingent events in the place itself (including, for example, the processes of weather; the use of technologies; the processes engaged in by other living things; and generationally mediated processes within work, schooling and leisure practices). In working within places across generational boundaries, we note that we do not feel it necessary for both generations to be physically co-present in one place at all stages of programming (see Mannion 2012) since places are often interpenetrated by other locations and connected through many lines of relation (for example, online).

From fixed self to becoming-in-relation

Another key consequence is that we are required to confront generational issues in our work (see Mayall 2000). Batsleer's (2013) summary of Irigaray's account of selfhood in relation to difference is of use here. Irigaray (2000) suggests we need to accept intergenerational difference as given and essential to the creation of some one in relation to others. The following principles (after Irigaray) provide a useful framework for making the premise on responding to difference consequential in intergenerational educational processes. These principles, we argue, will be pertinent whether one begins seeking to address the social inclusion of children (Mayall 2000), adults (Scharf and Keating 2012), or both (Mannion 2009). We require:

- A non-reducible commitment to the expression of intergenerational difference within the human and across the boundaries of the human with the animal and the human with the machine.
- A recognition of the non-reducibility of 'the other' to 'the same' and at the same time a recognition that it is in this way that speech comes to be possible.
- A foregrounding of a process of becoming subject in relation to generationally different others rather than a training of the subject by means of static knowledge.

- A respect for life and the existing universe rather than an education in the rule of the subject over places/the world.
- The learning of life in community rather than the acquisition of skills out of context.
- Construction of a liveable and more cultured future rather than submission to a tradition.

These principles refract two premises in different ways. They emphasize commitments to future performances rather than current conformities, the connection between knowledge and self-construction in relation to place and other, and the role of the non-human in reciprocal relation are all embedded in these principles. What learners need, therefore, in order for response-making to be possible, is some encounter or expectation of encounter with difference and the ongoing challenge to learn from and within traditions, and to change traditions through learning by altering existing habits when appropriate. Hultman and Taguchi (2010: 529) remind us that difference, in Deleuze's terms, is *positive* and arises through our 'connections and relations within and between different bodies, affecting each other and being affected'.

Conclusion

In this chapter we have outlined the sources of the rationales and consequences of two theoretical premises when used as a basis for intergenerational (and indeed many other kinds of) educational programming:

1   people and places are reciprocally enmeshed and co-emergent, and
2   people learn through making embodied responses to differences.

Taken together, we argue that our two premises are useful starting points for understanding, programming and researching intergenerationally lived experience in a relational manner (alongside other social signifiers such as gender, race and class). In a generationally responsive curriculum, we have argued that places are not backdrops to the social action, and that generational relations are linked to place–person relations. We have argued that humans are a constitutive part of places *and vice versa*, so they do not in*ter*-act with it *but rather in*tra*-act with it.* Our work suggests that schools and other educational settings can benefit from enrolling local community members from diverse generations *and* harnessing places into emergent processes of curriculum-making. By this view, learning is an all-age, emplaced process derived from encountering differences within social and material worlds. Because of the traditionally generationally niched and indoor nature of schooling, often separated off from people and places beyond the school, intergenerational educational practice is seen as particularly relevant.

However, our ongoing research points to the need to handle place and generation with care since the very differences that give rise to a viable curriculum also give rise to inherent tensions across generationally and place-based fault lines, for example, past/future, local/outsider, local/professional, school/community, and adult/child. By the same token, working within the contested zones of encounter across these fault lines is in fact the essential component of a vibrant and effective place- and generation-responsive educational intervention.

Place-responsive intergenerational pedagogies can be brought about through various forms of curriculum-making that involves intermingling the human and non-human, allowing the participating generations to be responsive to each other and to a changing and contingent environment. This work will involve educators, learners and their collaborators in actively seeking out *place-based intergenerational differences* and working with these in *non-representational ways*. The analysis here suggests this can be achieved through altering the boundaries for the participation of both the elements of place (materials, processes, other species) and the human in assembling curricula. In a place- and generation-responsive curriculum, differences can be found in our relations with place and with others through our embodied activities. On the basis of our two premises, *intergenerational learning* is said to have occurred when participants have responded to the generational differences. Becoming someone new through intergenerational education is tied to both place-based embodied material circumstances and to generational relations. Responding to the differences we notice in this work is core to generating the kind of knowledge necessary to create more intergenerationally inclusive, sustainable forms of eco-social flourishing 'in order to perhaps make it possible for others (humans and non-humans) to live differently in realities yet to come' (Hultman and Taguchi 2010: 540).

References

Ames, B.D. and Youatt, J.P. (1994) 'Intergenerational education and service programming: A model for selection and evaluation of activities', *Educational Gerontology*, 20(8): 755–776.

Barad, K. (2003) 'Posthumanist performativity: Toward an understanding of how matter comes to matter', *Signs*, 28(3): 801–831.

Barad, K. (2007) *Meeting the Universe Halfway*, Durham, NC: Duke University Press.

Batsleer, J. (2013) 'Youth work, social education, democratic practice and the challenge of difference: A contribution to debate', *Oxford Review of Education*, 39(3): 287–306.

Brown, R. and Ohsako, T. (2003) 'A study of intergenerational learning programmes for schools promoting international education in developing countries through the International Baccalaureate Diploma Programme', *Journal of Research in International Education*, 2(2): 151–165.

Casey, E. (1993) *Getting Back into Place: Toward a New Understanding of the Place-World*, Bloomington, IN: Indiana University Press.

Coull, Y. (2010) *Users Guide to Intergenerational Learning*, Glasgow: Generations Working Together, The Scottish Centre for Intergenerational Practice.

Deleuze, G. and Guattari, F. (1988) *A Thousand Plateaus: Capitalism and Schizophrenia*, trans. B. Massumi, London: Athlone Press.

Duvall, J. and Zint, M. (2007) 'A review of research on the effectiveness of environmental education in promoting intergenerational learning', *Journal of Environmental Education*, 38(4): 14–24.

Field, J. (2013) 'Learning through the ages? Generational inequalities and inter-generational dynamics of lifelong learning', *British Journal of Educational Studies*, 61(1): 109–119.

Edwards, R. (2010) 'The end of lifelong learning: A posthuman condition?' *Studies in the Education of Adults*, 42(1): 5–17.

Ellsworth, E. (2005) *Places of Learning, Media, Architecture, Pedagogy*, Abingdon: Routledge.

Gough, N. and Price, L. (2004) 'Rewording the world: Poststructuralism, deconstruction and the "real" in environmental education', *Southern African Journal of Environmental Education*, 21: 23–36.

Gould, E. (2012) 'Moving responses: Communication and difference in performative pedagogies', *Theory Into Practice*, 51(3): 196–203.

Granville, G. and Ellis, S.W. (1999) 'Developing theory into practice: Researching intergenerational exchange', *Education and Ageing*, 14(3): 231–248.

Grosz, E. (1994) *Volatile Bodies: Toward a Corporeal Feminism*, Bloomington, IN: Indiana University Press.

Grosz, E. (2005) *Time Travels: Feminism, Nature, Power*, Durham, NC: Duke University Press.

Hammad, S. (2011) 'Senses of place in flux: A generational approach', *International Journal of Sociology and Social Policy*, 31(9/10): 555–568.

Hekman, S. (2008) 'Constructing the ballast: An ontology for feminism', in S. Alaimo and S. Hekman (eds) *Material Feminisms*, Bloomington, IN: Indiana University Press.

Hultman, K. and Taguchi, H.L. (2010) 'Challenging anthropocentric analysis of visual data: A relational materialist methodological approach to educational research', *International Journal of Qualitative Studies in Education*, 23(5): 525–542.

Ingold, T. (2003) 'Two reflections on ecological knowledge', in G. Sanga and G. Ortalli (eds) *Nature Knowledge: Ethnoscience, Cognition, and Utility*, New York, NY: Berghahn.

Ingold, T. (2008) 'When ANT meets SPIDER: Social theory for arthropods', in C. Knappett and L. Malafouris (eds) *Material Agency: Towards a Non-Anthropocentric Approach*, New York, NY: Springer, pp. 209–216.

Ingold, T. (2011) *Being Alive: Essays on Movement, Knowledge and Description*, Abingdon: Routledge.

Irigaray, L. (2000) 'A two subject culture' in *Democracy Begins Between Two*, London: Athlone Press.

Jones, O. (2008) 'Stepping from the wreckage: Non-representational theory and the promise of pragmatism', *Geoforum*, 39(4): 1600–1612.

Kaplan, M.S. (2002) 'Intergenerational programs in schools: Considerations of form and function', *International Review of Education*, 48(5): 305–334.

Lorimer, H. (2005) 'Cultural geography: The busyness of being "more-than-representational"', *Progress in Human Geography*, 29(1): 83–94.

MacPherson, H. (2010) 'Non-representational approaches to body-landscape relations', *Geography Compass*, 4(1): 1–13.

Mannion, G. (2009) 'After participation: The socio-spatial performance of intergenerational becoming', in B. Percy-Smith and N. Thomas (eds) *A Handbook of Children's Participation: Perspectives From Theory and Practice*, London: Taylor and Francis.

Mannion, G. (2012) 'Intergenerational education: The significance of "reciprocity" and "place"', *Journal of Intergenerational Relationships*, 10(4): 386–399.

Mannion, G. and Adey, C. (2011) 'Place-based education is an intergenerational practice', *Children, Youth and Environments*, 21(1): 35–58.

Mannion, G., Fenwick, A. and Lynch, J. (2013) 'Place-responsive pedagogy: Learning from teachers' experiences of excursions in nature', *Environmental Education Research*, 19(6): 792–809.

Mayall, B. (2000) 'Conversations with children: Working with generational issues', in P.J.A. Christensen (ed.) *Advocating for Children: International Perspectives on Children's Rights*, London: Falmer.

Mulcahy, D. (2012) 'Affective assemblages: Body matters in the pedagogic practices of contemporary school classrooms', *Pedagogy, Culture and Society*, 20(1): 9–27.

Nespor, J. (2008) 'Education and place: A review essay', *Educational Theory*, 58(4): 475–489.

Hopkins, P. and Pain, R. (2007) 'Geographies of age: Thinking relationally', *Area*, 39(3): 287–294.

Perry, M. and Medina, C. (2011) 'Embodiment and performance in pedagogy research investigating the possibility of the body in curriculum experience', *Journal of Curriculum Theorizing*, 27(3): 62–75.

Peterat, L. and Mayer-Smith, J. (2006) 'Farm friends: Exploring intergenerational environmental learning', *Journal of Intergenerational Relationships*, 4(1): 107–116.

Quay, J. and Seaman, J. (2013) *John Dewey and Education Outdoors: Making Sense of the 'Educational Situation' Through More Than a Century of Progressive Reforms*, Rotterdam: Sense Publishers.

Ross, H. and Mannion, G. (2012) 'Curriculum making as the enactment of dwelling in places', *Studies in the Philosophy of Education*, 31(3): 303–313.

Sanchez, M. (2007, trans 2009) *Intergenerational Programmes Evaluation*, Documents Collection, Technical Documents Series, Madrid: Spanish National Institute for Older Persons and Social Services.

Scharf, T. and Keating, N. (2012) 'Social exclusion in later life: A global challenge', in T. Scharf and N. Keating (eds) *From Exclusion to Inclusion in Old Age: A Global Challenge*, Bristol: Policy Press.

Scottish Executive (2004) *A Curriculum for Excellence: The Curriculum Review Group*, Edinburgh: Scottish Executive.

Semetsky, I. (2013) 'Learning with bodymind: Constructing the cartographies of the unthought', in D. Masny (ed.) *Cartographies of Becoming in Education: A Deleuze-Guattari Perspective*, Rotterdam: Sense Publishers.

Somerville, M. (2010) 'A place pedagogy for "global contemporaneity"', *Educational Philosophy and Theory*, 42(3): 326–344.

Somerville, M., Davies, B., Power, K., Gannon, S. and de Carteret, P. (2011) *Place Pedagogy Change*, Rotterdam: Sense Publishing.

Thrift, N. (2003) 'Performance and...', *Environment and Planning A*, 35(11): 2019–2024.

Thrift, N. (2008) *Non-Representational Theories*, London: Routledge.

Thrift, N. and Dewsbury, J.-D. (2000) 'Dead geographies – and how to make them live', *Environment and Planning D: Society and Space*, 18(4): 411–432.

Vanderbeck, R.M. (2007) 'Intergenerational geographies: Age relations, segregation, and re-engagements', *Geography Compass*, 1(2): 200–221.

17 Moving from boats to housing on land

Intergenerational transformations of fisher households in southern China

Chen Fengbo and Samantha Punch

Introduction

Fishers on the Beijiang River in the Shaoguan district of the Guangdong province in the south of China traditionally tended not to own property on land and lived a life of uncertainty in boats on the water. Their fishing boats were their homes, as well as the tools with which they made their livelihoods. In the context of rapid economic change in China and environmental decline in the Beijiang River, in 2009 the local Shaoguan government introduced a programme of resettlement to move the fishers who lived on the river in the urban area to new, low-rent, urban housing. This chapter explores the process and impact of resettlement for fisher households that has resulted in the displacement of both older and young people from their local environment which in turn has led to key changes in livelihoods and intergenerational practices.

On the one hand, the transition to urban land-based housing offers young people new education and work opportunities yet, as in other Majority world contexts, the chances of securing better paid work are limited as formal urban employment is extremely competitive (Jones and Chant 2009). On the other hand, the housing and livelihood transitions add to the burden of the older generations who are losing the traditional labour contributions of young people whilst not being compensated with remittances from well-paid urban jobs (Punch and Sugden 2013). By considering patterns of intergenerational change this chapter contributes to recent debates around the importance of exploring age relations across the lifecourse (Hopkins and Pain 2007; Vanderbeck 2007) as well as relationally between the generations (Valentin and Meinert 2009; Tisdall and Punch 2012). In particular, with a focus on Chinese families undergoing resettlement from their boats to land, we explore the livelihood impacts on different generations and how, in turn, that effects intergenerational relations. The chapter outlines the background of fishing livelihoods, before discussing the methodology. It then considers the advantages and disadvantages of the housing transition for the fisher households as a whole, and finishes by exploring the different generational impacts of the move. This chapter uses the concept of 'intergenerational ambivalence' (Lüscher and Pillemer 1998) to understand how changing living arrangements can have differential social, economic and cultural effects with multigenerational households.

History of fishers in Shaoguan, South China

In the Pearl River area and along the coast of South China Sea, fishers have been a large and important social group since the Ming and Qing Dynasty. Known as *Danmin* (疍民), they see themselves as a different ethnic group, in terms of dialect and culture, from the people living on land. Living and working on their boats meant that they were also a marginal group as traditionally the only interaction with non-fishers was when they sold the fish (Ye 1991).

Up until the 1980s and early 1990s, fishing was still a feasible livelihood option for fishers as there were many fish in the river, especially during the fishing season. There were also employment opportunities in river transportation for coal and sand mining. However, by the early twenty-first century, the livelihood of fishers had become increasingly difficult as coal was transported by train instead of by river, and local sand mining now required government permission and a large amount of investment for mechanical dredging. The quantity of fish in the river also decreased as a result of the development of the hydro-electric dam, water pollution, and over-fishing using electric fishing tools (Luo *et al.* 2011). In 2011 in order to protect the depleting fish stocks, the government introduced a policy of 'closed season for fishing' on the Pearl River (China Agricultural Bureau 2010), prohibiting fishing between 1 April and 1 June, when most fish spawn. This has further exacerbated the livelihood crisis for fishers who began to realize they would no longer be able to rely solely on fishing and would have to find jobs on land. Liu *et al.* (2010) reported that the income of fishers still mainly comes from fishing, but that their livelihoods are becoming increasingly diverse.

It was only in the 1960s that former Premier Zhou Enlai set a policy to resolve the fishers' housing problem: 'the settlement on the shore of fishermen' (渔民陆上定居工程). However, land was very limited and, although housing was allocated to fishers in the coastal area, in Shaoguan City most fishers continued to live in the living boats known as *Zhujiachuan* (住家船) (Zan 2004). More recently in 2006, the local government of Shaoguan City set up a project to apply for the National Garden City (国家园林城市) from the central government, which focuses on the environment and landscape of the city. In 2008, some city deputies proposed a motion to remove the living boats and make the urban river cleaner and more attractive. The main purpose was to improve the environment and landscape of Shaoguan City, and a secondary benefit was to address the living conditions of the river residents. In 2009, two districts with low-rent housing and affordable housing were built: Shiliting and Tianziling. Shiliting is close to the Beijiang River, and Tianziling is further away, approximately 30 minutes by bus. The fishers pay low rent for the flats and also received some compensation from the government in return for no longer living on the river.

Before 2009, there were 124 large living boats and 156 fisher families (as some fishers lived in small fishing boats) in the watercourse of the urban area of Shaoguan City, which were distributed along three sites of the Beijing River

(Shaoguan Local Government 2009). The majority (111) of the large living boats were made of cement and iron with an area of 80–90 m², divided into several separate rooms for sleeping or storage. The remaining 11 living boats were made of wood, previously being used for transport but were later transformed to living boats without engines. In 2011, the 156 fisher families moved to the new low-rent government built flats: 61 households in Tianziling, and 95 households in Shiliting. Each flat is approximately 52 m² with two bedrooms, a dining room, a small bathroom and a kitchen. As part of the resettlement programme, 101 living boats were destroyed, 62 labourers received new skills training, and most house-holds secured compensation of CNY20,000 (US$3,175) for the resettlement and loss of the living boat. Although the government destroyed the living boats in 2011, most fishers still keep their small fishing boats and fishing licences. About a quarter of fisher households in Tianziling and about half of those in Shiliting still go fishing. Those who continue to fish are mostly older people, whilst the younger generation has sought alternative employment on land.

Methods

This chapter draws on data which are part of a larger multi-partner, multidisciplinary project: 'Highland Aquatic Resources Conservation and Sus-tainable Development' (HighARCS), funded by the European Commission (see www.wraptoolkit.org). The Beijiang River, as one of the upper branches of the Pearl River, was chosen as the field site in the Shaoguan Municipal Region, southern China. The overall aim of this research in China was to explore the threats and sustainability of the use of aquatic resources and the subsequent impacts on the life of fishers. The study has examined the social and economic factors involved, whilst looking for management and legislation issues related to the use of aquatic resources in order to suggest short-term actions as well as seek longer-term solutions. The resettlement project of fishers in Shaoguan was identified as one of the government actions for addressing the decrease in aquatic resources and diversifying household livelihoods.

In order to understand the impact of resettlement, the research team at South China Agricultural University conducted semi-structured interviews with a small sample of resettled fisher families in April and June 2012. With guidance from the Shaoguan fishery department and the Shaoguan public housing office, the team randomly selected nine households to take part from two districts: four from Tianziling and five from Shiliting. This district location is indicated at the end of the quotations used in this chapter. The interview included information on household characteristics, life before moving, and thoughts on resettlement prior to the move as well as afterwards, such as emotions and changes in lifestyle, consumption and working practices. Over two visits, the team endeavoured to speak to different family members, including older people, adults and children. In order to capture different generational perspectives we have broadly categorized each age range as: 'older people', typically the grandparents who were 60 years or more; 'adults' who were parents, mostly the son of the grandparents and their daughter-in-law; and 'children' who were studying, up to 18 years old. In total five grandfathers, five grandmothers, seven fathers, eight mothers, four boys and three girls were interviewed across the nine sample households. A qualitative thematic analysis was used to compare the life of fisher households before and after moving, whilst considering the gendered and generational differences of the impacts of the housing transformation.

The challenges and benefits of resettlement for fisher households

Housing has always been one of the major problems confronting fishing communities as they tend not to own any land (Sathiadhas *et al.* 1994). Housing is not merely a dwelling place but also a residence, which includes locality factors such as infrastructure, public and private facilities, and social, economic, political and relational factors that centre around housing (Kemeny 1992). For fishers the move from boats to new housing has wide social implications beyond the physical relocation. Housing has direct connection with the residents' livelihood space. For fishers, as Kraan (2009) points out, their livelihood space includes three elements: spatial, economic and social/cultural. Each space, connecting to the livelihood space, with its cultural, social, economic and geographical characteristics, produces a unique pattern of realignment between actors, processes and consequences. Thus, living in the on-land flats is a completely new life for the fishers. They are no longer a floating population but have a fixed space and sturdy house to live in which has a range of opportunities and constraints attached to it.

One of the key benefits mentioned by all of the fisher households is the increased safety and security of living in a house on land rather than a boat on the river which is vulnerable to the weather conditions, especially at night: 'We feel safer when we move to the house in the flats, and we are no longer afraid of flooding and heavy rain. The new house is firm, and we feel more comfortable and warm in the winter' (Daughter-in-law, Shiliting).

> When we lived in the boat, we always worried about the typhoon and flood. We would be afraid that the boat would overturn when it was shaking heavily. Mostly we don't sleep at night when the typhoon and floods are coming.
>
> (Grandfather, Shiliting)

The hydropower station also threatened the safety of the living boats docked near the bank of river, particularly when its downstream dam was opened. The flow would become rapid and the water level would fall very quickly, stranding or overturning the boats, which was very dangerous if people on board were sleeping. Some people living on the boat did not know how to swim, particularly women and children: 'My child and I were walking on the side of boat and we fell into the water. We can't swim; fortunately, my husband was nearby, and he saved us' (Daughter-in-law, Shiliting). Consequently very young children (up to about three years old) would be tied to the boat by a rope around their chests whilst their parents went fishing or when all family members were sleeping at night. Nearly all the adult fishers talked about similar experiences of being roped when they were a child to stop them falling into the water.

One of the greatest advantages of the move to new housing is the improved living conditions for fisher households. When they lived in their boat, they used wood rather than gas for cooking, they used water from the river for washing and cooking, they had no cable television, the washroom was small and dirty, and the bedroom was dark and small. In the new flats, they had a designated kitchen and living room, a toilet, gas for cooking and cable television. They were able to eat at a table instead of on the floor of the boat, and the house was cleaner with sanitation facilities and waste disposal:

> I have been fishing for 30 years, life in the boat is very hard and not restful. It is not a sustainable life. Even though the house is a little small, it is very clean and comfortable and it is definitely better than the boat.
>
> (Father, Shiliting)

> It is very easy to clean the room. We have bought a water heater to take a shower, but before, we had to use wood to heat the water to have a wash using a small wooden bowl. It is a big change for our life.
>
> (Mother, Shiliting)

> The ceiling of the boat is low, we can't put a table and chairs on the boat, we always had dinner just sitting on the ground. But now, we can have dinner together around the table.
>
> (Girl, Tianziling)

However, all of the new facilities need to be paid for, leading to an increased economic burden. Life on the boat did not require much cash income, and the unsold fish were eaten every day. The better living standards come at a cost as rent has to be paid along with utility bills and property charges. For most fishers financial concerns are their biggest worry:

> The big change is that we need to pay for living costs. Before moving, we didn't need to pay for gas, water, nor rent for the house, but all of these we need to pay for now. As well as food and transport, we need to pay CNY600 (US$95) for fixed living expenses, including: CNY113 for the rent, CNY260 for gas, CNY40 for water and CNY60 for electricity, CNY25 for property management fee, and some other fees such as parking bicycles.... The income of my son and daughter-in-law is about CNY2,000–3,000 (US$317–476) per month. We need that money just to sustain the daily life of the whole family, there is not so much money left after the basic expenditure. If the government increases the rent, I can't sustain the life of this family.
>
> (Grandfather, Tianziling)

It took quite a long period of adjustment for most of the fishers to adapt to their new lifestyles on land rather than on the river. As well as changing the main source of livelihood there was also the need to alter everyday mundane practices: 'When I just moved into the house, I don't like many things, because I don't know how to use them, such as the gas, the toilet and the microwave' (Grandmother, Tianziling). Some households also pointed out that for the larger fisher families the new flat is smaller than the living boat: 52 m^2 compared with 80–90 m^2 for their living boat. The size of the new housing is particularly problematic for families with several generations living together as they would only be allocated one house by the government. Nevertheless the new urban lifestyle offers greater opportunities for young people and adults to access work in factories or the service industry. It is also easier for children to attend school regularly. Thus there are a range of impacts attached to the housing transition and many fishers were ambivalent about the overall benefits and limitations as this quotation indicates:

> Living in the new house is better than living in the boat. It is better for young people to find a job and for children to go to school. It took a long time for us to adjust to the new life in the new house. The house is too small, and we need to climb the stairs every day. Sometimes, we miss the fresh air and freedom from when we lived in the boat. We need to buy everything we need. We don't know many people who live in the district, and we need to use a card when we enter into or go out of the residential lot.
>
> (Father, Shiliting)

One particular advantage of the housing transition is that it enables those fishers to acquire an urban household registration. Household registration, according to residential location, is very important in Chinese society (Wang and Murie 2000; Mallee 2008). Having a household registration in an urban area allows access to the city social security system and to schools in the district of the household registration. Thus, housing enables them to be less socially excluded (Somerville 1998). Whilst there is a similar social security system for rural household registration, the level of social protection is lower. Due to their mobility, traditionally most fishers have not been included in the rural or urban security systems. For some fishers who had rural household registration but migrated to live and work in the urban area, it was problematic for their child to attend school in the city. According to government policy regarding the housing resettlement of fishers, these households are able to have an urban household registration after their move into new housing, even for those who previously had a rural household registration. Thus these fishers can now buy medical insurance enabling access to government health subsidies, people over 60 years will receive a pension, and their child will be able to access a better education in the city. Nevertheless, some fishers still find that such benefits are limited: 'I can get the pension after 60 years old, but it is very low, about CNY70 (US$11) per month' (Grandfather, Tianziling). 'After we move into the new house, we can buy urban health insurance, but the reimbursement rate of health insurance is very low, about 30 per cent of medical payment. It's too expensive, we can't afford it' (Grandmother, Shiliting). By considering the household as a whole, the shift to new housing greatly improves the quality of fisher families' everyday lives. In particular they experience better material conditions even though these are accompanied by economic pressures.

Intergenerational change for fisher families

In contrast, if we look inside the household, the impacts of the transition vary for different family members. As already mentioned, traditional fisher households are a distinct social group living on the river relatively isolated from the rest of society and falling in between rural and urban areas. Before they moved to the new housing on land, fisher households tended to be solitary, private units, where the relationship between parents and their child was quite closed. After moving to an urban residential area, the nature of intergenerational relations within fisher households changed, creating some ambivalences and ambiguities for different family members. We use Lüscher and Pillemer's concept of 'intergenerational ambivalence' (1998) to illustrate tensions in relationships between parents and adult offspring that cannot be reconciled, such as contradictory emotions and motivations.

Change for older people

Similar to the processes of rural labour migration during the major transition of Chinese society from a rural economy to an industrialized and urbanized economy (Zhang and Song 2003), the young descendants of fishers also aspired to pursue more 'modern' employment within manufacturing or service industries. In contrast, the older generation continued to fish according to their traditional lifestyles. Thus the new housing offers more working opportunities for the younger generations seeking urban-based employment but many older people consider themselves to be lacking adequate skills for such alternative work and too old to re-train, particularly given that many of them are illiterate:

> I have been fishing for more than 50 years. I never went to school. You know, illiterate people can't find another job, so I have to go fishing and live in the fishing boat after the resettlement. The new house is so small that it is only enough for my son and his wife to live. The new house is hot in summer and too high, I always feel tired when climbing the stairs.
>
> (Grandfather, Shiliting)

This quotation suggests that a combination of factors lead to older men continuing to fish and live on their boats whilst their families live in the new urban housing. First, the house is too small for three generations of family members to live together; second, many of the older fishers have no pension and it is hard for them to find a job on land; and finally, most of the older fishers were so used to their traditional mobile lives in the boat that they did not like living in the fixed and closed urban environment. However, fishing alone on the river is not an easy task as it traditionally requires one person to steer the boat whilst the other fishes:

> I am still living in the fishing boat and go fishing every day, and I never sleep in the new house. My wife takes care of my grandson and lives with my son and daughter-in-law. When our family lived on the boat, my wife always went fishing with me, she steered the boat, I threw the net. But now, nobody helps me to control the boat, it is very hard for me to throw the net, so I just use the net cage to fish. I have to prepare food for myself and sell the fish by myself. I also think about finding a job on land, but the wage is very low and I will lose the fuel subsidy.... My son and daughter-in-law just have temporary jobs with low wages, I don't want to leave the financial burden to them. At the same time, I don't like doing the same thing and staying at the same place all day. I like the life of freedom, such as going fishing. When I get tired, I can have a rest. Moving to the new house is better for my grandson going to school, and for my son it is easy to find a job. But it is no good for the old people.
>
> (Grandfather, Tianziling)

The small size of the new house, the heavy financial burden and the desire to pursue their fishing lifestyle are the main reasons that many of the older men still live on their small fishing boats (the larger living boats were destroyed as part of the resettlement programme). They were also able to contribute to the fish consumption of their family when either they visited them in the urban area or their wives visited them on their boats. Most of the older fishermen told us that their working lives had not changed as they were still going fishing and living in the boat, but due to the separation from their families, some of their daily life was worse than before. However, they recognize that the main benefits are for the younger generations: 'living in the new house is better for my child and grandchild' (Grandfather, Shiliting). Thus, the move to new housing brings intergenerational ambivalence (Lüscher and Pillemer 1998) for young adults and their older fisher parents:

> I have gone fishing with my husband for more than 30 years. Now, I live with my son and daughter-in-law and am responsible for buying the food and taking care of our grandchild. I do not go fishing any more. The life after moving is good for my son and daughter-in-law. My husband still lives in the boat, I am willing to live with him, and it is also the wish of my husband, but my son doesn't agree: if I live with my husband, nobody cares for my son's child. If my husband lives in the new house, he will give up fishing and the fuel subsidy from the government. You know, we have no pension, if there is no income from fishing and fuel subsidy, our life will become very hard.
>
> (Grandmother, Tianziling)

Unlike their husbands, most of the older female fishers moved into the new house, usually living with their son and daughter-in-law. Living separately from their husbands, they worried about their husbands living alone on the boat, but felt obliged to undertake child care for their sons. Thus despite better living conditions, it was at the expense of their emotional well-being. This illustrates intergenerational ambivalence where mixed emotions merge around care provision, 'when warmth, tenderness, and delight coexisted with frustration, disappointment, and resentment' (Lüscher and Pillemer 1998: 416). Furthermore, as the older fishermen tend to have no savings or pension, they will have to rely on their adult child eventually by living with them when they are no longer able to fish. Thus the spatial constraints of this livelihood transition are temporarily resolved by older fishermen living in the boat and continuing to fish.

Change for young adults

For young adults who work or want to work in the city, moving to the new house is a key improvement as it accelerates the process of the young fishers merging into urban society and reduces their travel time and costs. Most young descendants of fishers already worked in urban areas before moving to the new house. For those young men who are not married, the new house is also an important asset for a future marriage. The younger adults adapt relatively quickly to the new urban lifestyle and for those without urban jobs, it is now easier for them to find better work on land.

> We are willing to move to the new house, but our parents are not happy to move. When we lived in the boat, all family members lived together, but since we move into the new house, we are divided into two parts: my father-in-law lives on the boat, my mother-in-law lives with us. If my mother-in-law lives with my father-in-law in the boat, I will stay at home to take care of my child instead of working. Our life will be very hard. My husband also worked on land before moving. I did temporary work in a hotel before, about CNY400 per month including meals, but now I am working in a factory, about CNY1,700 per month. When we lived in the boat, I spent one hour getting to work, but now, just 20 minutes. I work 12 hours per day, and always work at night. The work is very hard, but we can get more money, our economic status has improved.
>
> (Daughter-in-law, Tianziling)

However not all adults found suitable urban employment despite receiving some training from the government, as this man explains:

> The government supplied some skill training course for us, such as welder and garden worker, and they also recommended some jobs for us. But the wage is too low, only CNY600 per month for a city garden worker. I am

also going fishing. It is difficult for me to find a job, because I am illiterate and have no other skill.

(Adult male, Shiliting)

Whilst the urban employment opportunities offered an alternative to fishing, which was increasingly hard to survive on due to the reduced number of fish, the manufacturing and service sectors involved long hours with relatively low wages. Consequently for most young couples, both men and women had to work in order to meet the increased living expenses for the three-generational family living in the new housing. This in turn meant they had to rely on their older parents (mainly grandmothers) for child care. Hence the transition of livelihood space results in a key social change whereby grandmothers no longer fish but take on child-care responsibilities for their working children.

Change for children

As we have seen, for the children who lived on the fishing boat, they had a relatively sheltered upbringing, mostly staying on the boat surrounded by water. During early childhood, due to safety reasons, they were often tied by rope to the boat and had limited contact with people outside the fishing community. It was difficult for them to go to school regularly and study well. Even today, some young fishers are illiterate. Moving to the urban-based housing thus disrupts the traditional, rather closed, tie between parents and child. Thus the housing transition enables children to have broader social networks and to receive a better education in the city.

Our child is very happy, and they can play with their friends at a vacant area of the new district. When we lived in the boat, the child could not move much because they were tied by rope, they just cried to the child on the opposite boat. The young child mostly doesn't go to nursery school when living in the boat, but now, they should be sent to nursery school by paying CNY500 per month.

(Daughter-in-law, Tianziling)

The following quotations also indicate that children used to be stigmatized by their fisher identity and that they would be bullied by their peers at school for their marginalized social status:

When I lived in the boat, it was very difficult to make friends, my teacher and my classmates always laughed at me, and they said I was a son of fishers. After moving to the new house, I have many friends, I don't feel the discrimination any more. At least, I can tell them where I live now.

(Boy, Shiliting)

'I like life in the boat, because I can use the fishing rod to go fishing in the boat. But I like life in the new house too. I have made many new friends since I move to the new house.'

(Girl, Shiliting)

Hence, moving to the new house enhances the fishers' social status and enables children to mix more easily with their peers at school. Due to the decreased amount of fish in the river, nearly all the fishers said that they would prefer their children not to be fishers in the future. Many of them had decided not to even teach their children how to fish: 'I think the fishers will disappear in 10 years' time. The young people don't know how to repair the net. For example, even if I give the net to my son, he doesn't know how to fish' (Grandfather, Tianziling). And 'I never think my children will be fishers, they don't know how to fish. None of the young generation of fishers want to go fishing. Only old people go fishing' (Father, Shiliting). Thus this illustrates that there has been a cultural shift in terms of the transition of livelihood space as parents no longer want their children to pursue the traditional livelihood of fishing.

Conclusion

This chapter has shown that the housing transition from boats to land for fishers in southern China has had many collective benefits for fisher households. Yet the change in livelihood space has also had varied economic, spatial, social and cultural impacts on different generations. It has placed particular burdens on the older generation of fishers, as the older fishermen continue to fish by themselves partly due to the limited space of the new flats and partly in order to reduce the economic burden on the household by providing for themselves and contributing fish for household consumption. Their continued fishing livelihood is also because they consider themselves too old to develop new skills for the urban employment sector, many of them are illiterate which makes re-training difficult and for many it is a lifestyle habit which they do not wish to leave behind. The freer, more mobile life on the river is more appealing despite the limited income compared with the more enclosed lifestyle in fixed housing. Meanwhile, the older fisherwomen no longer fish alongside their husbands; instead they undertake the burden of child care for their grandchildren whilst the younger generation pursues dual-earner incomes in the city in order to meet the increased economic costs of the higher standard of urban living. Thus it is clear that the older generation is making substantial sacrifices for their children and grandchildren during this major shift in livelihood and housing conditions. It is less clear how the next generation will adjust to the urban, non-fisher livelihoods without such sacrifice and additional burdens being taken on by older people. Perhaps the increased levels of education pursed by the younger generation will enable the development of a wider skill base in order to secure higher paid employment, but evidence from other parts of the Majority world suggest that this is unlikely to be the case (Jeffrey 2010). Globally young people are becoming more educated yet failing to secure better paid urban employment which remains competitive with low wages and long hours (Jones and Chant 2009; Punch and Sugden 2013).

Thus whilst the material living conditions may have improved for fisher households who have moved to urban housing, the emotional and social well-being of the older generation has tended to decrease whilst their work burdens have increased by fishing or undertaking child care alone without their partner. Intergenerational relations thus involve tensions as ambivalences emerge in relation to the negotiated and constrained interdependencies which are worked out across the generations (see also Punch forthcoming). Intergenerational ambivalence (Lüscher and Pillemer 1998) develops as interdependencies are constrained by the economic position of fisher households in the city due to the higher costs of urban living, low wages, long hours and limited work opportunities. Yet these interdependencies are not fixed and are likely to transform as family members move through the lifecourse. Hence this study has emphasized the importance of including the perspectives of the older generation alongside those of parents and children when considering work, education and livelihood change for young people (Hopkins and Pain 2007; Vanderbeck 2007). Research within childhood studies has tended to focus more on either children's voices or those of children and their parents, but has yet to embrace a three-generational approach to understanding the nature of intergenerational and intra-generational relations across the lifecourse (Tisdall and Punch 2012). A consideration of generational issues across the lifecourse reveals varied consequences of the changing livelihoods as a result of environmental decline. These include economic, spatial, social and cultural changes regarding the livelihood space of fisher households which impacts on the nature of intergenerational relations. In particular, in this Chinese context of resettlement from boats to land, this had led to the increasing marginalization of older people who bear the greatest burdens regarding the transition of livelihood space.

Acknowledgements

The authors acknowledge financial support from the European Commission under Theme 6 of the Seventh Framework Programme (no. 213015). This publication reflects the authors' views, and the European Community is not liable for any use that may be made of the information contained herein. We also appreciate the generous support and contributions from South China Agricultural University: in particular Prof Luo Shiming, Dr Liu Yiming, Dr Shang Chunrong, Dr Cai Kunzhen, and the students: Zhuang Fengchi, Guo Shangyi, Mai Zhenyu for their inputs during data collection and data processing.

References

China Agricultural Bureau (2010) *Notice of the Ministry of Agriculture on the Implementation of the System of the Pearl River Fishing Ban Period* (in Chinese), Beijing: China Central Government.

Hopkins, P. and Pain, R. (2007) 'Geographies of age: Thinking relationally', *Area*, 39(3): 287–294.

Jeffrey, C. (2010) 'Timepass: Youth, class, and time among unemployed young men in India', *American Ethnologist*, 37(3): 465–481.

Jones, G.A. and Chant, S. (2009) 'Globalising initiatives for gender equality and poverty reduction: Exploring "failure" with reference to education and work among urban youth in the Gambia and Ghana', *Geoforum*, 40(2): 184–196.

Kemeny, J. (1992) *Housing and Social Theory*, London: Routledge.

Kraan, M.L. (2009) 'Creating space for fishermen's livelihoods: Anlo-Ewe beach seine fishermen's negotiations for livelihood space within multiple governance structures in Ghana', unpublished thesis, African Studies Centre, Leiden.

Liu Y., Shang C., Sugden, F., Luo S., Chen F., Wang W., Jiang B., Gao M., Li H. and Ye Y. (2010) 'Report on livelihoods dependent on highland aquatic resources: A case study of Shaoguan, China', *HighARCS Project Report WP4*, South China Agricultural University, online, available at: www.higharcs.org/download/output-matrix/Guandong%20China/Livelihood_Report_China.pdf (accessed 12 June 2014).

Luo S., Cai K., Liu Y., Jiang B., Zhao H., Cui K., Gan L., Fu J., Zhuang X., Tong X., Li H., He H. and Ye Y. (2011) 'Report on integrated action plan for conservation and sustainable use of aquatic resources in Beijiang River, China', *HighARCS Project Report*, South China Agricultural University, online, available at: www.higharcs.org/download/output-matrix/Guandong%20China/IAP_China_ Final.pdf (accessed 12 June 2014).

Lüscher, K. and Pillemer, K. (1998) 'Intergenerational ambivalence: A new approach to the study of parent–child relations in later life', *Journal of Marriage and the Family*, 60(2): 413–425.

Mallee, H. (2008) 'China's household registration system under reform', *Development and Change*, 26(1): 1–29.

Punch, S. (forthcoming) 'Youth transitions in the majority world: Negotiated and constrained interdependencies within and across generations', *Journal of Youth Studies*.

Punch, S. and Sugden, F. (2013) 'Work, education and out-migration among children and youth in upland Asia: Changing patterns of labour and ecological knowledge in an era of globalisation', *Local Environment: The International Journal of Justice and Sustainability*, 18(3): 255–270.

Sathiadhas, R., Panikkar, K.K.P. and Kanakkan, A. (1994) 'Traditional fishermen in low income trap – a case study in Thanjavur coast of Tamil Nadu', Marine Fisheries Information Service, Technical and Extension Series, 135: 5–10.

Shaoguan Local Government (2009) *The Treatment Scheme of River Living Ships in Shaoguan Urban Area* (in Chinese), Shaoguan: Shaoguan Local Government.

Somerville, P. (1998) 'Explanations of social exclusion: Where does housing fit in?' *Housing Studies*, 13(6): 761–780.

Tisdall, E.K.M. and Punch, S. (2012) 'Not so "new"? Looking critically at childhood studies', *Children's Geographies*, 10(3): 249–264.

Valentin, K. and Meinert, L. (2009) 'The adult North and the young South: Reflections on the civilizing mission of children's rights', *Anthropology Today*, 25(3): 23–28.

Vanderbeck, R.M. (2007) 'Intergenerational geographies: Age relations, segregation and re-engagements', *Geography Compass*, 1(2): 200–221.

Wang, Y.P. and Murie, A. (2000) 'Social and spatial implications of housing reform in China', *International Journal of Urban and Regional Research*, 24(2): 397–417.

Ye, X. (1991) 'The living customs and the geopolitical relations of Guangdong Danmin in Ming and Qing Dynasty', *Journal of Chinese Social and Economic History* (in Chinese), 1: 56–62.

Zan, J. (2004) 'The summary of resolving for the Guangdong fisher problem since the 1949', *Journal of Morden Chinese History* (in Chinese), 6: 91–100.

Zhang, K.H. and Song, S. (2003) 'Rural–urban migration and urbanization in China: Evidence from time-series and cross-section analyses', *China Economic Review*, 14(4): 386–400.

Part V

Intergenerationality and ageing

18 Exploring intergenerationality and ageing in rural Kibaha, Tanzania

Methodological innovation through co-investigation with older people

Gina Porter, Amanda Heslop, Flavian Bifandimu, Elisha Sibale Mwamkinga, Amleset Tewodros and Mark Gorman

Introduction

Increasing attention is being paid to the relationships between generations in Africa, not only among academic researchers but also in the development agencies. This is motivated principally by two interlinked trends: first, the relatively rapid population ageing being experienced in many developing countries; second, the prevalence of HIV and AIDS which has left many grandparents and grandchildren supporting and caring for each other as a result of a missing or incapacitated middle generation. This is well exemplified by Tanzania, where with an official HIV infection rate of 5.7 per cent and approximately two million orphaned and vulnerable children (UNICEF 2006: 29), around 50 per cent of orphaned children now live in households headed by older people, predominantly grandmothers and other older female family members.

Research studies have highlighted the reciprocal and symbiotic nature of such child–elder relationships in Tanzania and elsewhere, drawing on evidence gathered from those older people and children whose lives are so closely entwined together, often in difficult and impoverished circumstances (e.g. Whyte *et al.* 2004; Whyte and Whyte 2004; Ingstad 2004; Schatz and Ogunmefun 2007; HAI 2007; Kamya and Poindexter 2009; Skovdal 2010). Such marginalized population groups have commonly been engaged through methodologies such as action research and participatory appraisal. Far rarer, however, are studies in which children and older people have themselves played a major role in carrying out the research, through a process of direct co-investigation.

This chapter focuses on co-investigation as a means of exploring intergenerational relations, with particular reference to a research project where academic researchers and NGO staff (of varying ages) collaborated with a team of older people from a community in Kibaha district, Tanzania, to jointly investigate issues surrounding older people's access to transport and mobility and its wider implications. Our aim was to develop methods which would help assess not only the direct impact of older people's mobility on their own livelihoods and health-seeking behaviour but also the broader impacts on younger generations in their families and communities and related issues of intergenerational poverty transmission. Our investigations in ten villages in this district demonstrate how intimately the mobility of older people is tied up with that of their children, grandchildren and others in the communities where they reside.

The co-investigation methods employed in this study are documented below in detail. We are aware of only one earlier study where co-investigation has involved the recruitment of older people (a HelpAge study by Ibralieva and Mikkonen-Jeanneret 2009). To our knowledge this is only the second study using co-investigation methods which focuses specifically on intergenerational issues, and the first to consider this from an older people's perspective. (A previous study by Clacherty (2008) in north-west Tanzania worked with children who interviewed their peers living with grandparents.) Two of the authors had direct prior field experience of recruiting and working with community members as researchers (Heslop as facilitator with the aforementioned Ibralieva and Mikkonen-Jeanneret older people study and Porter with young people in their teens: Porter and Abane 2008; and Porter *et al.* 2010, 2012a). We both found the approach remarkably effective in improving outsider-generated research and development initiatives.

Background: how the study came about

The researchers who initiated this study came together from quite different but complementary backgrounds, conducive to the development of a co-investigative study focused on older people and intergenerational relations. Gina Porter is an academic researcher who has worked on mobility and transport issues in sub-Saharan Africa for many years, has a strong interest in developing participatory field methods and had recently piloted and led a large mobility study in Ghana, Malawi and South Africa in which 70 young people aged between 11 and 21 years were recruited as co-researchers. After training, these young people undertook in-depth research with their peers in the villages and towns where they resided or were at school, bringing to the fore issues and questions which the academic research team arguably might not have identified without their prior investigations. These questions formed the base on which the subsequent academic qualitative and survey research programme was based. The group of young researchers worked so successfully with their peers and the academic team that they built the confidence and enthusiasm to write their own booklet (available at: www.dur.ac.uk/child.mobility/ subsequently disseminated to ministries, schools, libraries and communities across Ghana and Malawi). In the course of that study, many interconnections between youth and older people emerged: in South Africa, Malawi and Ghana respectively, 14 per cent, 9 per cent and 9 per cent of the young people interviewed in our survey lived with grandparents (usually grandmother alone); overall, around 20 per cent in each country lived with other relatives/foster parents, many of whom were older people.

Amanda Heslop is a practitioner and researcher who has facilitated many studies with older people across the world including the work led by Ibralieva and Mikkonen-Jeanneret (2009). She has been involved in numerous studies with older people covering issues such as social protection in Tanzania and with migration and seasonality in Kyrgyzstan and Moldova. The other collaborators (FB, ESM, AT, MG) were involved with AH in the development of HelpAge's Cost of Love study in Tanzania (2004), their subsequent Building Bridges project (HAI 2007) or related HelpAge projects. These studies focused on the negative impacts of HIV and AIDS on older people, highlighting the caregiving responsibility for grandchildren that older people must take on after the death of children with AIDS and the importance of integrating older people in HIV and AIDS interventions. The investigations led to the development of an intervention model which is now being utilized by NGOs and government agencies in their work.

Gina Porter contacted staff at HelpAge International about a possible joint study because of the linkages between children and older people emphasized by the child mobility study and the wider potential implications for community mobility patterns. The ensuing discussions between us drew attention not only to growing concerns at HelpAge regarding the access constraints faced by older people, but also to our shared interest in co-investigation. Given these overlapping interests, the potential for a new study in which we would join forces to work with older people as co-investigators on a mobility study was evident. On the basis of the discussions we decided that the principal focus should be on older people's mobility and access to health and livelihoods.

Mobility, or lack of it, is implicated in many facets of older people's lives (Schwanen and Paez 2010). The linkage with health is an obvious one, as the need to access health care often increases in old age. However, income poverty is another common characteristic of Africa's older people, especially in societies like Tanzania where government measures do not provide universal social security coverage in old age (Apt 1997; van der Geest 1998; Heslop and Gorman 2002; Barrientos *et al.* 2003; Aboderin 2004). Given the lack of old-age social security, continuing access to livelihoods is frequently vital, not just for the elderly to support themselves, but also to support young orphans and others in their care (Clacherty 2008; Mudege and Ezeh 2009). Obtaining a livelihood also tends to require some degree of mobility.

We decided to base our field investigation in Tanzania, because this is a country where HelpAge already had strong experience of working with older people's groups, and where the in-country office had previously identified mobility issues as of significant concern to older people, especially in rural areas. Following identification of a suitable study area and research team, we planned to utilize a three-strand research methodology, as discussed in the next section.

Developing a field plan: field sites and methodology

Once a full literature review of published and grey literature had been undertaken and funding for the research (eventually) secured, careful planning was needed to select a suitable study area and a team of research assistants, and find a small number of older people prepared to invest time and energy in the project. Kibaha district was selected for the research because HelpAge International's Tanzania office had been involved in initiatives in the district for some years and had observed significant transport services issues for older people. In Kibaha district, HelpAge had worked with a local NGO, the Good Samaritan Social Services Tanzania (GSSST), which had established older people's groups in the town: thus GSSST was recruited to support our project too. We were also able to build on knowledge gained from another HelpAge study in the district with Ifakara Health Institute – a local health institute aimed at enabling health practitioners to assess and understand the burden of non-communicable diseases among older Tanzanians. This work had already pointed to the high cost, unsuitability, scarcity, irregularity and unreliability of means of public transport on routes where transport is available (but not affordable or unsuitable for older persons).

Within the district we elected to work in one core village, Vikuge, where our GSSST collaborator already had links, but also identified a further nine additional settlements with varying access conditions for research. One village was located on the paved road (Kongowe), and nine along poor unsurfaced roads (five with a clinic, four with no clinic). Five research assistants with experience of rural Tanzania and relevant language skills were then recruited through Research on Poverty Alleviation (REPOA), a Tanzanian research organization based in Dar es Salaam.

The project methodology was designed to incorporate three key strands in which co-investigation (as in the child mobility study) would occupy the first phase, setting the key issues for further investigation and analysis in the academic research component. The latter components would involve qualitative studies with older people and key informants, and then a survey questionnaire to older people to provide the third element of the triangulation.

The first element, *co-investigation through older people community peer research* forms the focus of discussion in this chapter and is discussed in some detail below. In brief, 12 older people peer researchers (henceforth OPRs) were selected for training as peer researchers. They attended an initial one-week training workshop (conducted in Swahili with English–Swahili translation as necessary) to develop some age-adjusted research methods and subsequently conduct qualitative research in the field with their peers. We had planned that they would work in their own village and use young Research Assistants (henceforth RAs) in the second and third research strands to extend the investigation to the additional nine villages. However, as we note below, the peer research team decided to extend their own research component across the full study area.

In addition to the OPRs' qualitative interviews, *check list qualitative interviews* were also conducted by our five young RAs (two women, three men, all in their twenties) using check lists prepared by the academic lead researcher. The check lists drew substantially on findings from exercises and discussions during the one-week training workshop with the OPRs (together with some additional preliminary field investigation). Following participation in the training workshop with older people and a further training week, the young RAs interviewed older men and older women in each of the ten study settlements, some key informants, notably clinic staff and motorcycle taxi (*boda-boda*) operators, and also some children who lived with their grandparents. Their interviews covered health-related and livelihoods-related transport issues. In total 194 in-depth interviews were conducted.

The third component of the study was the *questionnaire survey of older people* in the ten settlements. This drew on the preliminary qualitative research findings, and was administered by our young RAs to older people. Our aim was to have a minimum of 30 completed questionnaires per settlement but in some villages this was not possible since the total number of older people present at the time of the survey was under 30: 339 fully complete questionnaires were obtained.

Findings on transport issues from the three research strands are reported in Porter *et al.* (2013b). Here we focus on the research process in the study's most innovative component, co-investigation with older people from the study community, and the role this played in exploring intergenerational issues. We also reflect on intergenerational relations within the research team itself.

Selection of OPRs: issues of inclusion

The participation of older persons as researchers in the initial stages of the study was central to the design of the research. Given the development objective of the project it was important to enable this older person perspective to help shape the methodology. All 12 older researchers were residents of Vikuge, our core study village. In our selection of older researchers we aimed to include women and men of a range of ages 60 years and over, along the spectrum of able bodied to severely disabled. We subsequently added the criterion of literacy because we decided that it was important that older researchers were able to record their own discussions and fieldwork observations. Although it proved harder to recruit literate older women, an adequate gender representation was achieved with eight men and four women. Age composition ranged from 59 years to 69 years: we were unable to obtain any people older than this willing and able to participate in the work. Most of the older researchers described themselves as farmers and some engaged in additional occupations such as masonry and cattle husbandry. A number of women and men held positions on village committees concerned with water, pastoralists, the local courts and village leadership. When asked about physical considerations, all older researchers highlighted problems with sight (most used spectacles) and four reported mobility difficulties. Despite these conditions, all the older researchers travelled six kilometres twice a day from the village to our one-week training workshop on the back of hired motorcycles.

The workshop training process with the OPRs

The central purpose of the training was to build the skills of the OPRs and the RAs to carry out a series of interviews in the study focus village (trying out a number of research methods), to gather initial information about the transport and mobility problems of older residents (in order to build key research questions for the main research phase) and to agree a code of conduct. The older people's feedback on both methods and questions during the training week was vital to the design of subsequent research in the ten study settlements. The younger RAs worked with, supported and learned from the older researchers throughout the training and field-work, in preparation for their subsequent role in the research.

The one-week workshop included three half-days of fieldwork practice in the core village, during which the OPR research teams tested interview methods for generating qualitative information on the following health and livelihoods-related themes: daily livelihood and health journeys made by older persons; impacts of seasonal changes on transport and mobility; and means of transport-ing household produce, water and fuel by household members in which older people lived alone or with grandchildren. Four teams of four researchers were formed and team roles agreed (lead and second interviewers, observer and recorder). For the workshop week, each team comprised three OPRs and one RA. (Subsequently, however, the older person research teams operated in four groups of three, since they demonstrated sufficient confidence to undertake the work without any support from the young RAs and were happy to mostly work independently.) Initially one team was formed solely of women in order to capture gender dimensions that might be overlooked by mixed gender teams. Feedback from participants after two days' fieldwork suggested this was an unnecessary and potentially unhelpful measure, and teams were re-adjusted.

As well as practicing and testing out the methods, the whole team engaged in regular synthesis and discussion of information gathered: this was of crucial importance in focusing our attention on generational relations in the subsequent field research. On the final day, having expressed their desire to continue with the study and to extend their involvement to research across the ten villages, the OPRs planned further information-gathering activities for an additional few weeks. In order to extend their scope beyond their own village, they often trav-elled in small teams with the RAs to each of the other nine settlements, using project transport, but worked independently there.

Intergenerational relations within the research team: experiences from the workshop (and beyond)

The one-week training workshop, though focused primarily on the OPRs, was enriched by an extraordinary diversity of participant backgrounds, age variation and perceptions. The meeting was led by Amanda Heslop, the lead trainer from HelpAge International, but also included other local and international HelpAge staff, the lead academic researcher, the GSSST staff member, the five RAs and a national transport engineer, all of whom provided additional perspectives on emerging findings and on possibilities for subsequent practical interventions. Our five young local RAs worked alongside the older researchers during the workshop, at times supporting them, but in common with the rest of the group, mostly learning from the older people's experiences.

One of the interesting elements of this study was the relationship which evolved between generations within the larger research team itself, especially as observed in the workshop training week where the OPR and RA teams first came together. Although many of our wider team members were not much younger than the OPRs, our five young RAs provided a distinctly different generational element. During the training week the young RAs were present for two reasons: first to support the lead trainer, but second also so that they became familiar with key issues faced by older people and thought through how they would themselves relate in ensuing weeks both to the OPRs and to older people whom they would interview in strands two and three of the research.

The power dynamics involved in intergenerational research relations can be complex, as earlier work involving academics researching with children has indicated (Porter *et al.* 2012b). Co-constructed knowledge generated through a cross-generational process must raise questions as to the extent to which older people as social agents are really able to enter the research equation. It is possible that the OPRs initially perceived our young local RAs to be quite powerful, even intimidating, because of their urban, educated status and perhaps also because of their youth and evident physical energy, but over time the relationships between OPRs and RAs appeared to clarify into a seemingly different pattern, as we illustrate below.

We took careful note of how our individual young RA team members worked with older people (both OPRs and other older people in the villages) through the training week and were impressed by their courtesy and care in this respect. We asked the RAs to reflect on their experiences too, since none had worked with older people as colleagues or informants before: they were positive but pointed out the need to make adjustments for 'When you see an older person for interview, it's different – they get tired'.

Over the week, the RAs got to know the OPRs better and to understand how to help support them. Our OPR team also began to come forward with their own observations on age-related characteristics. The need for adjustment was noted when they interviewed some of their older and less literate peers in the villages: 'We had to give her time to speak slowly' (male OPR) and 'We had to slow down, he was repeating himself' (male OPR). At the same time, the RAs also started to learn from, and to be advised by, some of the OPRs, about how to approach and pace interviews with older people in the villages:

MALE OPR, ADVISING RAS: Use your experience – you need to know your respondent – read the facial expressions, responses.

The following conversation took place between a group of OPRs and RAs reflecting on their latest field experience in a feedback session one afternoon. They were reviewing a joint interview with an elderly widow in the village:

YOUNG MALE RA: She didn't want us to rush – it was a slow interview. We did that but the time [it took] was difficult. We had to wait for her to chew it and respond.

MALE OPR IN THE SAME TEAM: It was difficult for her to understand the issues and she had bad hearing. With older people you may just need to take more time and explain things.

WOMAN OPR IN THE TEAM: Yes, many old people are trying to call for support so they may have a lot to say. Some of the women may not be used to being interviewed. They get tired. You can see the fatigue.

Both RAs and OPRs accompanied their observations by examples of how they adapted to issues of pace, hearing and understanding. Rather than seeing this as insoluble, both age groups showed the capacity to adjust and to be open to learning from some of the oldest and most vulnerable people in the communities where we worked.

The confidence of the OPR team grew over time, as they built up a strong knowledge of the difficulties and disadvantages faced by their peers in settlements across the district. One (male) member of the team, on the basis of the information the team had collected, had no hesitation in getting up at our final national review workshop in Dar es Salaam and lecturing the meeting – including the young Chief Medical Officer for Tanzania, who had been invited to open the meeting – about the needs of older people. Both male and female OPRs were observed to participate actively in the small group discussions with senior ministry and NGO staff which followed. We hope that the confidence they have built through the project will help them to take a stronger role in advocating for change in their own community in the future.

Methods and findings from the training workshop

This final section considers methods in which the OPRs were trained during the one-week workshop and some reflections on issues relating to intergenerational relations which were raised through the associated exercises and thus formed the base for subsequent investigation in the main phase of work.

A variety of methods were demonstrated and tried out by the OPRs during the training workshop. These included visual diagramming, interviewing (avoiding closed questions, using 'helper' questions, starting interviews with open questions and avoiding leading questions), mobility mapping and mobile interviews (to gather information about daily journeys, including those made for livelihood and health-care reasons) and seasonal calendars to further explore accessibility and associated livelihood and health issues. Participants also practised using a timeline to develop a conversation about daily carrying activities, gathering information on the types of loads carried, who carried them, what method they used, and how often various loads were carried during the day or a week. Although pedestrian load-carrying is widespread across Africa, its potentially important implications both for health and livelihoods are little studied (Porter *et al.* 2012a, 2013a, 2013b). The OPRs were able to work with most methods, with the exception of the mobile interviews. These proved difficult to arrange at the time of community visits, so instead the OPRs kept a journal for two weeks of their own journeys. The OPRs continued with this work after the workshop, electing to travel with the RAs to all ten villages, where they worked in their own teams. In total they conducted 74 interviews.

In terms of findings about older people's lives and relations with other generations, the preliminary interviews in the training workshop and associated field trials in Vikuge provided important context and indicated many issues that we would need to explore in depth in the main field research component. The daily timeline interview proved of particular importance in exploring generational relations and indicated the frequency of symbiotic relationships in which older people support children in their care, while their young charges (beyond the age of about six years) are active on behalf of their grandparents and elders, carrying their messages, collecting medicines, going to the grinding mill, helping carry water and firewood and so on. (Though children may not be expected to reciprocate, Skovdal and Campbell (2010) suggest children often reciprocate in a conscious effort to nurture potentially protective social relationships from the wider social recognition these efforts may generate.) Feedback from field exercises, for instance, drew particular attention to problems experienced by older people in fetching water because of the weight of the load and distances involved: some were unable to carry it by head; some used smaller containers, increasing the number of journeys. They had to carry water even when unwell and many depended on help from family and neighbours and timed their water collection accordingly. Interviews suggested how important help from grandchildren and younger neighbours was in this respect. One woman for instance carried a bucket on her head while her young grandchildren hauled a wheelbarrow with five 20-litre cans twice a day (in this case they need water for their two cows); in another case an old woman carried a ten-litre can on her head three times a day, accompanied by her small granddaughter carrying three litres. Grandchildren also participate in carrying firewood; they often walk two to three kilometres to collect larger amounts about twice a week (older women reported carrying 2–15 kilos – their grandchildren were reported to carry more). Older women without this help had to manage with small amounts of biomass debris collected from the farm or around the home daily. To explore carrying practices further, one of the main interviews was adapted to focus on how older people in 'vulnerable' households (where older people lived alone or lived with grandchildren without younger adults) obtained their water, firewood and food (farm produce) for the home and became a key focus of interviews in the main study.

The mobility mapping also showed how many older people walked up to one hour to their farms every day to cultivate the land and to bring home food – mainly cassava – for family consumption. Older people cultivate a fraction of their land, often just half an acre, due to lack of family labour and ability to pay for labour and other inputs. Some older people with little land, or inadequate resources to farm and children too young to help farm in their care, also told how they had to undertake paid day work as labourers to obtain sufficient food for the family. Table 18.1 shows the range of issues which emerged from the initial co-investigation with the OPRs during the training week.

Table 18.1 Findings from the preliminary training workshop with older people researchers

Findings	Themes for further inquiry
• Fetching water is a problem for older people (OP) • Many OP look after grandchildren with little help (most of the 97 orphans in Vikuge are cared for by OP) • Casual labour is part of livelihood strategy for some OP • Nearest road is generally passable but: potholes and mud in wet season, sand in dry season • High cost of transport affects OP's ability to access health services; health emergencies a big problem • Health insurance (TSh5,000 (US$3.30) per household) often denied to OP even when covered by household insurance • Journeys on foot, often to farm, very hazardous in wet season; snakes, fallen trees, mud	*Casual labour:* How common is it among OP, which OP do it, who do they labour for, How much do they get? *Income:* How do OP get cash income (pension, remittance, sale of produce); what are their travel needs for this? *Health insurance:* And implementation of free health care for OP. Why do hospitals ask OP for insurance cards? *Household composition:* How are older people in vulnerable/skip generation households supported? *Carrying:* How do OP transport water, firewood and farm produce to their homes? *Boda-boda:* As the main local transport, how do OP use this form of transport and what are their views on it? Source: Mobility map interviews: daily/health and livelihood journeys
• Main crops planted: maize, cassava, peas, lentils, cashew, rice, pumpkin • Most OP grow only for household consumption, larger volumes, when produced, is sold at farm • Farm related carrying often hard for OP as mainly by head-loading. Mainly tools and seeds to the farm and food for household consumption and firewood from the farm • Most OP cultivate a small proportion of their land due to physical decline and lack of cash to hire labour and buy inputs • Few people keep animals	*Older farmers:* What prevents OP farming all their land; do older farmers receive any government support? What is the definition of a 'farmer'; what proportion of farmers are older people? *Transport to market:* If OP could get their produce to market would they get a better price than selling at the farm gate? What are the barriers to selling at market? *Methods:* What proportion of farmers use fertilizer; what prevents them? *Cash income for OP:* What are the main sources of cash for OP; of these, which are the most valuable and why? Source: Seasonal calendar interviews: seasonality of livelihood journeys

- Main water source is wells, most run dry in dry season; some houses have piped water to the compound and pay a monthly fee but flow is sporadic – one or two times a week or a month (different views)
- Family remittances to OP quite common and mainly delivered by hand; some send by phone requiring journey to agent in town
- One example of mobile (cell) phone use for business (ordering tobacco supply from 120 km away)
- Grandchildren play key role carrying water and firewood in OP households

Alternative haulage: How common is use of hand carts/wheelbarrows for hauling water/firewood?

Animal keeping: How common for OP?

Care of grandchildren by OP: How do OP ensure children attend and continue in school; how do OP manage (carrying etc.) when children leave home?

Remittances to OP: How common is this; how is money delivered?

Phone use: What proportion of OP own their own mobile phones or have access to one? What do they use them for; how does this affect their transport needs?

Source: Timeline interview: carrying water, firewood and food

Conclusion

We set out to show in this chapter how co-investigation with older people over a one-week training period substantially aided the identification of key issues for further research regarding older people's lives and mobility patterns – including the extent to which their lives are bound up with other generations. Initial work by and with the OPRs showed how many older people gain access to services indirectly through both adults and children in the community: young people carry their messages, collect medicines, go to the grinding mill and help carry water and firewood. At the same time older people often have to care for and in large part support young grandchildren (sometimes with little or no material or financial support from other family members). The OPRs' work, and the subsequent qualitative and survey research which built on it, suggest that the symbiotic relationships which develop in these difficult contexts, for the most part, benefit all concerned. However, a complementary study, with young people as co-researchers, focused on exploring intergenerational relations and associated mobility patterns through the eyes of children resident with grandparents in the same communities, could add substantially to this picture and would form a logical extension of the work presented here.

A subsidiary theme of the chapter was intergenerational relations within the project research teams. The workshop week enabled us to observe the development of relations between our OPRs and a group of young RAs who had been appointed to help support the OPRs in their training week and subsequent studies, but also to then work on the larger scale (qualitative and survey) academic research study. This drew our attention to the power dynamics where research involves different generations and, in particular, to the power relations which may be at play when co-constructed knowledge is generated through a cross-generational process. We observed how, during the training week, there was a subtle change as our village OPRs grew in their confidence as researchers and repositories of significant local age-related knowledge and began to advise their more educated young urban co-investigators: subsequently this translated into interventions beyond the village context at our national workshop.

The involvement of the OPRs as co-investigators has certainly improved knowledge of older people's mobility issues and the extent to which the mobilities of older and younger generations are intertwined in Kibaha district: the findings are likely to have relevance for much of rural Tanzania. At the same time, the skills the OPRs developed during the research process have the potential to help raise the voice and profile of older people in their community and beyond. Both aspects potentially have important implications for future development interventions: the task now is to take these forward.

References

Aboderin, I. (2004) 'Decline in family material support for older people in urban Ghana', *Journal of Gerontology, Series B*, 59(3): S128–S137.
Apt, N. (1997) 'Aging in Ghana', *Caring*, 16(4): 32–34.

Barrientos, A., Gorman, M. and Heslop, A. (2003) 'Old age poverty in developing countries: Contributions and dependence in later life', *World Development*, 31(3): 555–570.

Clacherty, G. (2008) *Living With Our Bibi: Our Granny is Always Our Hope*, report for World Vision, Tanzania.

HAI (HelpAge International) (2004) *The Cost of Love: Older People in the Fight Against AIDS Tanzania*, Dar es Salaam: HelpAge International.

HAI (HelpAge International) (2007) *Building Bridges: Home-Based Care Model for Supporting Older Carers of People Living with HIV/AIDs in Tanzania*, London: HelpAge International.

Heslop, A. and Gorman, M. (2002) *Chronic Poverty and Older People in the Developing World*, CPRC Working Paper 10, Manchester: IDPM, University of Manchester.

Ibralieva, K. and Mikkonen-Jeanneret, E. (2009) *Constant Crisis: Perceptions of Vulnerability and Social Protection in the Kyrgyz Republic*, London: HelpAge International.

Ingstad, B. (2004) 'The value of grandchildren: Changing relations between generations in Botswana', *Africa*, 74(Special issue 1): 62–75.

Kamya, H. and Poindexter, C.C. (2009) 'Mama Jaja: The stresses and strengths of HIV-affected Ugandan grandmothers', *Social Work in Public Health*, 24(1/2): 4–21.

Mudege, N.N. and Ezeh, A.C. (2009) 'Gender, aging, poverty and health: Survival strategies of older men and women in Nairobi slums', *Journal of Aging Studies*, 23(4): 245–257.

Porter, G. and Abane, A. (2008) 'Increasing children's participation in transport planning: Reflections on methodology in a child-centred research project', *Children's Geographies*, 6(2): 151–167.

Porter, G., Hampshire, K., Bourdillon, M., Robson, E., Munthali, A., Abane, A. and Mashiri, M. (2010) 'Children as research collaborators: Issues and reflections from a mobility study in sub-Saharan Africa', *American Journal of Community Psychology*, 46(1/2): 215–227.

Porter, G. Hampshire, K., Abane, A., Munthali, A., Robson, E., Mashiri, M., Tanle, A., Maponya, G. and Dube, S. (2012a) 'Child porterage and Africa's transport gap: Evidence from Ghana, Malawi and South Africa', *World Development*, 40(10): 2136–2154.

Porter, G., Hampshire, K., Dunn, C., Hall, R., Levesley, M., Burton, K., Robson, S., Abane, A., Blell, M. and Panther, J. (2013a) 'Health impacts of pedestrian head-loading: A review of the evidence with particular reference to women and children in sub-Saharan Africa', *Social Science & Medicine*, 88: 90–97.

Porter, G., Tewodros, A., Bifandimu, F., Gorman, M., Heslop, A., Sibale, E., Awadh, A. and Kiswaga, L. (2013b) 'Transport and mobility constraints in an aging population: Health and livelihood implications in rural Tanzania' *Journal of Transport Geography*, 30: 161–169.

Porter, G., Townsend, J. and Hampshire, K. (2012b) 'Editorial: Children and young people as producers of knowledge', *Children's Geographies*, 10(2): 131–134.

Schatz, E.J. and Ogunmefun, C. (2007) 'Caring and contributing: The role of older women in rural South African multi-generational households in the HIV/AIDS era', *World Development*, 35(8): 1390–1403.

Schwanen, T. and Paez, A. (2010) 'The mobility of older people: An introduction', *Journal of Transport Geography*, 18(5): 591–668.

Skovdal, M. (2010) 'Children caring for their "caregivers": Exploring the caring arrangements in households affected by AIDS in western Kenya', *AIDS Care*, 22(1): 96–103.

Skovdal, M. and Campbell, C. (2010) 'Orphan competent communities: A framework for community analysis and action', *Vulnerable Children and Youth Studies*, 5(Supp. 1): 19–30.

UNICEF (2006) *Africa's Orphaned and Vulnerable Generations: Children Affected by AIDS*, New York, NY: UNICEF.

Van der Geest, S. (1998) 'Reciprocity and care of the elderly: An anthropological comparison between Ghana and the Netherlands', *Tijdschr Gerontol Geriatr*, 29(5): 237–243.

Whyte, S.R. and Whyte, M.A. (2004) 'Children's children: Time and relatedness in eastern Uganda', *Africa*, 74(Special issue 1): 76–94.

Whyte, S.R., Alber, E. and Geissler, P.W. (2004) 'Lifetimes intertwined: African grandparents and grandchildren', *Africa*, 74(Special issue 1): 1–5.

19 The intergenerational help desk

Encouraging ICT use in older adults in England

Irene Hardill

Introduction

In this chapter we focus on intergenerational geographies by critically reflecting on participation in research funded in the UK by the New Dynamics of Ageing research programme. The Sus-IT project (shorthand for Sustaining IT use by older people to promote autonomy and independence) was concerned with understanding the problems and circumstances which might cause people to 'disengage' or give up using information and communication technologies (ICTs),[1] such as computers, the internet or mobile phones. ICTs are fundamentally altering the spatial and temporal organization of economic and social life. A recent report on internet use and older adults in Ireland highlighted the utility of the internet as both a communication tool and an information resource (CARDI 2012). For older people, ICTs can be powerful assistive technologies, helping them to maintain their independence, social connectedness and sense of worth in the face of declining health or limited capabilities, but they can also offer new and empowering opportunities to improve an individual's quality of life. Old age and space are entwined processes (Schwanen *et al.* 2012), and using ICTs can alter the spatial and temporal contours of everyday life: for example, ICTs can be used to undertake tasks from home that once needed to be undertaken outside the home, such as online shopping and paying bills. ICTs offer people the possibility of keeping in touch with others via mobile phones while outside the home. Intergenerational support from family members and non-kin in the community play an important role in supporting older people building and maintaining ICT *savoir faire*, technologies increasingly used by older people to stay in touch with children and grandchildren (Hardill 2013). In order to understand the challenges of sustaining digital engagement this chapter examines the ways in which digital technologies are embedded and embodied into the practices of everyday life, highlighting the role of intergenerational linkages with particular reference to England. After this introduction, the chapter is divided in three parts. The first examines older adults and ICT use. This is followed by a section highlighting the Sus-IT project, and the final section is the conclusion.

Older adults, old age and digital technology

In this section we begin by examining older adults and public policy in the UK, with specific reference to England, and then turn to the use of digital technologies by older adults. In common with other advanced capitalist countries England is experiencing population ageing. Older adults in England[2] are identified in public policy as being over 50 years of age, and they make up over 34 per cent of the population (ONS 2011), ranging in age from 50 to over 100 years.

Chronological age, therefore, is used as a public policy marker, and the over-50s have been the focus of research into demography and ageing by the UK Research Councils (Walker 2007). In 2001, New Labour published the national service framework for older people (Department of Health 2001) and in that report three broad groups of the 'over-50s' were identified:

- Entering old age: These are people who have completed their career in paid employment and/or child rearing. This is a socially constructed definition of old age, which, according to different interpretations, includes people as young as 50, people who are active and independent and many remain so into late old age.
- Transitional phase: This group of older people are in transition between a healthy, active life and frailty. This transition often occurs in the seventh or eighth decades but can occur at any stage of older age.
- Frail Older People: These people are vulnerable as a result of health problems such as a stroke or dementia, social care needs or a combination of both. Frailty is often experienced only in late old age.

So we can see that within the framework of the modern lifecourse, old age is defined as one of a number of discrete phases each delineated by changes in social and economic roles (Cole 1992). Growing older according to the Department of Health (2001: 107) was seen as representing a period of increased dependency, as physical strength, stamina and suppleness decline, and the individual has to cope with chronic and long-term conditions. In the book *A Fresh Map of Life*, Peter Laslett (1989) portrayed four distinct 'ages' of a person's lifecourse. He represented the 'First Age' as the period of childhood dependency, the 'Second Age' as the time of independence, employment and maturity, the 'Third Age' as the period during which people are freed from work and family constraints and have time to pursue a good quality of life. Finally, he saw the 'Fourth Age' as characterized by dependence and declines in health. Peter Laslett offered a view of old age that was not just inactivity, declining health and mobility, loneliness and poverty. In some ways the Department of Health report resonates with Peter Laslett's (1989) conceptualization of the lifespan, with old age divided into a third and fourth age.

The de-standardization of the lifecourse in post-industrial societies means that lives are less predictable, less collectively determined, less orderly and more flexible, indeed working life is considered to be more precarious (Featherstone

1991). Retirement, for example, is identified as less of an event and more as a 'zone' through which people pass, making adjustments over time to their commitment to paid work, including adjusting the amount of paid work undertaken, against a changing policy context with the abolition of the default retirement age in the UK (Stockdale and MacLeod 2013). Everyday life for the over-50s is now more complex in the current economic downturn, concerns about pensions appears to be resulting in rising levels of economic activity, and 'unretirement' (Kanabar 2012).

In a recent review article (Schwanen *et al.* 2012) argued for a more nuanced understanding of growing older pointing to a blurring of lifecourse identities which increasingly characterizes life in advanced liberal democracies. Such blurring makes it increasingly difficult to compartmentalize childhood, adulthood and old age into distinct life stages or to use chronological age indicators (i.e. age in years) to identify people as 'young' or 'old'. Schwanen *et al.* (2012) also highlighted that independence, mobility and experimentation have become the ideals for later life under neoliberalism in advanced liberal democracies, which has, in many ways, refracted and complicated – but not undone – industrial era connotations of old age with dependency, decline, passivity and obsolescence. Neoliberal discourses of 'active ageing', according to which older people are to enhance their well-being through participation in social, economic, cultural, spiritual and civic affairs (WHO 2002).

In terms of intergenerational relations the growing importance in Western countries of grandparenting as an identity construction in later life has been highlighted (Tarrant 2010; see also Tarrant, this volume). Tarrant (2010), for example points to the increase in the number of children growing up in single- or dual-parent families with complex employment and caregiving arrangements. Caregiving roles of grandparents to grandchildren appears to be increasing in the economic downturn of the late 2000s and early 2010s (Kanabar 2012).

Today it is commonplace to say that ageing is embodied, emplaced, as well as relational. These ideas are heavily indebted to the writings of Graham Rowles in his 1978 book *Prisoners in Space*, and to the writings of the late Glenda Laws (1993, 1995). In Sus-IT we have focused not on the body one has but on the body *one is*, the body through which one participates in the world and that emerges from the interactions between the physiological body and the people, artefacts and other forms of materiality it encounters. Bodies are seen as objects onto which values, ideas and discourses can be inscribed and as material entities representing and reproducing those values, ideas and discourses.

In our work we have focused on the ways in which digital technologies – computers, the internet, mobile (cell) phones – have become embedded and embodied into the practices of everyday life, and the challenges people encounter in staying engaged (Hardill and Olphert 2012; Hardill and Mills 2013). We have been concerned with the context of digital technologies, and how the spatial and temporal patterns of everyday life are changed by the use of ICTs, as digital technologies become spatially and temporally embedded in everyday relational practices. Folding together places and people separated by time and space via a 'connected presence', digital technologies, for example, can help reduce social isolation by linking people together.

Older people did not learn about ICTs when at school, nor did all use ICTs when in paid employment; ICT skills have been acquired in later life, some as part of paid work, or through accessing support in the community. As a result the nature and quality of ICT training and support is critical in supporting older people's development of 'internet self-efficacy' (Hardill and Olphert 2012; Hargittai 2008). As part of the ESRC E-Society programme in the UK Lindsay *et al.* (2008) worked with 108 older adults who were provided with free home computers and a one-year broadband subscription. Half the sample received ICT support in the form of 'facilitated learning' while the other half received no support. Lindsay *et al.* (2008) found that many of their participants first needed to overcome their fear of the technology before they could learn how to use it effectively. Significant differences were recorded between those who received ICT support and those who did not six months after being involved in the project. Over half of those who received ICT help subsequently searched for information online on public authority websites, compared with 28 per cent of those who had not received help (ibid.: 326). They concluded that older adults first needed some form of encouragement and then a compelling proposition linked to their own lives to help overcome their fear of turning a computer on and going online (ibid.: 323). In summary, once older people have access to, and acquire the skills to use ICTs, ICTs can become part of everyday life, as is illustrated by US research that shows that 70 per cent of people aged 65 or over who had started using the internet stated that they typically use it every day (Zickuhr and Madden 2012).

Turning first to data derived from the special Eurobarometer (Eurostat 2012) on ICT infrastructure (computer, internet and broadband) and ICT use across the member states of the EU, 76 per cent of UK households owned a computer in 2011, but the proportion is less, at 41 per cent for households with people aged 60+ years. When it comes to access to the internet, 74 per cent of UK households have access, but the proportion is only 38 per cent of households with adults aged 60+ years (ibid.). For older people the most common reasons for using the internet were searching for information and emailing (Hardill 2013). Turning to mobile phone use, in 2011 91 per cent of individuals in the UK possessed one, but only 51 per cent of people over 75 years did so (Ofcom 2012) As with ICT infrastructure, mobile phone ownership levels varied by income group, with 92 per cent of the highest decile owning a mobile phone in 2009, compared with 67 per cent of households in the lowest decile (ONS 2011). Mobile phones are generally rated as the most vital digital technology (Pfaff 2010), equally as vital as landlines (Crang *et al.* 2006: 2562), and are no longer merely devices for sending and receiving telephone calls; rather, they now embody different communication technologies (text, visual, video, web browsing as well as audio) (Kwan 2007).

Shove and Pantzar (2005) argue that artefacts and forms of competence only have meaning and effect when integrated into practice, and thus that it is through the integrative work of 'doing' that elements are made animate, sustained and reproduced. When that stops, fossilization sets in. In their study of ICT use (mobile phone and the internet) in two contrasting communities in Newcastle upon Tyne, Crang and Graham (2005) found that for some people using the internet and mobile phone was episodic while for others use was pervasive, the internet and mobile phones had become an integral part of everyday life.

Even within households with a networked computer not all household members may value the technology or be proficient users. There must be a compelling reason to make people use them. Among retired people in the UK, 'just not interested' was cited as the main reason for not using the internet and giving up using it; 'do not know how to use' was the second most common reason for non-use, while for retired people 'not for people my age', 'computer not available' and 'too expensive' were also presented as reasons for non-use (Dutton and Blank 2011). In recent academic work it has been reported that providing ICT access alone is not always enough to bridge the digital divide as social capital (such as skills and education) is also needed to engage with technology (Lindsay *et al.* 2008). This view is echoed by Sinclair and Bramley (2011) whose work emphasized the ways in which technology is socially embedded in the tasks of everyday life. In the following section we turn to the Sus-IT project.

Context and approach to the research

In the remaining part of this chapter we draw on data collected as part of Sus-IT, which as we mentioned above was funded in the UK as part of the British New Dynamics of Ageing (NDA) Research Programme (2009–2012). In 2010 a team based at Memorial University, Canada secured funding from the Canadian Institutes of Health Research (CIHR) to undertake related work on sustaining digital engagement in Canada. As a result Sus-IT has undertaken cross-national comparative work into the challenges of sustaining digital engagement. In this chapter I draw on the work undertaken in the UK.

The New Dynamics of Ageing (NDA) Research Programme was established by five of the UK Research Councils – ESRC, EPSRC, BBSRC, MRC and AHRC[3] – to fund multidisciplinary research to better understand the way in which older people's lives may be changing as a result of social, economic and technical developments (Walker 2007). In common with the other research programmes NDA identified people over 50 years as older people, and placed their involvement high on its agenda (Hennessey and Walker 2011; Walker 2007).

Sus-IT was scoped to address the lacuna identified in the NDA call on ICT use (Damodaran and Olphert 2010). Indeed the fundamental research question Sus-IT posed was inspired by lived reality, the challenges the father (an elderly widower, who lived alone) of a member of the UK project team was facing as he struggled to remain digitally engaged in the face of declining eyesight. He relied on his children and grandchildren to sustain his digital engagement. So at the heart of the Sus-IT project was a research problem being experienced at first hand. Sus-IT aimed to understand the challenges faced by older people using digital technologies, to identify ways to help older people to be confident and competent users of computers and other digital products, and to explore how older people's use of these technologies could be maintained in the event of declining capability and/or changed circumstances, events fused with emotion. Much previous research has focused on the differences between those who have access to the internet and those who do not (Light 2001; Warf 2001; Selwyn 2004). While long-term adoption of internet use provides an important indicator of online engagement, however, little is known about the factors that support sustained use and those that discourage it.

Sus-IT was designed to fill this gap in the literature. Sustainability of internet use, we argue, represents another disparity, since there are barriers to its use that go beyond issues of access. The methodological approach we applied in Sus-IT was interactive rather than extractive, undertaking research *with* older people, rather than *on* them (Damodaran and Olphert 2010; Hardill and Olphert 2012). In reaching out to diverse groups of older adults a range of methods were employed through interactive sessions using images, producing a participatory video, photo diaries and personal stories of digital engagement to stimulate discussion and debate on ICT usage from a lifecourse perspective (Hardill and Olphert 2012). As a concept, a lifecourse is defined as 'a sequence of socially defined events and roles that the individual enacts over time' (Giele and Elder 1998: 22). A lifecourse approach affords researchers the possibility to examine an individual's life history using a variety of data-gathering tools. Engaging qualitatively or quantitatively with time enables a more finely grained understanding of everyday life, and the uncovering of how the personal is interlinked with the immediate and wider social context.

We worked through 'gatekeepers' responsible, for example, for the provision of computer support, formal computer classes, University of the Third Age (U3A), Older People's Fora and community groups across England. We first approached gatekeepers to explain the purpose of Sus-IT, the modus operandi of the project team and to invite participation of the groups they represented in Sus-IT. The gatekeepers offered advice as to the most suitable methods of engaging with their members, and some involved intergenerational working. A key theme to emerge from these discussions was the importance older people attached to a relational approach for sustaining digital engagement, this centred on accessing one-to-one support, either formally through IT clinics, or less formally by calling upon support from kin and friends, or at IT drop-in centres. Our Sus-IT events offered participants new linkages especially for those older adults unable to access IT support from grandchildren or children.

Each interactive event was therefore planned with a specific group of older adults in mind; their needs were foremost, throughout. This led to a variety of events at which presentations about Sus-IT were given, other speakers invited (e.g. from digital outreach teams, local authorities and older people's charities), and a variety of methods (described above) for capturing ICT use were employed. Moreover, in response to demand from different groups we hosted special events (e.g. Festival of Social Science 2010 and 2012, Silver Surfer

Day 2010), and delivered IT taster sessions, mobile-phone clinics, and held joint events with Youth Assemblies and Older People's Assemblies. Our participatory research approach also encouraged some groups to be very proactive and engage in actively co-creating a research agenda with us (Hardill and Olphert 2012).

About 750 older adults participated in Sus-IT events in England and Scotland, and completed a questionnaire-based survey of Digital Engagement, which included quantitative data about factors such as extent, frequency and scope of use of digital technologies, and qualitative data about attitudes and experiences with technology. Within what seemed to be digitally engaged households, there was a spectrum of 'onlineness' with one partner, often the male partner in heterosexual households, being the most intensive internet user. We noticed, therefore, the importance of coupledom in supporting ICT use within households, and the consequential impact of its cessation through bereavement. The widowed partner in such households often struggled to sustain ICT use. We also found that the degree to which digital technologies were integrated into everyday life did not correlate smoothly with chronological age. A recurring theme was the importance of kin (children and grandchildren) in supporting and encouraging older adults to engage with digital technologies (cf. Tarrant 2010). Finally, ICT support also came from non-kin in the community, with young and older volunteers offering one-to-one ICT support via classes and ICT drop-in clinics provided by community groups (Hardill 2013). A number of the Older People's Assemblies we worked with had previously undertaken projects whereby members of young people's groups (local schools or youth assemblies) provided one-to-one help and support with computers and mobile phones. Such intergenerational work was highly valued by the groups, and older people reciprocated by visiting local schools and giving talks about the local community.

With a subset of participants, life-history interviews were conducted to explore ICT use, and we analysed these data using a framework based on competency of use, which looked at the ways in which computers, the internet and mobile phones are spatially and temporarily embedded in everyday relational practices. We have built on the work of Shove and Pantzar (2005) and Crang and Graham (2005) discussed in detail above to develop our framework, which captures a spectrum of onlineness and includes:

- *Pervasive use*: confident ICT users who have developed 'internet self-efficacy' (Hargittai 2008); ICTs used daily forming an integral part of the architecture of everyday life; networked PCs/laptops used with confidence to undertake a wide range of everyday tasks; such as for communicating with other people, including children and grandchildren [via Skype/email]; as a source of information; for organizing everyday life, including searching for information and services; in some cases mobile phones are used to access the internet. Upgrading ICTs does not pose a challenge indeed it is enjoyed, choosing new ICTs is almost a hobby.

- *Episodic use*: sporadic use of ICTs, while some said they 'coped' using them, others were 'scared' of using them; ICTs not 'always on'; limited range of applications used, and not confidently.
- *Fossilization*: episodic ICT usage declines to complete cessation.

As the particular focus of Sus-IT was on understanding the problems and circumstances which might cause people to 'disengage' or give up using the internet and ICTs, *fossilization* captures the process by which ICT usage, for a variety of reasons – social, health-related, economic – declines, to the point of complete cessation. In the following section the focus shifts to ICTs and intergenerational linkages.

ICTs and intergenerationality

In this section we use indicative case studies to illustrate five dimensions of ICTs and intergenerationality. As reported earlier, in their study of older people and ICTs, Lindsay *et al.* (2008: 323) concluded that older adults first needed some form of encouragement and then a compelling proposition linked to their own lives to help overcome their fear of turning a computer on and going online (ibid.). A second UK study by Sykes *et al.* (2008) found that older adults tend to access information through personal contacts, especially their social networks. While some of our Sus-IT participants received ICT support from family members, not all felt that their children were the best teachers, because of limited patience and/or not explaining things simply, which is examined in more detail below.

A number of participants had attended computer classes; in some cases more than one as they had forgotten what they had learned on previous courses. There were mixed views on the merits of courses, generally those who had been using computers longer had no interest in a course, while novice users did express some interest in a tailored course that would help them, but there was a clear preference for one-to-one learning, especially task/problem-based to deal with a specific issue. The preference was, therefore, for task-related classes, with a focus on everyday applications of ICT functions, and for informal support from family and friends to supplement the one-to-one training. There was a feeling that group settings re-enforced previous poor learning experiences, and tutees can end up feeling left behind. ICT drop-in classes were popular, where people receive one-to-one help to solve a particular problem.

Turning now to ICTs and intergenerationality, a recurring theme in the in-depth interviews was that new linkages occurred through the use of Skype and email, and mobile phones, for example. An indicative example is provided by Judy (divorced, early sixties, working part-time). Judy is a confident ICT user. Most of her family live nearby, and she is in close touch with her children and grandchildren. Digital technologies have been an integral part of both her home and work life for some time. Indeed, she still works part-time in a job that requires the daily use of a computer. She told us that she helps work colleagues with their computer tasks, and when at home she loves using her networked laptop, especially for surfing the internet, and checking up on her daughters via Facebook. Judy told us that the only way she can keep in touch with her adult daughters is via text messaging. They never reply to her phone calls or answer her phone messages, but they respond immediately to her text messages. So text messaging enables Judy to communicate with her children. She also follows them on Facebook, so she knows what is happening in their lives. The issues Judy raised were repeated by other research participants, if they wanted to keep in regular touch with children and grandchildren they had to engage with the digital technologies younger people use to communicate with others, young or old.

A second dimension of ICTs and intergenerationality is through the provision of ICT equipment. A number of our research participants indicated that they had acquired ICT equipment (mobile phones and computers) from their children. In some cases it was their old equipment, in other cases new equipment was bought with the parent's needs in mind, but the older user may not necessarily have been included in the decision-making. While the artefacts were provided, this did not necessarily extend to giving older people the support they needed to sustain usage. The members of a number of community groups we worked with asked us to run mobile-phone clinics and IT help desks to help their members get to grips with the artefact they had been given. We now examine the case of Bally, a widow who is in her mid-fifties. Her three children bought her a networked computer and they continue to offer Bally the help and support she needs to sustain her ICT usage, they are her 'help desk'. Bally cited the example of Skype: her children helped her to learn how to use Skype so she can keep in touch with them.

> My daughter lives in Nottingham, my son lives in Bristol and my eldest son is coming and going to Holland all the time. So it is nice for me to use my Skype and I can see them when we talk.

Unlike other participants, Bally's children really help her to sustain her ICT usage, they 'troubleshoot' her problems, at the end of a telephone rather than face-to-face as they live some distance from her.

A third intergenerational linkage is through the ICT support provided by grandchildren, which was highly valued. Indeed some participants preferred to ask their grandchildren as they were more patient. An illustrative example is provided by Ruth, a 75-year-old married, retired grandmother who told us about Bradley, her 11-year-old grandson, who lives nearby, who is her 'ICT help desk' for both her mobile phone and computer. Bradley helps his grandfather, too. He sits and explains things slowly and so clearly, much better than Ruth's daughter, Bradley is very patient, and he does not make Ruth feel stupid. Some were supported by wider kin, such as Betty, an 83-year-old widow, whose only daughter lives abroad. Betty has enlisted the help of her niece, whom she meets weekly, who has provided the one-to-one help that Betty needed to learn how to use her mobile phone. She has written down simple instructions and practices every day, and now Betty has the confidence to turn to mastering an iPad, which her daughter bought her so they can email each other. Again Betty's niece is providing the one-to-one support she needs.

A fourth intergenerational linkage involves non-kin within neighbourhoods and communities. The gatekeepers of two Older People's Assemblies, for example, told us of digital engagement work they had successfully undertaken with local schools and a Youth Assembly. This included running mobile-phone and internet clinics. These two groups requested we help organize more intergenerational events. One group specifically wanted us to run mobile-phone clinics; most of their members possessed a mobile phone, often given to them by their children, and a number were not confident users. So at four consecutive meetings (August 2010–June 2011) we worked together to develop a programme of activities, which included mobile-phone 'clinics' offering one-to-one help (Hardill and Olphert 2012).

Another group asked us to enlist the help of the Youth Assembly for an internet workshop (Hardill and Mills 2013).We also involved the local library, which has a 'community hub' as part of the GO ON UK network (www.go-on.co.uk). GO ON UK is a network of community-based UK online centres that amongst other things help people get online and make the most of online life; they have a local focus, and rely on the commitment of volunteers to help others integrate technology into their lives. The GO ON UK hub we worked with has a suite of networked computers and undertakes community development work to improve the use of digital technologies. Staff at the hub were keen to develop intergenerational linkages to support older people's skills development. Members of the Youth Assembly worked with staff at the hub to produce a guided work programme to improve older people's access to three key public service websites. This was then delivered by members of the Youth Assembly, who worked one-to-one at a speed dictated by the older person offering him/her the help and support required to access information from the key websites.

Conclusion

In this chapter we have explored intergenerational geographies by focusing on research undertaken in the UK, which was funded by the New Dynamics of Ageing research programme. The Sus-IT project was concerned with understanding the problems and circumstances which might cause people to 'disengage' or give up using ICTs such as computers, the internet or mobile phones. We found no correlation between age and the propensity to engage with ICTs. Indeed people embedded digital artefacts into the tasks of everyday life, acquiring a specific skill set for internet use, 'internet self-efficacy' (Hargittai 2008), when they had a specific reason to use such technologies. The compelling reason to invest time and effort into learning how to use them was often to keep in touch with children and grandchildren. Sustaining digital engagement is linked to the significant, indispensable and crucial ICT support role of (extended) family members, who provide intergenerational support, along with motivational factors, acting as drivers for digital engagement. Through ICTs, new linkages and connections between the generations can be formed (cf. Tarrant 2010). But not all older adults received the help and support needed to become confident users from family members, and for such older adults support from the community, often involving young people, organized formally by neighbourhood and community groups is providing a vital resource supporting older people sustain their use of digital technologies.

Acknowledgements

This chapter draws on research funded by the New Dynamics of Ageing programme, *Sustaining IT use by older people to promote autonomy and independence* (Sus-IT) (RES-353-25-0008). I wish to acknowledge the contribution of the Sus-IT team, in particular L. Damadoran, J. Sandhu, S. Keith, M. Heeley and W. Olphert, and our research participants. The views expressed here are those of the author alone.

Notes

1 ICTs include telecommunications technologies, such as telephony, cable, satellite and radio, as well as digital technologies such as computers, information networks and software (Damodaran and Olphert 2006: 6).
2 In England older adults are identified as people over 50 years of age by policy makers (Department of Health 2001) and in recent research programmes on demography and ageing, such as the New Dynamics of Ageing research programme (Walker 2007).
3 Economic and Social Research Council, Engineering and Physical Sciences Research Council, Biotechnology and Biological Sciences Research Council, Medical Research Council and Arts and Humanities Research Council respectively.

References

CARDI (2012) *Focus on Internet Use and Older People*, Belfast: Centre for Ageing Research and Development in Ireland.

Cole, T.R. (1992) *The Journey of Life: A Cultural History of Aging in America*, Cambridge: Cambridge University Press.

Crang, M. and Graham, S.D.N. (2005) *Multispeed Cities and the Logistics of Living in the Information Age*, ESRC Project Report: Swindon, online, available at: http://dro.dur.ac.uk/1035/.

Crang, M., Crosbie, T. and Graham, S. (2006) 'Variable geometries of connection: Urban digital divides and the uses of information technology', *Urban Studies*, 43(13): 2551–2570.

Damodaran, L. and Olphert C.W. (2006) *Informing Digital Futures: Strategies for Citizen Engagement*, Dordrecht: Springer.

Damodaran, L. and Olphert, C.W. (2010) 'Sustaining digital engagement: Some emerging issues', *Proceedings of Digital Futures 2010*, Nottingham, 10–12 October 2010, online, available at: www.horizon.ac.uk/images/stories/f50-Damodaran.pdf.

Department of Health (2001) *The National Service Framework for Older People*, London: Department of Health, online, available at: www.dh.gov.uk/en/publicationsandstatistics/publications/publicationspolicyandguidance/DH_4003066.

Dutton, W. and Blank, G. (2011) *Next Generation Users: The Internet in Britain*, Oxford: Oxford Internet Institute.

Eurostat (2012) 'Individuals using the Internet for Internet banking', database table, online, available at: http://epp.eurostat.ec.europa.eu/tgm/table.do?tab=table& plugin=1 &language=en&pcode=tin00099.

Featherstone, M. (1991) *Consumer Culture and Postmodernism*, London: Sage.

Giele, G.H. and Elder, J.Z. (1998) *Methods of Lifecourse Research: Qualitative and Quantitative Approaches*, London: Sage.

Hardill, I. (2013) *E-government and Older People in Ireland North and South*, Belfast/ Dublin: CARDI.

Hardill, I. and Mills, S. (2013) 'Enlivening evidence-based policy through embodiment and emotions', *Contemporary Social Science*, 8(3): 321–332.

Hardill, I. and Olphert, C.W. (2012) 'Staying connected: Exploring mobile phone use amongst older adults in the UK', *Geoforum*, 43(6): 1306–1312.

Hargittai, E. (2008) 'Whose space? Differences among users and non-users of social network sites', *Journal of Computer-Mediated Communication*, 13(1): 276–297.

Hennessey, C.H. and Walker, A. (2011) 'Promoting multi-disciplinary and inter-disciplinary ageing research in the United Kingdom', *Ageing and Society*, 31(1): 52–69.

Kanabar, R. (2012) 'Unretirement in England: An empirical perspective', *Royal Economic Society*, online, available at: www.webmeets.com/res/2013/m/ viewpaper. asp?pid=241.

Kwan, M.-P. (2007) 'Mobile communications, social networks and urban travel: Hypertext as a new metaphor for conceptualising spatial interaction', *Professional Geographer*, 59(4): 434–446.

Laslett, P. (1989) *A Fresh Map of Life: The Emergence of the Third Age*, London: Weidenfeld and Nicolson.

Laws, G. (1993) '"The Land of Old Age": Society's changing attitudes toward urban built environments for elderly people', *Annals of the Association of American Geographers*, 83(4): 672–693.

Laws, G. (1995) 'Theorizing ageism: Lessons from postmodernism and feminism', *The Gerontologist*, 35(1): 112–118.

Light, J. (2001) 'Rethinking the digital divide', *Harvard Educational Review*, 71: 709–734.

Lindsay, S., Smith, S. and Bellaby, P. (2008) 'Can informal e-learning and peer support help bridge the digital divide?' *Social Policy and Society*, 7(3): 319–330.

Ofcom (2012) 'Communications market report 2012', online, available at: http://stakeholders.ofcom.org.uk/market-data-research/market-data/communications-market-reports/cmr12/.

ONS (Office for National Statistics) (2011) *2011 Census, Population and Household Estimates for England and Wales*, Table P04: Usual resident population by five-year age group, local authorities in England and Wales, London: ONS, online, available at: www.ons.gov. uk/ons/publications/re-reference-tables.html?edition=tcm%3A77-257414.

Pfaff, J. (2010) 'Mobile phone geographies', *Geography Compass*, 4(10): 1433–1447.

Rowles, G.D. (1978) *Prisoners of Space? Exploring the Geographical Experience of Older People*, Boulder, CO: Westview Press.

Schwanen, T., Hardill, I. and Lucas, S. (2012) 'Spatialities of ageing: The co-construction and co-evolution of old age and space', *Geoforum*, 43(6): 1025–1032.

Selwyn, N. (2004) 'Reconsidering political and popular understandings of the digital divide', *New Media and Society*, 6(3): 341–362.

Shove, E. and Pantzar, M. (2005) 'Fossilisation', *Ethnologia Europaea – Journal of European Ethnology*, 35(1/2): 59–63.

Sinclair, S. and Bramley, G. (2011) 'Beyond virtual inclusion – communications inclusion and digital divisions', *Social Policy and Society*, 10(1): 1–11.

Stockdale, A. and MacLeod, M. (2013) 'Pre-retirement age in-migration: Opportunities and challenges for remote rural areas', *Journal of Rural Studies*, 32: 80–92.

Sykes, W., Hedges, A., Groom, C. and Coleman, N. (2008) *Opportunity Age Information Indicators Feasibility Study*, London: Department for Work and Pensions.

Tarrant, A. (2010) 'Constructing a social geography of grandparenthood: A new focus on intergenerationality', *Area*, 42(2): 190–197.

Walker, A. (2007) 'Why involve older people in research?', *Age and Ageing*, 36(5): 481–483.

Warf, B. (2001) 'Segueways into cyberspace: Multiple geographies of the digital divide', *Environment and Planning B: Planning and Design*, 28(1): 3–19.

WHO (2002) *Active Ageing: A Policy Framework*, Geneva: World Health Organization.

Zickuhr, K. and Madden, M. (2012) *Older Adults and Internet Use*, Washington, DC: Pew Research Centre.

20 (Grand)paternal care practices and affective intergenerational encounters using Information Communication Technologies

Anna Tarrant

Introduction

This chapter considers the role of communication technologies in grandparenting, specifically grandfathering, with a focus on empirical data collected from men who are grandfathers. The changing spatialities of grandfather–grandchild relations, and relationships across generations more generally, are explored as they facilitate gendered care and practices of intimacy (referring to 'the quality of close connection between people and the process of building this quality' (Jamieson 2011: para. 1.1)). Excerpts from qualitative, in-depth interview data collected from 31 men are presented, revealing that grandfathers actively seek to maintain intimate intergenerational relationships with grandchildren through the use of technology. This reveals the affective nature of doing care in disembodied ways, challenges gender and age biases in understandings of familial care and reveals tensions created by extending care beyond the homespace.

Recent scholarship reveals that grandparents are active in seeking ways to communicate with their family members and they are a significant resource in families, often providing informal child care and emotional and financial support (Wellard 2011; Wheelock and Jones 2002). However, this is taking place against a complex backdrop of increasing mobility that is altering distance and proximity between grandparents and grandchildren, the altering of family formations, and transformations to the spatial, relational and political characteristics of caring and loving. Taking distance and proximity as an example, analysis of longitudinal data from the British Household Panel Survey demonstrates that proximity to grandchildren can strongly increase the exchange of care between grandparents and grandchildren (Chan and Ermisch 2011). As a result of increased mobility, however, families are now more likely to be geographically spread than ever before. According to Holdsworth (2013), a family decline narrative is often assumed as a result of this increased mobility. This suggests that living at a distance might have an erosive impact on intergenerational relations between grandparents and their grandchildren, reducing the frequency of involvement as well as the quality and character of relations between young and old familial generations. However, reduced frequency of involvement with grandchildren and assumptions of decline in intergenerational exchanges of care between grandchildren and grandparents are not realistic in a technologically advanced society and deserve further scrutiny.

Despite developments in thinking about how family life and its spatialities are changing (see Duncan and Smith 2006), intimate relations across generations (and consequent intergenerationalities) have varying implications for individuals across time and space (Hopkins *et al.* 2011; Valentine 2008; Vanderbeck 2007) and how cultural constructions of ageing might have different implications for people in the lifecourse (Pain *et al.* 2000), social geographers have rarely examined the spatialities of contemporary grandparenting (Tarrant 2010). Driven by this position and Vanderbeck's (2007: 202) question about how space can 'facilitate and limit extrafamilial intergenerational contact and the nature of intergenerational relationships', this chapter first puts modern grandfathering in context within literature from demography, sociology and social geography before describing the research's design. The main focus of the chapter is how traditional familial and intergenerational intimacies, as well as the processes of loving, caring and knowing are being transformed by communication technologies. Grandfathering via ICTs challenges the ideal that proximity is an important feature of family life and highlights alternative ways in which men are engaging in intergenerational care in contemporary societies. The chapter concludes by highlighting how the space opened up by ICT is facilitating new kinds of familial engagement but has also become a site in which intergenerational difference and tension may be exacerbated.

Grandparenthood in context

Uhlenberg (2004) argues that an examination of the demographic forces structuring intergenerational relationships between grandparents and their grandchildren is a useful starting point, not least because dislocating them from their temporal and spatial context results in a superficial understanding of the dynamics that shape the relationship. Indeed, grandparenting, or the practices integral to facilitating intergenerational familial relationships, is both active and dynamic (Arber and Timonen 2012). This has been strongly attributed to global social, economic and demographic change, framed by diverse welfare and cultural contexts (Arber and Timonen 2012).

Existing literature about grandparents from several disciplines including demography, geography and sociology suggests that diversity in contemporary grandparenting can be understood and explained by interconnected processes occurring at both local and global scales. At a macro-social scale, a number of interrelated social changes associated with globalization and late modernity, including demographic change, geographic mobility, trends in paid employment and the communication revolution (Arber and Timonen 2012), are shaping the micro-level practices and social relationships that grandparents experience and engage in across generations.

At the macro-social scale, leading sociological theorists such as Beck (1992) and Giddens (1992) explain that we have entered a 'late modern' epoch of 'individualization and 'detraditionalization' thought to have weakened the social ties of kinship and marriage (Duncan and Smith 2006). This weakening, often related to 'increasing mortality, decreasing fertility and shifting patterns of divorce, remarriage, lone parenthood and women's labour force participation' (Kemp 2007: 855) impacts on how grandparents negotiate their family relationships (Kemp 2007). Growing numbers of fragmented and reconstituted families require the renegotiation of relationships across several generations beyond the often-studied parent–child relationship.

On the other hand, a key paradox of these powerful trends is the contention that personal love and intimacy have actually become all the more important (Duncan and Smith 2006). Family is still considered significant even though it has become much more diverse, fluid and constructed through choice. Theoretical debate has consequently moved away from conceptualizing the 'family' as a noun, to a verb (Morgan 1996), drawing attention to the diverse ways in which people 'do' family and family practices, as well as practices of intimacy (Jamieson 2011). Drawing on a spatial metaphor appropriate to geographical knowledges about families and intergenerational spaces, Jacqui Gabb (2008: 64) conceptualizes families as 'affective spaces of intimacy within which meanings and experiences are constituted by family members in an historical socio-cultural context'.

Human geographers are contributing to these more recent debates by theorizing about the interconnectedness of local and global processes as they impact on family and intergenerational ties and how relations materialize through everyday spatial practices (Valentine and Hughes 2012). As Plummer (2001) highlighted over a decade ago, the connections between intimate life and globalization are frequently forgotten (Plummer 2001). Valentine (2006), for example, argues that while globalization, as a complex process of movement and flows, might be thought to undermine and threaten intimacy, which is often assumed to require physical proximity, it has also created new global spaces that facilitate intimacy and intimate practices over distance. New ways of 'doing' family and intimate relationships that are emotional and affective are emerging and being maintained between different homes, over large geographical distances (Valentine 2008). Empirical examples of transnational families (Bacigalupe and Lambe 2011) and couples 'living apart together' (Holmes 2004) exemplify these alternative ways of 'doing' intimacy. The internet in particular plays an important role in facilitating these sometimes intergenerational relationships and has become a key space in which intergenerational love and caring are enabled and different practices of intimacy can be explored and developed (Bell and Binnie 2000; Valentine 2008). Ames et al. (2010) argue that Skype and iChat technologies have become part of a constellation of activities that families now perform to enact, create and reinforce family values and love across distance. Indeed communication technologies are becoming increasingly enmeshed in everyday caringscapes (Bowlby et al. 2011).

The use of technologies by grandparents, including communication technologies, have rarely been considered, perhaps because it is commonly assumed that new technologies characterizing the global society have been considered as exclusionary to older populations. The lower use of technologies among older people is thought to reflect the structural marginality of older people, super-imposed on existing social inequalities (Gilleard and Higgs 2008). For those who have taken up technology use, it has been thought to have implications for relations between generations by altering the power dynamics between older and younger generations (Wilk 2000) through shifting knowledge to younger generations. As a consequence, how grandparents enact loving and caring in global contexts, and the spatialities of these enactments, remains invisible.

Why grandfathers? gendered care and intimacy

Up to this point, the chapter has considered 'grandparent' as a gender-neutral identity in order to explain the context in which grandparenting and intergenerational relationships with grandchildren are currently enacted and negotiated. Grandparenting is a gendered role, however, and while it may be unusual to only focus on grandfathers, the feminization of what is a multidisciplinary literature about grandparenthood (Arber and Timonen 2012) has resulted in a very limited understanding of grandfathering and paternal familial practices beyond father-hood. Men's voices as grandfathers have been less well represented in the various multidisciplinary literatures. The identities and practices of older hetero-sexual men, particularly in the family, remain mostly invisible and marginalized in academic thought (Leontowitsch 2012) largely because the association of care and the domestic with women (particularly mothers, but also grandmothers) remains deeply entrenched. Men's identities and care practices, particularly later in life, have either been forgotten or are not problematized. However, as men age, questions of masculine competence are raised in both public and private arenas of their lives (Reich 2007) and men are likely to experience distinct challenges to their identities (Calasanti 2004; Hearn 2007).

One social arena in which older men renegotiate their identities is grand-fatherhood (Tarrant 2012, 2013), a paradoxical role in which men are thought to enact masculinity through acting the 'sage' and wise man, but are also afforded the opportunity to become more nurturing in their relationships, particularly with grandchildren (Davidson *et al.* 2003). Grandfathering, however, as a gendered experience has received much less consideration than grandmothering. Indeed grandparent is often equated with grandmother. This chapter remedies this by considering grandparent practices and intergenerational relationships from the perspectives of men.

Methods and methodology

The data presented in this chapter come from a larger research project that examined the contemporary geographies of grandfathering and the influence of inter-generational relations on the construction of grandfather identities (Tarrant 2012). While the use of communication technologies was not the main focus of the research, this topic emerged unexpectedly as one of the ways in which several of the participants described their involvement with their grandchildren and the 'doing' of family and grandfathering, opening up interesting ideas about the evolving spaces of intergenerational contact between grandfathers and grandchildren.

For the study, 31 in-depth, semi-structured interviews were conducted with British grandfathers living in the Lancaster district in the north-west of England, between June 2009 and July 2010. This choice of method facilitated the examination of the embodied practices of intimacy and family relationships grandfathers engaged in, as they were experienced in particular spaces, such as the home. As mentioned, only men were recruited for this study in response to the 'feminized' bias of existing research on grandparenthood outlined earlier (Mann 2007) and the desire to understand men's constructions of grandparenting and care from their own perspectives. Previous research has argued that grandfathers are simply less involved and interested in family than women and that they are also more reticent about talking about family matters (Cunningham-Burley 1984) and taking part in research. Mann (2007), however, attributes this to a lack of consideration about the role of masculinity in research settings and the ways in which grandfathers might present themselves.

I found that the men who had agreed to take part were highly involved in relationships with their grandchildren and most were comfortable with talking about family life, saying they welcomed the opportunity to be asked to take part and countering some hegemonic ideas about masculinity (Barker 2008). Nonetheless, the language they used did focus more on their embodied practices, rather than the emotional experiences of grandfathering, and as others have found (Kornhaber 1996; Scraton and Holland 2006) they often only talked at a fairly superficial level. This required some creativity during the interviews on behalf of the interviewer in order to understand how men felt and why they answered the questions the way they did.

The grandfathers were asked a range of questions in the interviews. These were loosely structured around key themes to ensure that relevant information was gathered but also so that the men could discuss what was important to them. They were first asked to provide a brief biography (in which they often discussed paid employment before family) and then to describe their personal and familial circumstances. This was then followed up with questions structured around topics relating to the expectations and realities of grandparenting. The use of computers and communication technology emerged frequently in these discussions as the men reflected on the practicalities and dynamics of their relationships with their children and grandchildren.

The characteristics of the participant sample reflected the diversity of circumstances in which individuals are grandfathering in Britain today, particularly in relation to their personal and familial circumstances. There was great variation in the number and ages of grandchildren and the men's relationship status. Most were retired professionals, although three of the participants were still employed. Professional status is an important factor in examining the use of communication technologies because access and ability to use computers and telephones was not problematic for these men. For others, this may not be the case and any future study of grandparents with more limited access to information and communications technologies (ICTs) would enhance the findings presented here. Of the 31 men interviewed, 19 of the men had grandchildren who lived outside of the Lancaster district. Five of those men had grandchildren whose parents had emigrated to places such as Canada, Holland and Australia. The men's ages ranged from 51 to 88; all were white and able bodied, except one who was Jamaican and another who was partially sighted, and all identified as heterosexual.

The data was analysed using software program ATLAS.ti and a grounded theory approach (Charmaz 2006) that involved a two-tier coding system in which axial and open codes were used. The themes generated from the coding system allowed both common themes and divergences to emerge, each of which related more generally to how grandfathers 'do' grandparenting, where they do it, and how this made them feel. It was from this analytical approach that the importance of using communication technologies emerged as significant to the conduct and affective nature of contemporary grandfathering.

(Grand)paternal care: uptake of new technologies

Twelve of the participants in this study discussed their use of different forms of communication technology, highlighting the ways in which they were 'doing' grandfathering. While it is not appropriate to generalize qualitative data, the use of computers complemented the diverse range of different practices of intimacy that they engaged in with their grandchildren. These practices varied depending on the age of grandchildren but included telling stories, reading, playing and helping with homework, and babysitting. Those with older grandchildren, or grandchildren old enough to use computers were more likely to include some kind of interaction that either included the direct use of a computer, or that involved contacting grandchildren who lived distantly.

Like the men in Sorenson and Cooper's (2010) study of contemporary grandfathering, many of the grandfathers had computers in their homes and referred to telephone calls, emailing and texting as ways of keeping in contact with their children and grandchildren, revealing the ways in which ICT provides an additional space for intergenerational intimacy (Valentine 2006):

> Well we, occasionally we, send them e-mails or receive e-mails.... Mostly birthday celebrations and things like that. The rest of the family, the children we communicate with every week, not every day, but every week, by computer, by telephone mostly.
>
> (David, age 86, grandchildren live in Canada)

And 'I text and that, they [grandchildren] also send me photographs. I can receive a photograph on mine [mobile phone] but I can't send one, but it's such an easy one it's pointless upgrading it (laughs)' (Jonah, age 84, grandchildren live locally). Interestingly, Jonah lives locally to his grandchildren, and even great-grandchildren, yet he maintains contact with them via this means, as well as face-to-face. Distance between grandfathers and their grandchildren was not always a dependent factor in relation to frequency and type of contact. David's comments were also typical of the narratives as a whole (although not necessarily more widely representative) in that in discussing contact and interaction, they often did so with reference to their wives and partners.

The telephone in particular provides a space for communicating with grandchildren and showing care for them in situations where family breakdown has occurred and there is fear of losing contact. Paternal grandparents in particular have been found to fear a loss of contact with grandchildren when their sons get divorced (Timonen *et al.* 2009), reducing the amount and frequency of face-to-face contact that grandparents can have with their grandchildren. As a consequence the men's personal and familial circumstances produced particular gendered caringscapes that determined the use of ICTs. These technologies facilitate a range of alternative grandfathering care practices in these instances:

> Because of the domestic situation, I do not go and visit them [grandchildren], which is unfortunate. They sometimes text me or are on the phone to me, and the same with the granddaughters. I'm in contact with them, they're just at the end of the phone, if I do ... want to see them or, or they need me you know.
>
> (Andrew, age 70)

The grandfathers did not always instigate contact between themselves and their children and grandchildren, as Roger's (age 74) narrative indicates: 'My oldest granddaughter often sends me texts and things. My daughter rings me probably, twice a week every week, 'cause, 'cause my wife died 'bout, well exactly ten years ago, so my daughter looks after me (laughs) yeah.'

Widowhood and living alone has been found to be a more depressing experience for men than for women and one that predisposes men to being particularly vulnerable to social isolation later in life (Yetter 2010). In this instance ICTs are used by his daughter and granddaughter to care for him and to check on him since his wife died. This reveals that women are still primarily enacting caring practices, albeit more distantly. Technologies such as the telephone enable caring but the gendered division of care has become superimposed on technology spaces. Care via technologies is evidently a multi-directional process, one in which women can continue to care for their (grand)parent and vice versa. This points to the ways in which technology spaces facilitate interdependent intergenerational care and intimacy (Holdsworth and Morgan 2005).

Technological intimacy: affordances and tensions

Several quantitative studies suggest that grandparents significantly value contact and communication with grandchildren and are using new media (telephones, computers, text messaging services) to stay in touch (see Hurme *et al.* 2010 for a review) with them. These figures say very little however about the emotional and affective nature of using these technologies, other than to state a preference for a particular media or to show that face-to-face contact is more valued. For those living at a distance in particular, the use of technologies for contact was highly valued by the men. Percy (widower, age 72), whose daughter and grand-daughter live in Canada, discussed his qualitative experiences of seeing them both via a webcam. Most notable was his use of highly emotive language, describing it as both *marvellous* and *a miracle*:

> As regards seeing them physically, the last time I saw [daughter] and [granddaughter], was last March.... That was the last time I saw them, sort of physically, and I'm going across to Canada in June this year. Since [granddaughter's] been born I've seen her twice ... I think it's twice in Canada and once here so three times in three years. I see her, and they send me email, photographs, so that helps and, we have a webcam on the computer and, it's marvellous really, it doesn't always work, sometimes it er, the char ... the people don't move but ... you know it's a miracle really, you have this picture, this instant picture through your computer, the little webcam isn't it? And that does help.

Building on the conclusions of previous studies (Bell and Binnie 2000; Valentine 2008), the internet (and the use of a webcam here particularly) facilitates intimate, familial relationships across distance and beyond the domestic, creating new global spaces where different familial intimacies take place. Percy talks about the webcam as 'helping' and while he is not explicit about why, his practices here enable and sustain a sense of closeness and a special relationship between him, his daughter and granddaughter, despite problems with the technology. Most notably, Percy talks about 'I', despite living with his second partner. Percy spends a great deal of time face-to-face with his step-granddaughters. It is his biologically related grandchild who lives distantly. Being related and living at a distance in this example contributed to his desire to 'see' his granddaughter and daughter online.

Reg, age 66, has two sets of grandchildren, one living in Oxford and another in Aberdeen. For him and his wife, being able to talk to their grandchildren on the telephone is extremely valuable because he feels it supports 'good' relationships with them:

> Being a grandparent ... it's very special, because we hope that we'll have a good relationship with our grandchildren, that they will see that we offer, almost a second home, that there's somebody else there to care, and to love them ... because instinctively we were thrilled when the, it's quite special being a grandfather, or grandparents ... if I tell you, when we speak to [grandson, age 2] on the phone. It's a nice warm glow. It sounds pretty, pretty doesn't it? It's just a feeling ... you're important to someone.

Here the telephone facilitates the doing of love and intimacy and provides an emotional experience and one in which feeling important is valued. Interactions between generational groups have consequence for a range of social issues that impact on individuals' lives (Hopkins *et al.* 2011) and in this instance, technology facilitates intergenerational cohesion and the notion of being special. In contrast to Percy's narrative discussed above, Reg talks about 'we'. Both he and his wife use the telephone to speak to their grandson suggesting that gender is not always a factor that influences who uses these technologies to care. In this instance technology breaks down the notion that women are key kin-keepers and facilitates the doing of caring by both men and women.

While these stories are positive about the use of technologies, men were also concerned about its use and described a range of tensions. Bill, age 70, discusses his opinion of using webcams and other technologies for contacting his grandchildren:

> It will never be a substitute for face-to-face, for body language and all that but, I'm experimenting with it, it's a bit radical but, yeah I want to extend if I can, this time of real contact before they become monsters, you know into, into boyfriends and girlfriends yeah, yeah.... I think it's a new world, there are dangers in it, it's like everything else, it can be used for harmful ways or positive ways and I tend to think ahead, I think one day we'll have screens, up there, and we'll switch it on and say 'hi, how are things?'
>
> (Bill, age 70)

Here Bill privileges face-to-face involvement and intimacy with his grandchildren but also demonstrates how public discourses and fears around the use of technologies can blur public/private boundaries in the mediation of intergenerational relationships. Ambivalence around technology is shaped by wider cultural fears that construct children as in need of protection and at risk. These fears are not unfounded especially given that potentially dangerous activities most associated with public spaces can now be accessed more easily in the home (see Valentine and Hughes's 2012 exploration of the impact of internet gambling on family relationships as an example). Bill's comment demonstrates that while technologies can transform intergenerational relations for the good – by staying in contact with grandchildren more regularly – it is also a potential site of risk for grandchildren and Bill is concerned that his grandchildren use the technology for safe contact. This example also indicates that the surveillance of children and young people at home is extended by communication technologies opening up new, complex power relations between generations.

Another tension that emerged in relation to the use of communication technologies appeared to be the result of generational difference and a lack of 'knowing' grandchildren. Having asked about his relationships with his grandchildren living in the USA, Arnold, aged 65 explains:

> It's interesting, very, very, very rarely [do we contact them], and that's not because we don't want to it's because, our son will say 'do you want to talk

to grandad?' and they'll say 'no', because they're doing something else, but I think it's as much to do with ... they don't know us, they don't know us.

Communication technology use, while facilitating a space for intergenerational cohesion as the previous examples indicate can also exacerbate the emotional distance between generational groups as well as generational difference. Hopkins *et al.* (2011) describe several intergenerationalities that might also be conflicting and challenging as well as cohesive. If grandchildren are resistant to speaking on the telephone and communicating with grandparents then these technologies potentially reinforce the physical and emotional distance that might be felt by grandparents. This highlights the superficiality of some ICT interaction between generations and suggests that ICTs are spaces in which the doing of intimacies can be false and lacking 'real' connections.

Conclusions

This chapter has explored the complexities of knowing, loving and caring for grandchildren in a global society in which key demographic trends and processes such as increased mobility, family fragmentation and reconstitution and individualization are considered to be weakening kinship and family ties and intergenerational relations. Alongside accounts of the different practices and new technologies that grandfathers are experimenting with, as well as reflection on their personal and familial circumstances, these men evidently value their intergenerational relationships with grandchildren despite the fact that global processes of social change undoubtedly impact on the realities of everyday life for grandfathers.

The physical distances they live from grandchildren and the constitution of their families across multiple generations do impact upon the doing of grandfathering yet rather than automatically resulting in decline, communication technologies provide alternative spaces for facilitating loving and caring relationships. They influence the doing of grandfathering across distance and proximities and in age-segregated societies (Vanderbeck 2007) have a transformative power that demands a more critical understanding of the impact of social change on families and on intergenerational intimacies and interactions more generally.

Most evident is that the facilitation of emotional and affective relationships across multiple familial generations remains significant in explaining what 'doing' grandfathering entails and what it means, adding weight to the relational, connected and embedded (Mason 2004) nature of individual generational identities. The empirical findings presented in this chapter therefore support ongoing criticism of individualization as a driver of change in personal life (see Jamieson 2011). In seeking and utilizing alternative spaces of contact and intimate practice, it is evident that grandparents, including men, continue to value the interdependence of kin, genders and generation; they just conduct themselves, and their relationships in evolving ways.

In support of previous studies (Quadrello *et al.* 2005; Hurme *et al.* 2010), the grandfathers described using and experimenting with a range of communication technologies including telephones, mobile phones and webcams that compress spatial distances. Perhaps more useful though, especially in terms of advancing spatio-temporal approaches to intergenerationality, is that the qualitative nature of the study provided insight into the affective nature of using such technologies to facilitate intergenerational relations. Visual technologies such as webcams particularly enable the 'seeing' of children and grandchildren beyond embodied interaction, while technologies that rely on verbal or textual communication (such as mobile phones, telephones and emails) support the often mutual, interdependent relations of care between grandfathers and their children/grandchildren.

As Valentine (2006) points out, however, the use of ICTs and the incorporation of new technologies in the home is much more complex than simply providing positive or negative experiences of intimacy. It can shift the power dynamics between children and grandparents, as knowledge and competence between both alters. Problems with the technology, such as pauses in Skype, create a range of emotional tensions that may not be encountered in face-to-face interaction, and there are ongoing fears over children's exposures to dangerous materials that have the potential to increase the familial surveillance of children to family members beyond parents. There is also evidence that the availability of new technologies sometimes exacerbates generational difference between grandparents and their grandchildren, as Arnold's narrative indicates. It can then become a source of tension and reinforce an intergenerational care gap. Consequently new technologies, while potentially transformative, may also intensify age segregation and complicate generational power relations in the family.

The narratives of grandfathers were central in this chapter because (grand) paternal care practices in families are often invisible and marginalized. The gendered nature of the research recruitment therefore allowed consideration of older men's spatial engagement in intergenerational relations from their point of view. Nonetheless, the gender bias in the recruitment was fruitful in highlighting the ways in which the intersections of age/generation and gender continue to influence experiences of intergenerational relationships and care. While the men sometimes spoke of 'we' (i.e. them and their wives, partners, etc.) in maintaining contact and relationships with grandchildren, computers and technologies were a significant part of their carescapes as grandfathers. However, it is difficult to ignore the evidence that daughters and granddaughters were also described as instigating communication with grandfathers. This especially occurred in situations where the men were widowed or there had been divorce or separation in the family. This might indicate that technologies support interdependency in multigenerational familial relationships where physical distance is involved. These findings also support the notion of an ongoing gendered division of labour and consolidation of gendered identities, despite the possibilities technologies provide men in becoming more nurturing in their continuing familial and social relationships. While technologies were used by the men to renegotiate family practices, indicting potential for increasing their engagement in familial relation-ships, there is evidence of a distinct continuity to the gendering of family life extending to technology spaces.

Finally, it is evident that technologies have the potential to both maintain, but also transform intergenerational relations and the politics of the spaces in which they are used. The use of communication technologies between grandparents and grandchildren has far-reaching implications for considering how positive inter-generational relations across multiple generations might be facilitated and har-nessed. The Sus-IT project, discussed in Chapter 19 of this book, goes some way towards understanding these implications in order to develop ways to encourage the digital engagement of older ICT users.

References

Ames, M.G., Go, J., Kaye, J. and Spasojevic, M. (2010) 'Making love in the network closet: The benefits and work of family videochat', *Proceedings of the 2010 ACM Con-ference on Computer Supported Co-operative Work*, Savannah, Georgia, 6–10 Febru-ary, pp. 145–154.

Arber, S. and Timonen, V. (2012) *Contemporary Grandparenting: Changing Family Relationships in Global Contexts*, Bristol: Policy Press.

Bacigalupe, G. and Lambe, S. (2011) 'Virtual intimacy: Information communication tech-nologies and transnational families in therapy', *Family Process*, 50(1): 12–26.

Barker, J. (2008) 'Men and motors? Fathers' involvement in children's travel', *Early Child Development and Care*, 178(7/8): 853–866.

Beck, U. (1992) *Risk Society: Towards a New Modernity*, London: Sage.

Bell, D. and Binnie, J. (2000) *The Sexual Citizen: Queer Politics and Beyond*, Cam-bridge: Polity Press.

Bowlby, S., McKie, L., Gregory, D. and MacPherson, I. (2011) *Interdependency and Care Across the Lifecourse*, Abingdon: Routledge.

Calasanti, T. (2004) 'Feminist gerontology and old men', *The Journals of Gerontology: Series B, Psychological Sciences and Social Sciences*, 59(6): S305–S314.

Chan, T.W. and Ermisch, J. (2011) *Intergenerational Exchange of Instrumental Support: Dynamic Evidence from the British Household Panel Survey*, working paper, Depart-ment of Sociology, University of Oxford, online, available at: www.sociology.ox. ac.uk/working-papers/intergenerational-exchange-of-instrumental-support-dynamic-evidence-from-the-british-household-panel-survey.html (accessed 16 June 2014).

Charmaz, K. (2006) *Constructing Grounded Theory: A Practical Guide Through Qual-itative Analysis*, London: Sage.

Cunningham-Burley, S. (1984) ' "We don't talk about it…": Issues of gender and method in the portrayal of grandfatherhood', *Sociology*, 18(3): 325–338.

Davidson, K., Daly, T and Arber, S. (2003) 'Exploring the social worlds of older men', in S. Arber, K. Davidson, and J. Ginn (eds) *Gender and Ageing: Changing Roles and Relationships*, Maidenhead: Open University Press.

Duncan, S. and Smith, D. (2006) 'Individualisation versus the geography of "new" fam-ilies', *Twenty-First Century Society*, 1(2): 167–189.

Gabb, J. (2008) *Researching Intimacy in Families*, Basingstoke: Palgrave Macmillan.

Giddens, A. (1992) *The Transformation of Intimacy*, Cambridge: Polity Press.

Gilleard, C. and Higgs, P. (2008) 'Internet use and the digital divide in the English longitudinal study of ageing', *European Journal of Ageing*, 5(93): 233–239.

Hearn, J. (2007) 'From older men to boys: Masculinity theory and the life course(s)', *NORMA: Nordic Journal for Masculinity Studies*, 2(2): 79–84.

Holdsworth, C. (2013) *Family and Intimate Mobilities*, London: Palgrave Macmillan.

Holdsworth, C. and Morgan, D.H.J. (2005) *Transitions in Context: Leaving Home, Independence and Adulthood*, Maidenhead: Open University Press.

Holmes, M. (2004) 'An equal distance? Individualisation, gender, and intimacy in distance relationships', *Sociological Review*, 52(2): 180–200.

Hopkins, P., Olsen, E., Pain, R. and Vincett, G. (2011) 'Mapping intergenerationalities: The formation of youthful religiosities', *Transactions of the Institute of British Geographers*, 36(2): 314–327.

Hurme, H., Westerback, S. and Quadrello, T. (2010) 'Traditional and new forms of contact between grandparents and grandchildren', *Journal of Intergenerational Relationships*, 8(3): 264–280.

Jamieson, L. (2011) 'Intimacy as a concept: Explaining social change in the context of globalization or another form of ethnocentricism?', *Sociological Research Online*, 16(4): art. 15, online, available at: www.socresonline.org.uk/16/4/15.html.

Kemp, C. (2007) 'Grandparent–grandchild ties: Reflections on continuity and change across three generations', *Journal of Family Issues*, 28(7): 855–881.

Kornhaber (1996) *Contemporary Grandparenting*, London: Sage.

Leontowitsch, M. (2012) 'Interviewing older men', in M. Leontowitch (ed.) *Researching Later Life and Ageing: Expanding Qualitative Research Horizons*, London: Palgrave Macmillan.

Mann, R. (2007) 'Out of the shadows? Grandfatherhood, age and masculinities', *Journal of Aging Studies*, 21(4): 281–291.

Mason, J. (2004) 'Personal narratives, relational selves: Residential histories in the living and telling', *Sociological Review*, 52(2): 162–179.

Morgan, D. (1996) *Family Connections: An Introduction to Family Studies*, Cambridge: Polity Press.

Pain, R., Mowl, G. and Talbot, C. (2000) 'Difference and negotiation of "old age"', *Environment and Planning D: Society and Space*, 18(3): 377–393.

Plummer, K. (2001) 'The square of intimate citizenship: Some preliminary proposals', *Citizenship Studies*, 5(3): 237–253.

Quadrello, T., Hurme, H., Menzinger, J., Smith, P.K., Veisson, M., Vidal, S. and Westerback, S. (2005) 'Grandparents use of new communication technologies in a European perspective', *European Journal of Ageing*, 2(3): 200–207.

Reich, J. (2007) 'Unpacking "the pimp case": Aging black masculinity and grandchild placement in the child welfare system', *Journal of Aging Studies*, 21(4): 292–301.

Scraton, S. and Holland, S. (2006) 'Grandfatherhood and leisure', *Leisure Studies*, 25(2): 233–250.

Sorensen, P. and Cooper, N. (2010) 'Reshaping the family man: A grounded theory study of the meaning of grandfatherhood', *The Journal of Men's Studies*, 18(2): 117–136.

Tarrant, A. (2010) 'Constructing a social geography of grandparenthood: A new focus for intergenerationality', *Area*, 42(2): 190–197.

Tarrant, A. (2012) 'Exploring the influence of intergenerational relations on the construction and performance of contemporary grandfather identities', PhD thesis, Lancaster University.

Tarrant, A. (2013) 'Grandfathering as spatio-temporal practice: Conceptualizing perform-ances of ageing masculinities in contemporary familial carescapes', *Social and Cultural Geography*, 14(2): 192–210.

Timonen, V., Doyle, M. and O'Dwyer, C. (2009) *The Role of Grandparents in Divorced and Separated Families*, Family Support Agency and Social Policy and Ageing Research Centre, Trinity College Dublin.

Uhlenberg, P. (2004) 'Historical forces shaping grandparent–grandchild relationships: Demography and beyond', in M. Silverstein and K. Warner Schaie (eds) *Focus on Intergenerational Relations across Time and Place*, New York, NY: Springer.

Valentine, G. (2006) 'Globalizing intimacy: The role of information communication tech-nologies in maintaining and creating relationships', *Women's Studies Quarterly*, 34(1/2): 365–394.

Valentine, G. (2008) 'The ties that bind: Towards geographies of intimacy', *Geography Compass*, 2(6): 2097–2110.

Valentine, G. and Hughes, K. (2012) 'Shared space, distant lives? Understanding family and intimacy at home through the lens of internet gambling', *Transactions of the Institute of British Geographers*, 37(2): 242–255.

Vanderbeck, R.M. (2007) 'Intergenerational geographies: Age relations, segregation and re-engagements', *Geography Compass*, 1(2): 200–221.

Wellard, S. (2011) 'Too old to care? The experiences of older grandparents raising their grandchildren', *Grandparents Plus Report*, London, online, available at: www. grandparentsplus.org.uk/wp-content/uploads/2011/03/GP_Older GrandparentsOnline.pdf (accessed 16 June 2014).

Wheelock, J. and Jones, K. (2002) '"Grandparents are the next best thing": Informal childcare for working parents in urban Britain', *Journal of Social Policy*, 31(3): 441–463.

Wilk, L. (2000) *Intergenerational Relationships: Grandparents and Grandchildren*, European Observatory on Family Matters at the Austrian Institute for Family Studies, Family Issues Between Gender and Generations: Seminar Report, European Commis-sion, online, available at: https://getinfo.de/app/Intergenerational-relationships-Grandparents-and/id/BLCP%3ACN035754398 (accessed 16 June 2014).

Yetter, L.S. (2010) 'The experience of older men living alone', *Geriatric Nursing*, 31(6): 412–418.

21 Making sense of middle-aged gay men's stories of ageing

Paul Simpson

Introduction

The city of Manchester (England) has been a mini-laboratory for research on the relations of sexual and gender difference. The city's internationally known 'gay village', consisting of three streets, containing 36 bars and several gay-directed services (sex shop/sauna and barbers), has provided a focus for several empirical studies which consider expression of identities and social relations in urban 'homospaces'. Moran *et al.*'s (2004) interdisciplinary study concluded that queer experiences of public space especially at/just beyond the village's periphery are structured by symbolic and (fear of) actual violence. Other studies have focused on how relations of inclusion/exclusion in these spaces are related to patterns of difference based on gender and class. Binnie and Skeggs (2004) have demonstrated how gay users of this space themselves express symbolic violence towards 'gangs' of working-class women thought to represent less affluent areas. Their brash, 'excessive' sexuality is thought to embody the wrong kind of cultural capital (classed knowledge/tastes/display) in putatively cosmopolitan space. In terms of age and intergenerational relations, Whittle (1994) has portrayed the village as thoroughly ageist space where younger men are dupes of a consumerist, youth and body-obsessed gay culture and older men are rendered abject by it. In contrast, Haslop *et al.* (1998) have shown how gay men's uses of village bars reflect 'communitas, individualism and diversity ... identity and mood'. Whilst these accounts of conflict and commodification and of agency and belonging serve as resources to think with/against, they overlook ambivalences in how the relations of difference are negotiated, especially those concerning gay ageing/ageism.

This chapter attempts to address the above-identified theoretical gap and add nuance to debates concerning gay intergenerational relations/conflict. It draws on interviews and observations conducted in Manchester's gay culture. Specifically, I highlight the value of an analytical framework, adapted from the work of Rachel Thomson (2009) (on youth and gendered biographical development), which strategically combines tools from Foucauldian constructionism (more a genus of critical humanism) and theorizing of Bourdieu that invokes critical realism. The former assumes that subjects construct and are capable of critical appreciation of their social worlds using categories of thought, symbols and language. The latter assumes that underlying structures animate/shape thought and interaction involving reflexivity, i.e. when habitual thought/practice is not in sync with situated norms. Such an approach offers several related methodological and theoretical advantages. First, it affords understanding of social reality as heterogeneous. It enables demonstration of how the resources/narratives of ageing, apparent in generationally shaped 'technologies of the self' (Foucault 1987) and 'ageing capital' that forms an age habitus (see Simpson 2013a) can be situated within 'fields of existence' (Bourdieu 1984). Second, it treats ageing subjectivities and men's ways of relating as shaped by discourses but grounds them within specific fields of existence. This allows conceptualization of subjectivity and ways of relating as outcomes of the dialectic between structural/discursive constraints and everyday choices/opportunities for agency. Third, it enables exploration of different responses to gay ageism (spoken and/or enacted) and specifically how participants differentiate themselves from younger gay men. This approach is similar to that of Heaphy (2007) who demonstrates the uneven character of (classed and raced) resources of gay ageing (though aspects of gay ageing could transcend class differences (Simpson 2013a)). Men's accounts of differentiation, influenced by ageing capital, were implicated in capitulation to gay ageism and reverse ageism, others involved complex forms of negotiation with ageism whilst still others involved expression of an 'authentic' ageing self in ways that challenged gay ageism without derogating younger gay men. In sum, and like the work of Burnett (2010), I contribute to understanding of the under-theorized issue of negotiation by middle-aged subjects of ageing identities in particular (liminal) intergenerational spaces.

Manchester's gay culture

Manchester is the third largest city in the United Kingdom. Based on 2011 census data, the city of Manchester has a population of 503,000 and Greater Manchester, consisting of ten local authorities, has a population of 2.68 million (Office for National Statistics 2012). Its 'gay scene' is historically mutable: it has not always been concentrated in one 'gay' district. According to one interviewee, Leo (61), there had been a vibrant, partly integrated 'gay scene' (spread more across the city centre) prior to the decriminalization of male homosexuality in 1967. Nowadays the village's 36 bars (that mushroomed as a consequence of urban regeneration in the 1990s) and its nearby social/support groups are magnets for men living in the north-west region. The village is the largest night-time leisure zone in Manchester and popular with heterosexual people (Binnie and Skeggs 2004), attracting 20,000 visitors every weekend (Parkinson-Bailey 2000). All informants frequented it to some degree and it emerged as the structuring presence in their stories of ageing. Further, whilst informants are old enough to have been affected by gay liberation discourses in the 1970s, their self-consciousness was also shaped by Section 28 of the Local Government Act 1988 (that forbade local authorities to promote gayness as an equally valid lifestyle/relational choice and form of kinship) and hysterical media responses, public hostility and government cautiousness concerning HIV/AIDS (Watney 1987). But, the Beacon of Hope (an HIV/AIDS monument) installed in Whitworth Gardens in 1997 just across the canal symbolized growing official tolerance of sexual difference. The glass-fronted bars now proudly proclaim themselves as spaces for sexual difference and, in continental style, many of them put out tables and chairs overlooking the canal. Informants who remembered the village prior to its current state gave accounts of it as more cohesive space where men were obliged to rub along together and negotiate across differences in just a few bars. The village's present spatialized character – where bars are associated with certain 'types' of people by age, class and gendered sexuality such as older men, younger men, the 'trendy' and 'rough and ready' and 'bears' (proclaiming the viability of the older, fatter, hairier body) – was thought to signify fragmentation of gay community. Having sketched the research, historical and geographical context of the study on which this chapter is based, I now explain the methods used to generate participants' stories of ageing in gay cultural space.

Methods

My research is based on in-depth interviews with 27 men aged 39–61 (conducted largely in their homes) and 20 observation sessions conducted in Manchester's gay village. The two methods share a view of the workings of subjectivity (epistemology) and the constitution of culture/society (ontology). This mixed methods approach afforded comparison and contrast between different kinds of story/orders of data. The plausibility or salience of cultural events might be enhanced if data from both methods suggest convergence (Mason 1996). Besides, each method compensated theoretically and practically for the limitations of the other. Whilst the former elicited detailed, spoken narratives difficult to tell/hear or probe in village bars, the latter was used to generate detailed accounts of embodied display and interaction *in situ*. On its own terms, however, a semi-structured interview schedule elicited detailed, verbal reconstruction of events that could be probed. It was designed to explore: responses to the signs of ageing and body management with age (dress, grooming, exercise, diet); and patterns of kinship/ways of relating past and present. An advantage of participant observation is that it recognizes the 'multivocality' of situated experience (Brewer 2000). My role as an observer was of necessity covert and a semi-structured observation schedule was used to record interaction. This focused on: dress/grooming; peer-aged and intergenerational interaction; behaviour alone/in pairs; and interaction in/between small groups. The method helped identify the salience of events and distinctiveness of cultural practices (Brewer 2000) e.g. 'cruising' styles/modes of approach/interaction. Both methods brought me closer to the 'object' of study (Maxwell 1996) – midlife gay male ageing – and enable qualified claims to the transferability of the knowledge produced. Stories of agency and age and gender subversion through disco dancing or stories of bodily discomfort/shame could be expressed in gay villages/scenes in other post-industrial cities.

Moreover, a purposive sampling strategy used in relation to both methods was designed to accommodate key dimensions of difference in terms of: men's class, ethnicity and relational circumstances; and different spaces (those for younger men, those for older men and mixed-age ones) where modes of inter-action differed according to time/day of the week (and season when men mingled outdoors). This involved observing different actors and use of various foci. Interviewees were recruited through project publicity – a leaflet and poster. Sixteen interviewees (60 per cent of the sample) were recruited via three social/support groups. Six men (just over one-fifth of the sample) emerged in response to publicity in mixed-age bars and the same proportion came from informants' and acquaintances' personal networks. One respondent contacted me through publicity displayed in the village sauna. Fourteen of the 27 men (52 per cent) were aged 50–61 and 13 men (48 per cent) were aged 39–48. Twenty-four respondents (89 per cent) described themselves as 'white British', one respond-ent self-defined as 'mixed race', another as 'oriental' and another as 'Irish-European'. The two 'non-white' men spoke of the village as 'white space' where racial difference could be exoticized. In addition, 17 respondents (63 per cent) were single and the remainder partnered. Informants were allocated to both socio-economic and cultural class categories. This strategy, which recognizes the multidimensional character of social class, enabled analysis of consistencies and discrepancies between the economic and cultural resources that subjects were able to mobilize. In terms of class incongruity, an individual might be highly educated but on a lower income or conversely, economically comfortable but able to access a limited range of cultural pursuits. Socio-economic class cat-egories were based on employment/income-related data (i.e. whether men were in paid employment, working part-time/on minimum wage or in work that held out few prospects for development/promotion) and cultural class categories were based on the notion of 'cultural capital' (Bourdieu 1984). This refers to embod-ied knowledge required to access/conduct certain cultural pursuits. However, tastes for pop music or pulp fiction were probed to discern whether informants were able to read them at a meta-level (structure or social/political resonances). The actual sample was evenly spread across two socio-economic classes – 14 men (52 per cent) were working class and 13 men (48 per cent) were middle class. Twenty-two respondents (81 per cent) described a range of interests/capa-cities suggesting cultural omnivorousness, though nine of these men originated from working-class backgrounds.

Analytical framework

My research was undergirded by interpretivist methodology, which recognizes reflexivity as an accomplishment of everyday social engagement (Lynch 2000). However, such approaches risk uncritical adoption of institutionalized narratives (Bourdieu 1990). For example, accepting subjects' first-order constructs would have involved homogenizing younger gay men as shallow dupes of gay/consumer culture. I also wanted to avoid the binary thinking that suffuses the literature on gay male ageing, which casts middle-aged/older gay men either as victims of the relations of gay ageism (Hostetler 2004) or as defiantly continuing to cruise the gay scene for socio-sexual opportunity (Berger 1982). Indeed, my approach addresses several conceptual/theoretical gaps in the literature on gay male ageing, which is dominated by the notion of 'accelerated ageing' – being considered middle-aged/old before one's time (Bennett and Thompson 1991). Ageism might operate more acutely in gay male culture especially when compared with heterosexual men but accelerated ageing fails to illuminate study participants' varying, contextually shaped, ambivalent responses to gay ageing/ageism as explained below.

To avoid the above-identified problems, I adapted a framework for analysis developed by Thomson (2009), which involved selecting conceptual tools from two methodologies – Foucault's 'technologies of the self' (constructionism) and Bourdieu's concept of 'field' (1984), the latter invoking critical realism as does my innovation 'ageing capital' (see below). Technologies of the self are described as 'discursive resources used to construct an identity' (Thomson 2009: 23). They indicate resources for self-direction through which subjects can set their own rules/goals, free from unwanted/constraining forms of self-governance (Foucault 1987). Ageing capital (Simpson 2013b) is my term to encompass participants' references to a set of multivalent, context-dependent (often transferable) emotional, cognitive and political resources. It was manifest in claims to accumulated emotional strength, self-acceptance, age-appropriate bodily display/performance and growing awareness of/competence in managing the relations of gay culture and wider society. Such knowledge/competencies could interact with class-inflected 'cultural capital' (Bourdieu 1984) – capacities to deploy knowledge of social relations. Whilst ageing and cultural capitals are mutually influencing, the former can act independently of the latter. Knowledge gained through life experience could compensate for 'deficits' in educational experience and supplied opportunities to develop critical capacities. For instance, Les (53) who had been a 'kept boy' from his late teens to his late twenties was the only interviewee to question the characterization of cross-generational relationships as quasi-prostitution where the physical/sexual capital of youth is traded for the economic and/or emotional capital of older men. He spoke of how the financial/emotional largesse of an older lover can deny the younger party opportunities for autonomy.

Ageing capital was also implicated in intergenerational conflict over reputational resources (symbolic capital), specifically in the struggle for dominance over authoritative interpretation and representation of legitimate gay subjectivity, relationality and culture. In a more critical, empowering register, it invoked the notion of an authentic, midlife sexual subjectivity on which informants drew to distinguish themselves from the need to produce fashionable forms of self-presentation (associated with younger gay men and younger selves). This was commonly expressed in the normative idea of a more 'natural', less sculpted/elaborated body. This indicated a holistic self where appearance that involves 'dressing for comfort' should be a faithful reflection of a more 'real' inner self consisting of values and personality (prioritized over individualized projects of the body). Ageing capital could be used to re-aestheticize and legitimate the midlife/ageing body-self as still desirable and creative. Such thinking values the ageing process, suggesting that middle-aged gay men might be freer from the discursive pressures of gay/consumer cultures whilst contradicting stereotypes of them as obsessed with preserving youth and sexual marketability. In a more constraining register, differentiation narratives, as expressed through ageing capital (generationally inflected knowledge), could result in limitations on self-expression and relating to others. Specifically, they were implicated in reverse ageism that constructs younger men as insubstantial, underdeveloped and self-obsessed. Ageing capital could be thwarted or go awry when we consider that informants' efforts at self-recuperation were undermined by the narratives they drew on to assert *superiority* for their generationally informed ways of knowing, doing and being. In addition, the resources of ageing occur within 'fields of existence' (Thomson 2009) – cultural arenas with their own norms. These situated rules of the game become so entrenched within the body that they animate habitual thought/practice or 'habitus'. The latter also indexes a generationally inflected 'collective consciousness' shaped by distinct historical influences (Edmunds and Turner 2002). As already intimated, this generational habitus was influenced by endemic homophobia. Further, working-class informants indicated exclusion from the economic, cultural and reputational resources required to develop connections on the more middle-class gay social/support group scene. Conversely, middle-class informants indicated a failure of their cultural capital when they reported feeling incongruous when on the village bar/club scene where socializing revolves around a working-class and youth-coded taste for alcohol consumption and 'celebrity gossip'.

The above strategic 'pick and mix' framework enables one conceptual tool to compensate for problems with the other and allows explanation of different aspects of multiform social reality where structure and discourse are imbricated. Age-inflected technologies of the self sidestep the 'reproductionist' notion of 'habitus' (deeply embodied habits) as eternally fated to repeat itself whilst the idea of field avoids reducing (inter-)subjectivity to free-floating discourse (Thomson 2009). This conceptualization of reflexivity would allow consideration of how diverse subjectivities and modes of relating emerge from the untidy dialogue between constraint and choice (Thomson 2009). I demonstrate below that such thinking takes us beyond analysis of midlife gay men's responses to gay ageism not just as either conformist or voluntarist but, again, to register the ambivalent ways in which midlife gay men narrate/express their ageing identities, differentiate themselves and conduct social relations.

Stories of alienation: the multiple, hidden injuries of ageism

Common in *interview* stories were accounts registering ageing as constraint and exclusion. Subjects recounted partial, contingent tolerance of the older body-self within a culture viewed as obsessed with visual/surface appearance (Martins *et al.* 2007), especially in the village. There was a sharp sense of alienation where men might be physically 'on the scene' but not feel part of it.

Middle-aged gay habitus could be experienced as out of sync with prevailing norms in the youth-oriented sub-field of a nightclub. During observation in the early hours of Saturday morning, it was noticeable that men occupying the various podia/dance floor were in their twenties and early thirties. Many were stripped to the waist, with shaved, muscled torsos. In response to this sensorium of dry ice, dizzying lighting, loud, fast, thumping music, sweat and the energy of those dancing, I noticed a man of about mid-forties, carefully picking his way along the periphery of the dance floor as if battling against a storm. He appeared almost fixed to the spot under the onslaught of the 'high energy' multisensory experience. His facial expression and body posture communicated that his whole being felt embattled. When midlife gay men are enticed into spaces associated with younger men or more mixed generational venues, they could then find them internally spatialized along the lines of age. The protagonist in the above scenario appeared peripheral to the main event. Like the few other men of his age who were present, he was a solitary figure and stayed well back from the edge of the dance floor, which suggests a micro-politics of gay cultural space where discourse sets considerable constraints on mobility, self-expression and relating within it. Midlife gay men here appeared literally reduced to the status of immobilized onlookers. The club featured in the above description was visited a second time on 'retro' night (on Mondays). On this occasion too, the older men observed were mostly in their forties and none appeared older than about mid-fifties. They tended to survey the dance floor from a distance or corner, rarely moving in order to 'cruise' and even if in peer-aged groups, middle-aged patrons tended to remain internally focused, indicating habitual social distance and civil indifference between the generations.

Accounts of social distance from younger men were particularly commonplace in interviews. For example, Sam (45) described the village and the wider commercial gay scene in the following terms:

> It's completely worked out bodies.... It's all youth, body, perfect white shining teeth.... Yes, people might *admit* that they've got a little bit more wisdom but who needs wisdom when you can get a nice young, firm pair of buttocks?

Typically, Sam uses his experience of ageing as a gay man (ageing capital) to draw attention to the over-focus on the surface self, which could violate middle-aged gay men's normative sense of 'authenticity' (where the exterior should reflect the more 'real' interior self of personality and values). In this field of existence men's subjectivity is narrated as if reduced to fragmented body parts and over-production of appearance – 'nice firm buttocks' and 'perfect white shining teeth' which are privileged over the whole person, including 'wisdom' accruing from life experience. Indeed, the more culturally middle-class informants spoke at times in distinctly classed ways of how wisdom (part of habitus) might be something one has to own up to ('admit') or even suppress in cultural space where 'celebrity gossip', *Big Brother* and idolization of Madonna and Kylie represent the lingua franca. But, in referring to the labour that goes into producing appearance – of youthful, 'worked out bodies' – Sam registers the socially constructed character of age and youth itself. However, his statement indicates the operation of a form of discourse in relation to the village, which invokes moral and epistemic differentiation from younger gay men. The latter are thought to over-rely on the visual, which constrains them into seeing subject-ivity only in objectified ways and were, by implication, depicted as incapable of achieving authenticity. This form of differentiation was narrated independently of differences of class and race and involves the use of age-related knowledge in an attempt to reclaim self-worth in a hierarchized arena that denies midlife gay men's bodily and personal value. But, simultaneously this narrative 'resource' relies on stereotypes expressing ageism towards younger gay men. Subjects' attempts to recuperate an ageing-body self could then draw on discourses that undermined their generational claims to know/represent a more 'authentic' and valid form of gay embodiment, subjectivity and mode of relating. Gay ageism is a two-way street and can work multi-directionally against younger and old gay men – the latter commonly thought to signify a state beyond socio-sexual sub-jectivity, as representative of morbidity and closer to death, social and actual (Simpson 2013a). But the narratives outlined above are also strongly suggestive of discursive constraints on men's abilities to mobilize the resources of ageing, which appear to operate more sharply in the commodified, youth-oriented expressions of Manchester's gay male culture.

Stories and spaces of ambivalence

Whilst the above narratives of capitulation to gay ageism and reproduction of reverse ageism appear to support the depiction of Whittle (1994) mentioned above, men's stories were more diverse and complex than this homogenizing analysis that invokes a form of false consciousness. There were also more ambivalent responses, often occluded in the extant literature on gay ageing involving negotiation with (gay) ageism.

Such narratives are broadly suggestive of reflexive capacity: ageing capital and 'technologies of the self' might pave the way for personal transformation and resistance to gay ageism or at least a more questioning stance in relation to it. This, however, was far from guaranteed. But, above all, this form of nar-rative pointed up subjectivity and ways of relating as the contingent outcomes of the dialectic between structural/discursive constraints and everyday choices/ opportunities for agency. This positioning between constraint and agency or complicity and resistance (Nelson 1999; Lovell 2003) was particularly appar-ent in subjects' mediation of habitual body management regimes (related to diet and exercise) to stave off weight gain as a sign of growing older and the more questioning shadow narratives that permit giving into sensual pleasures related to food, alcohol and relaxation etc. For instance, Jonathan (42), who lived a largely domestically oriented gay existence with his long-term partner invoked strenuous thinking of the relationship of the self to the body when he described himself as being 'realistic about my body'. Whilst he saw his body changing (weight gain) he wanted to 'combat it ... through exercise'. In the same narrative episode, he also declared that: 'I'm quite happy with it ... though I can be critical of it.... But, when I am critical, because there is a bit too much fat there, it's in an idealistic way.' Jonathan indicates negotiation between two aspects of an interior self in flux: one 'idealistic', the other 'realistic'. Ambivalence is visible in narration of the ageing body as something to be controlled. The desire to 'combat' a changing, thickening body here highlights a struggle between an internalized critic, the product of gay ageism amply buttressed by consumer culture, which requires a 'better', trimmer body as a sign of ethical and aesthetic self-care. But this demand is tempered by a more forgiving 'realistic' and agentic self that entailed succumbing to and enjoying sensual pleasures.

Contradictory experiences of an ageing gay self were also available on the village scene, which also suggested a breakdown or frustration of men's abilities to deploy the resources of ageing, especially in more overtly sexualized spaces. During a midweek, mid-evening observation in a bar associated with middle-aged gay men, I noticed a South-East Asian man of about late forties sitting alone behind a group of seven men of varying ages. The man in question sported well defined biceps and torso, significant in the gay male erotic imaginary, and was wearing a tight-fitting, rugby-style, short-sleeved t-shirt that accentuated his toned physique. For about half an hour, he occupied himself by leafing through a magazine, leaflets, switching between a mobile phone and pocket diary, snatching glances at the group and around the bar. He made frequent subtle adjustments to the nap of his t-shirt. At one point, he was practically hugging himself; right hand holding his left shoulder with one leg crossed over his knee and left hand on right ankle, foot twitching. The subject's surface appearance as a viable (gym-toned) body and the subtle monitoring of self in relation to unknown others may have suggested openness to socio-sexual opportunity and implicit claim for viable sexual subjectivity. Yet this is contradicted by the protective body language of a 'defended self' (Hollway and Jefferson 2000), which suggests the power of the gay gaze to operate simultaneously as governance in ways that can discomfort/constrain an ageing, differently raced self in a field understood as 'white space'. Such body glossing techniques, which carve out 'private' territory in public space (Goffman 1971), could disguise that the subject is looking for sex but does not want to break the cardinal rule that enjoins avoidance of looking desperate. There are then ways in which some midlife gay men might experience the gay commercial scene as a site of visual pleasure but equally and contradictorily as a potential locus of hostility involving risks that could compromise the expression of a dignified, authentic, middle-aged, gay male sexual self.

Stories of agency

Accounts of agency and resistance to gay ageism were generated more often in observations, though all informants offered some critique in this regard despite differences of race and class. Non-white and working-class informants were equally adept at articulating excoriating critique of gay ageism and, in some cases, racism, though working-class informants often reprised middle-class critique of the fashion industry and its young gay followers. But, in more positive mode, men's critiques of gay ageism suggest ability to disrupt the dominant idea that youth represents the benchmark of desirability and is representative of Manchester's gay bar scene/culture. An 'authentic' middle-aged socio-sexual self loomed large in these more critical accounts and suggested deployment of ageing capital and age-inflected 'technologies of the self' to reclaim the value of the midlife body and avoid constraints on the expression of identity and ways of relating. Whilst stories of agency were often expressed in freedoms from/resistances to exercise regimes and youthful, homonormative forms of self-expression, the resources of ageing could be used to recuperate the midlife body in ways that do not enjoin derogation of younger others. Further, one informant's narrative stood out for its ethical questioning of the very notion of 'impotence' in a culture thought to place primacy on sexual prowess. Besides, according to the government, 'erectile dysfunction' affects over 40 per cent of men aged over 40 (National Health Service 2013). Bill (55) reported that his experience of this meant:

> I can't fuck ... most of the time.... And if you can't do one thing, well, we can do something else. Although this is a bit limiting ... it doesn't stop me doing other stuff.... What you're up against here is the stereotype of what sex should be.... Well, I'm not too bothered about what it should be.... This is something I associate with when I was a lot younger.... It has forced me to think about sexual satisfaction in much broader terms.... It's about discovering what's pleasing between people ... I'm not a performing seal. I'm not a machine. I'm a real, flesh and blood human being ... with real feelings, a real story to tell.... So, it's about ... putting the humanity back into the sexual situation.

This account of turning 'tragedy' into triumph from a highly educated informant indicates the use of age- and class-inflected technology of the self to negotiate the kinds of sex that do not involve penetration or orgasm as the defining endpoints. The informant has unravelled age-inflected, homonormative discourse that sanctions what 'real' gay sex should consist of. The informant's thinking challenges the default position in gay culture (and beyond) that 'proper' sex is penetrative and should involve a literal pay-off (Hawkes 1996). Instead, Bill adverts to sexual practices and pleasures that he finds intrinsically satisfying. He describes a form of sexual ethics that entertain the idea that the whole body is a field of erotic possibilities, enjoin mutual pleasure and re-establish him as a viable sexual subject. Bill's resistance to the youthful athleticism and machine-like, porn-star sexual efficiency contains the claim that his experience of sex in midlife has been characterized by attempts to make it more convivial.

In contrast to the interpretations of Binnie and Skeggs (2004) and Whittle (1994), the commercial gay scene is not uniformly alienated, commodified space. It is also the locus of aural, visual and kinaesthetic pleasures where the notions of age/ageing can be blurred and ageism challenged (consciously and subconsciously) or else forgotten through performance. The village is the site of many different displays of gendered sexuality such as drag performance, cross-dressing, trans, fetish, indie, retro, skinhead and the midlife staple of 'dressing for comfort' rather than for fashion. There are many colourful, spectacular, the-atrical forms of self-presentation (suggestive of physical and sexual capital) and opportunities for middle-aged men to look/be looked at. Such performances indi-cate continuing enjoyment and productive negotiation of discourses constituting the village scene in terms of how to make it work rather than as an oppressive experience. The possibilities of sexual opportunity, letting go of inhibitions and for respite from harsh ageist scrutiny on the scene that are available in its more age-friendly sub-fields might explain why middle-aged men continue to frequent the village despite the risk of rejection/erasure as socio-sexual subjects. Parts of the gay district then can figure as a space where midlife gay men might proclaim continuing validity of an authentic (sexual) self and one that is more elaborately styled and presented. The gains of ageing might also liberate midlife gay men from pressures to justify their ageing presence. In short, they may consider they 'no longer have much to prove' (Vince, 49), which could also signify relative freedom from inhibitions over creative use of the body that informants reported experiencing when younger. The rules of the game in some age-friendlier spaces and even late-night, mixed-age venues in the village can enable self-expression without fear of ageist censure or scrutiny and sanction ludic practices between men of different ages. In one observation episode (that lasted nearly 30 minutes), I noticed two men – one about early sixties and the other mid-twenties – leap onto the floor with determined facial expressions to dance to Cher's *Are You Strong Enough?* Their routine involved energetic windmilling of arms and suit-able gurning as they lip-synced the words to each other. During Christina Aguil-era's *Cause I'm a Fighter*, they began to attract an audience as they punched their fists in the air/towards each other mock aggressively. Other dancers cleared the floor to watch, as the two began spontaneously to mirror each other's move-ments/gestures and at one point shook their imaginary though ample showgirl breasts at each other. Their routine covered the whole of a sizeable dance floor and segued into dancing side-by-side, moving backwards and forwards in step, waving an index finger in front of them to the histrionic refrain of *One Night Only!* The audience was transfixed; enthusiastic applause followed and prompted the older man to imitate a pole-dancing routine.

The description above, developed from field notes, was a particularly eye-catching instance of many displays of dancing and shared campery observed in the mixed spaces (by age, gender, class and sexuality) which contributed to the village's variegated sensual geography. The men's mesmerizing dance moves, routines and lip-syncing referenced gay disco camp classics and musicals, which suggest a form of gay habitus consisting of complex inter-articulations of physical and cultural capital. The men's shared interest here suggests the possibility for unity around embodied knowledge and forms of storytelling which could help overcome age barriers and ageism at least temporarily if not on a more permanent basis. The code-switching of the two men between forms of self-expression understood as 'masculine' and 'feminine' and parodying these polar opposites is also suggestive of technologies of the self. In exercising these situated bodily freedoms, the protagonists point up the performativity of ageing, gender and sexuality.

Conclusion

I have drawn attention to the theoretical and methodological advantages of a mixed qualitative methods research design central to which is an analytical framework that strategically adapts conceptual tools from Foucauldian constructionism and Bourdieusian critical realism. If technologies of the self represent the former, ageing capital/field represent the latter. First, I have shown the capabilities of this framework for illuminating lived experience as heterogeneous and contradictory – the latter often missing from extant literature on sexuality and ageing. This framework has been used to point up multiform gay identities where (gendered) sexuality can inter-articulate with uneven resources of age, class and race to reinforce social distance between different gay men. Second, my framework illuminates how subjectivities and men's modes of relating are shaped by the dialogue between structural and discursive constraints and the opportunities that men create for agency. Third, it has produced a plurality of stories of ageing that show how the resources of ageing are implicated in capitulation to, negotiation with and resistance to gay ageism. Although largely located in the village as a particular field of existence, they are stories which might transcend the local contexts in which they were generated – they could be expressed in urban contexts with gay scenes/villages similar to Manchester. Finally, I have not only complicated realist notions of objectivity (suggesting that the methods/ analytical framework brought me closer to the object of study) but have contributed to debate around a 'politics of method' that involves careful knitting together of 'different lines of enquiry and different ways of seeing' (Mason 2011: 77). Putting methods into dialogue as elaborated above indicates a productive agenda for research on a situated politics of age and other, interacting forms of differentiation.

References

Bennett, K. and Thompson, N. (1991) 'Accelerated ageing and male homosexuality: Australian evidence in a continuing debate', *The Journal of Homosexuality*, 20(3/4): 65–75.

Berger, R. (1982) *Gay and Gray: The Older Homosexual Man*, Chicago, IL: University of Illinois Press.

Binnie, J. and Skeggs, B. (2004) 'Cosmopolitan knowledge and the production and consumption of sexualized space: Manchester's gay village', *Sociological Review*, 52(1): 39–61.

Bourdieu, P. (1984) *Distinction: A Social Critique of the Judgement of Taste*, London: Routledge.

Bourdieu, P. (1990) *The Logic of Practice*, Cambridge: Polity Press.

Brewer, J. (2000) *Ethnography*, Buckingham: Open University Press.

Burnett, J. (2010) *Contemporary Adulthood: Calendars, Cartographies and Constructions*, Basingstoke: Palgrave Macmillan.

Edmunds, J. and Turner, B. (2002) *Generations, Culture and Society*, Buckingham: Open University Press.

Foucault, M. (1987) *The History of Sexuality, Volume Two: The Uses of Pleasure*, trans. R. Hurley, London: Penguin Books.

Goffman, E. (1971) *Relations in Public: Micro-Studies of Public Order*, Harmondsworth: Pelican Books.

Haslop, C., Hill, H. and Schmidt, A. (1998) 'The gay lifestyle: Spaces for a subculture of consumption', *Marketing Intelligence & Planning*, 16(5): 318–326.

Hawkes, G. (1996) *A Sociology of Sex and Sexuality*, Maidenhead: Open University Press.

Heaphy, B. (2007) 'Sexuality, gender and ageing: Resources and social change', *Current Sociology*, 55(2): 193–210.

Hollway, W. and Jefferson, T. (2000) *Doing Qualitative Research Differently: Free Association, Narrative and the Interview Method*, London: Sage Publications.

Hostetler, A. (2004) 'Old, gay and alone? The ecology of well-being among middle-aged and older single gay men', in De Vries, B. and Herdt, G. (eds) (2004) *Gay and Lesbian Aging and Research: Future Directions*, New York, NY: Springer, pp. 143–176.

Lovell, T. (2003) 'Resisting with authority: Historical specificity, agency and the performative self', *Theory, Culture & Society*, 20(1): 1–17.

Lynch, M. (2000) 'Against reflexivity as an accidental virtue and privileged source of knowledge', *Theory, Culture & Society*, 17(3): 26–54.

Martins, Y., Tiggeman, M. and Kirkbride, A. (2007) 'Those Speedos become them: The role of self-objectification in gay and heterosexual men's body image', *The Personal and Social Psychology Bulletin*, 33(5): 643–647.

Mason, J. (1996) *Qualitative Researching*, London: Sage Publications.

Mason, J. (2011) 'Facet methodology: The case for an inventive research orientation', *Methodological Innovations Online*, 6(3): 75–92.

Maxwell, J. (1996) *Qualitative Research Design*, London: Sage Publications.

Moran, L., Skeggs, B. with Tyrer, P. and Korteen, K. (2004) *Sexuality and the Politics of Violence and Safety*, London: Routledge.

National Health Service (2013) 'Erectile dysfunction (impotence)', online, available at: www.nhs.uk/conditions/Erectile-dysfunction/Pages/Introduction.aspx (accessed 1 December 2013).

Nelson, L. (1999) 'Bodies (and spaces) do matter: The limits of performativity', *Gender, Place & Culture*, 6(4): 331–353.

Office for National Statistics (2012) 'Manchester population and religion', online, available at: www.ons.gov.uk/ons/about-ons/business-transparency/freedom-of-information/what-can-i-request/previous-foi-requests/population/greater-manchester-population-and-religion/index.html?format=printk (accessed 14 May 2013).

Parkinson-Bailey, J.J. (2000) *Manchester: An Architectural History*, Manchester: Manchester University Press.

Simpson, P. (2013a) 'Differentiating the self: The kinship practices of middle-aged gay men in Manchester', *Families, Relationships and Societies*, 2(1): 97–113.

Simpson, P. (2013b) 'Alienation, ambivalence, agency: Middle-aged gay men and ageism in Manchester's gay village', *Sexualities*, 16(3/4): 283–299.

Thomson, R. (2009) *Unfolding Lives: Youth, Gender and Change*, Bristol: Policy Press.

Watney, S. (1987) *Policing Desire: Pornography, AIDS and the Media*, London: Comedia/Methuen.

Whittle, S. (1994) 'Consuming differences: The collaboration of the gay body with the cultural state', in S. Whittle (ed.) *The Margins of the City: Gay Men's Urban Lives*, Aldershot: Ashgate Publishing Limited, pp. 27–41.

22 Negotiating urban space

Older people and the contestation of generational and ethnic boundaries

Tine Buffel and Chris Phillipson

Introduction

Two major forces are set to shape the quality of daily life in the twenty-first century: population ageing and urbanization. Both have become dominant concerns for public policy with the interaction between them raising issues for all types of communities – from the most isolated to the most densely populated. By 2030, two-thirds of the world's population will be residing in cities, with the major urban areas in the developed world likely to have 25 per cent or more of their populations aged 60 and over. Katz *et al.* (2008: 474) view the present century as the 'urban age', one that is unfolding at a 'dizzying pace and with a scale, diversity, complexity, and level of connectivity that challenges traditional paradigms and renders many conventional needs and practices obsolete'. Cities are now regarded as central to economic development, attracting waves of migrants and supporting new knowledge-based industries (Burdett and Sudjic 2011). However, the extent to which the new 'urban age' will produce 'age-friendly' communities, creating opportunities for older people as well as strengthening ties across generations, remains uncertain. Reflecting this, demographic transitions, economic restructuring and technological innovation may have led to increased segregation between generations (Izuhara 2010). This may also be linked with concerns about conflicts within urban space arising from inequalities in access and influence of different social and generational groups.

A number of initiatives have been launched in attempts to build more cohesive communities and to bridge generational and related divisions. These include: first, the European Year of Older People and Solidarity between Generations in 1993; second, the World Assembly on Aging (UN 2002) and actions linked to working on a 'Society for All Ages'; and third, the establishment of the annual European Day of Intergenerational Solidarity. The branding of 2012 as the European year of 'Active Ageing and Intergenerational Solidarity' further emphasized the importance of enhancing solidarity between generations. Such activities, despite their importance, have rarely addressed the pressures arising from the context within which ageing takes place, notably (especially for high income countries) those within cities and city regions.

The aim of this chapter is to explore the engagement of older people in urban settings, and examine evidence for the operation of processes of inclusion and exclusion. The chapter begins by highlighting the importance of using perspectives from urban sociology and allied disciplines to understand social changes affecting older people. Linked with this discussion is a review of changes affecting cities, notably those relating to globalization, economic inequalities, cultural divisions and their associated impact on relations between generations. The chapter will then develop a framework for exploring these changes, followed by an illustration drawn from two empirical studies which examine older people's negotiation of generational, ethnic, class and other boundaries in changing urban neighbourhoods. The chapter will conclude by discussing the basis for establishing the rights of older people to secure spatial justice and participation in the daily life of cities.

Urban theory and age-friendly cities

Although ageing and urbanization can rightly be viewed as major trends for the present century, the two developments have tended to be kept apart both in respect of policy and research. This has happened despite urging from pioneer scholars of urban society such as Lewis Mumford (1956) that we should be seeking 'age integration' rather than 'age segregation' in our cities. Against this, the direction of policy over the post-war period appears to have been in the opposite direction, with the focus on developing age-segregated provision associated with sheltered housing (especially in the 1950s and 1960s) through to retirement communities and the more recent evolution of urban retirement villages. An alternative approach, however, has stressed the importance of developing what have been termed by the World Health Organization (WHO 2007) as 'age-friendly communities'. Accordingly:

> It should be normal in an age-friendly city for the natural and built environment to anticipate users with different capacities instead of designing for the mythical 'average' (i.e. young) person. An age-friendly city emphasizes enablement rather than disablement; it is friendly for all ages and not just 'elder-friendly'
>
> (WHO 2007: 72; see also Fitzgerald and Caro 2014).

This approach may be viewed as consistent with new perspectives influencing urban development over the course of the 1990s and early 2000s, notably ideas associated with 'sustainable' (Satterthwaite 1999) and 'harmonious cities' (UN-Habitat 2008: xi). The former raised questions about managing urban growth in a manner able to meet the needs of future as well as current generations. The idea of harmonious development emphasized values such as 'equity, gender parity, inclusiveness and good governance' (UN-Habitat 2008: x), regarded as essential principles of urban planning. Such themes were also influential in the elaboration of ideas associated with 'lifetime homes' and 'lifetime neighbourhoods' (DCLG

2008; Atlanta Regional Commission 2009), which emerged alongside recognition of the need for more systematic interventions to support population ageing at a community level. An additional influence was recognition of the development in many localities of what have been termed 'naturally occurring retirement communities' (NORCS), i.e. neighbourhoods that, with the out-migration of younger people, have effectively evolved into communities of older people (Scharlach and Lehning 2013). The key issue behind the 'lifetime' concept was an understanding that effective support for older people within neighbourhoods would require a range of interventions linking different parts of the urban system – from housing and the design of streets to transportation and improved accessibility to shops and services.

On the one side, then, policy initiatives such as those from the WHO appeared to open-up the possibility of stronger cross-generational links within cities, together with support for reducing class, gender and ethnic divisions within urban areas. On the other hand, much of the debate remained disconnected from the pressures on vulnerable groups given changes affecting cities. Whilst the dominant approach was towards encouraging what came to termed 'ageing in place' (Wiles *et al.* 2012), the places in which older people were ageing often proved hostile and challenging environments (Buffel *et al.* 2013; Kelly-Moore and Thorpe 2012). Moreover, whilst much of the discussion has been around how to secure meaningful places for older people, changes affecting the character of urban space were largely ignored (see, further, Peace 2013). Here, the debate in urban geography and sociology focused on a variety of characteristics associated with the political economy of cities, many of these linked with economic and social inequalities arising from urban development (Soja 2010). Tonkiss (2005: 46), for example, discussing what she refers to as the 'spatiality of the urban problem', refers to a variety of challenges which arise from the impact of urban processes: 'Most notable here are the ways in which cycles of economic and spatial change make certain spaces and certain people redundant'. Drawing on the work of Robert Park, she goes on to argue that: 'Just as the growth of cities junks the material culture of urban life as it becomes obsolete, so these processes "scrap" those individuals who are resistant to the demands of progress' (Tonkiss 2005: 46).

The key point here concerns the impact on daily life of those in control of developing and managing urban space. Logan and Molotch (cited in Zukin 2010: 227) make the point that: 'city dwellers want to enjoy the use-values of their communities and homes, but developers are interested in maximizing exchange values – in making money'. This disconnect has become especially important in the context of meeting the needs of groups – notably older people – for whom the neighbourhood plays a central role in shaping the quality of daily life (Gardner 2011). Tonkiss (2005: 74) argues that: 'The political economy of the city is not confined ... to questions of who owns what, but with how this spatial economy is regulated in terms of access, exclusion and control'. Sassen (2000) observes that cities have become a strategic terrain for a series of conflicts and contradictions – amongst which the management and support of vulnerable populations is one of the most acute.

An important question concerns the extent to which the construction of the modern (or late-modern) city as the 'site for the new consumerism' (Savage *et al.* 2003: 149) results in social exclusion for groups such as older people. Rodwin *et al.* (2006: 7) suggest that while world cities offer extensive cultural and entertainment opportunities, they are expensive places in which to live. They illustrate this by citing a study of New York City that found that only 1-in-20 older households had sufficient money to take full advantage of the quality of life offered by the city. Comparable data is unavailable for British cities, although a relevant finding from the English Longitudinal Study on Ageing (ELSA) for the Social Exclusion Unit was that a larger percentage of older people living in London than in the rest of the country experienced multiple types of poverty and deprivation (ODPM 2006). This is consistent with an analysis of global cities that emphasizes the increasing divergence of the lifestyles and opportunities of wealthy and poor residents, itself a manifestation of growing inequalities linked to social class, ethnicity and, in some respects, age and generation.

Spaces of consumption and processes of generational change

Following from the above review, the argument developed in this chapter is that older people can be seen as an important group affected by the new 'spaces of consumption' opening up within cities, but these run in parallel with 'spaces of exclusion' within which the daily life of many groups is maintained and experienced. On the former, Zukin (2010), in her study of New York, analyses developments such as the gentrification of neighbourhoods which make claims on space which subsequently displace or marginalizes long-term residents. She views such spaces as involving a mixture of cultural as well as economic power, illustrated by new aesthetic tastes characteristic both of wealthier groups and/or younger generations supplanting those of long-term residents:

> The tastes behind these new spaces of consumption are powerful because they move long-time residents outside their comfort zone, gradually shifting the places that support their way of life to life supports for a different cultural community. Bistros replace bodegas, cocktail bars morph out of old-style saloons, and the neighborhood as a whole creates a different kind of sociability. Against the longtimers' sense of origins newcomers pose their own new beginnings.
>
> (Zukin 2010: 4)

Kelly-Moore and Thorpe (2012) link this process to the operation of 'cohort replacement' as a feature of neighbourhoods undergoing social and demographic change. Accordingly:

> [t]his cohort replacement can represent shifts in neighborhood socio-economic status, shared values, or culture, creating tensions between cohorts

of older long-term residents and younger cohorts of recent transplants. Once such transition reaches a tipping point, many older adults can feel disenfranchised in their own neighborhood.

(Kelly-Moore and Thorpe 2012: 499)

Reflecting the above argument, Savage *et al.* (2005: 44) comment in relation to one gentrifying Manchester locality that 'there is no sense of a past, historic, community that has moral rights on the area: rather the older working-class residents, when they are seen at all, are seen mainly as residues': the use of the term 'residues' illustrating divisions between older residents and more recent (younger) arrivals. But the issue of difference is almost certainly one of age *and* social class: older working-class residents lacking the resources to match the lifestyles of younger middle-class residents.

Community studies involving older people suggest that they may be especially affected by the kind of changes analysed by Zukin (2010) and Savage *et al.* (2005). These suggest that older people derive a strong sense of emotional attachment from both their home and the surrounding community (Townsend 1957; Phillipson *et al.* 2000). Indeed, Rowles (1978: 200) makes the point that 'selective intensification of feelings about spaces' might represent 'a universal strategy employed by older people to facilitate maintaining a sense of identity within a changing environment'. While this may be possible in relatively secure and stable neighbourhoods, some residential settings may impede the maintenance of identity in old age. This may be especially the case in the 'zones of transition' marked by a rapid turnover of people and buildings, and in unpopular urban neighbourhoods characterized by low housing demand and abandonment by all but the poorest and least mobile residents (Rogers and Power 2000; Newman 2003). Disadvantaged urban neighbourhoods, and the people who reside in them, may also be prone to 'institutional isolation' (Gans 1972), as services and agencies withdraw and access worsens to basic facilities such as food shops, telephones and banking (Scharf *et al.* 2003).

Although the evidence reviewed points to significant spatial inequalities affecting the lives of older people, further research is needed to capture the experiences of those living in areas characterized by high levels of deprivation, in particular those most affected by the cycles of economic and social change referred to above. The next section of this chapter aims to examine this aspect through drawing on a study of older people living in similar types of deprived urban communities across two European nations, with particular emphasis on spatial aspects of exclusion in the different areas.

Methodology of the study

The data for the present research were derived from two studies of older people living in England and Belgium which aimed to explore perceptions of the neighbourhood and experiences of social exclusion among people aged 60 and over (Buffel *et al.* 2013). The English study was conducted between 2000 and 2003;

the data from the Belgian study were collected in Brussels between 2007 and 2009. In both countries, the study areas were selected on the basis of criteria of urban deprivation, using the Index of Local Deprivation (DETR 1998) for those in England and the Atlas of Deprived Areas (Kesteloot *et al.* 1996; Kesteloot and Meys 2008) for those in Belgium. The study areas in England were: Club-moor, Granby and Pirrie in Liverpool; Park, Plashet and St Stephens in the London Borough of Newham; and Cheetham, Longsight and Moss Side in Man-chester. The research areas in Belgium were all located in the Brussels-Capital Region, and included: Marollen in Brussels, Brabantwijk in Schaerbeek, and Old-Molenbeek in Molenbeek. Each of the English communities was ranked in the 50 most deprived neighbourhoods in England (out of more than 8,000 neigh-bourhoods) at the time of the study. The research areas in Belgium were ranked in the 20 (of 178) most deprived areas in Brussels (Jacobs and Swyngedouw 2000). While the areas differ in their population profile, socio-economic struc-ture and their proximity to their respective city centres, they share an accumula-tion of features associated with intense urban deprivation. These include above-average rates of unemployment and low-income households, and relatively poor housing conditions (Kesteloot and Meys 2008; Social Exclusion Unit 1998).

The findings presented in this study are based on 124 semi-structured inter-views in England and 102 semi-structured interviews in Belgium with people aged 60 and over. In England, interviews were undertaken with 85 older people of English origin; 19 older Somali people in Liverpool and 20 older Pakistani people in Manchester. Data collection in Belgium comprised 59 interviews with older people of Belgian origin; 20 interviews with older Moroccan people; and 23 interviews with Turkish elders. Recruitment of respondents ranged from the more formal to the fully informal: through relevant community organizations, including social service centres, voluntary and religious organizations, as well as through informal gatherings or meeting places in the neighbourhood. This pur-posive sampling strategy sought to target a heterogeneous group of people in order to reflect a broad spectrum of neighbourhood connections.

The participants in England were aged between 60 and 87 years with an average age of 72 years. The majority of older migrants in England were in their sixties, although exact birth dates among the migrant population were often uncertain, mainly due to the absence of birth certificates in the case of respon-dents from Somalia. The participants in Belgium were aged between 60 and 97 years with an average age of 69 years.

In both countries, interviews were undertaken in the language of the respon-dents' choice by members of the research team, or by interviewers recruited from the relevant ethnic group. All interviews were audio-recorded, transcribed verbatim and, where necessary, translated into either English (for the English study) or Dutch (for the Belgian study). Both studies employed a similar topic-list, and included such issues as participants' perceptions of their neighbourhood, the nature of their social relationships with neighbours and experiences of urban daily life.

Community change, space and social exclusion

The interviews indicated that a focus on neighbourhood and place identifies both the daily challenges and exclusionary pressures facing older people, as well as the way they strive to create a sense of home and belonging. The interviews were carried out in the context of environments that had undergone major population change and industrial decline: factors which may contribute to an experience of rejection or exclusion from the locality (Scharf *et al.* 2003). This was especially significant given the length of time respondents had lived in each locality – 45 years for white British and 52 years for those of Belgian origin. Many commented on the changes they had experienced over this time and how this had affected their sense of 'home'. Whilst some reported significant improvements in their lives, others compared present aspects of their neighbourhood unfavourably with earlier times. Typical observations included: 'We were always together, day and night.... Everyone was watching out for each other ... We never thought we would end up like this [feeling unsafe] ... never' (71-year-old white woman, 71 years in the neighbourhood, Old-Molenbeek, Brussels). And 'Well you don't congregate like same as like on bonfire night. In the old days all the neighbours used to be sat outside with chairs.... Having treacle toffee and roasted potatoes.... Nobody cares about you now' (68-year-old white man, 68 years in the neighbourhood, Cheetham, Manchester).

Respondents often reflected on difficulties in coming to terms with economic and social changes brought about by the decline of industries supporting the various communities – especially in the decades from the 1970s onwards. Many experienced financial insecurity over this period, with loss of work affecting themselves, family members and friends. A 69-year-old man who lived in Old-Molenbeek, also known as 'the little Manchester'[1] of Belgium, reported how this had affected his social network:

> Everyone who used to live here were workers, but when they lost their job they had no reason to stay.... Anyone who had the choice moved away. They were fed up with all the misery. Nobody wanted to raise children here. One of my mates who lived across the street was a plumber ... he installed our bathroom. He moved to New-Molenbeek. Our neighbour was a cabinet-maker, he moved to Dilbeek. I haven't heard from them for a long time.

High population turnover is often a feature of deprived urban areas, one that may have a detrimental effect on social networks and relationships (Smith 2009). In both studies, people commented on the changing composition of their locality, with the moving away of family, friends and acquaintances. As a result, some respondents commented upon the absence of people of a similar age (and generational background) to themselves. The impact of population change on the experience of exclusion from local social relationships was reflected in the comments of a 65-year-old white woman in Cheetham, Manchester:

I only live here because my house is here.... When we moved here first we had the school, we had the church, we had my husband's work and it was different. The population has all changed you know.... There's more of a transient population now.

Equally, the replacement of 'old style' cafes with new types of bars seemed also to favour younger over older generations:

Everything has become more expensive around here. People have invested [purchased property] in this neighbourhood, so the prices are going up, which makes it even more attractive to invest here. All the old traditional cafes and pubs have been replaced by expensive restaurants and antique shops. Lots of 'Maroliens' [a word to describe the residents of this neighbourhood] complain about this issue. It has just become far too expensive here for ordinary people.

(67-year-old woman, 39 years in the neighbourhood,
Marollen, Brussels)

It's a neighbourhood of passers-by. People don't stay here. In the olden days there used to be loads of cafes and pubs but they have all disappeared now. They have been replaced by chic shops and restaurants. Yes, it's definitely a neighbourhood in transition. New people are coming in, but that's not to everyone's advantage. There are still a lot of poor people living here who can't afford to do their shopping here anymore. There are both rich and poor people in this area but they don't seem to interact with each other.

(71-year-old white man, 41 years in the neighbourhood,
Marollen, Brussels)

References to what has been termed a 'loss of togetherness' (Blokland 2003) figured prominently in such comments. Many people who had 'aged in place' felt that the 'community spirit' that once characterized their area had been lost. This was evident, for example, in the comment below:

Most people who I knew around here are gone now. They either died or they moved away.... In the old days, the neighbourhood was much more sociable. Now, there are a lot of new arrivals and we don't socialize with them. It's difficult ... I feel like a stranger in my own neighbourhood.

(79-year-old white woman, 51 years in the neighbourhood,
Brabantwijk, Brussels)

For this and other respondents there was a discrepancy between the present reality of their lives and a 'lost community' to which they expressed attachment. Some long-term residents made reference to 'new arrivals', people belonging to different ethnic backgrounds from themselves, as if they presented a threat to the 'safe' sense of community associated with earlier times:

There aren't many of us [white people] down here now because they're nearly all Asians.... All right, some of them are quite nice – some of them. I speak to them and I'm sociable to them, but you still feel as if you're being taken over.

(75-year-old white woman, 46 years in the neighbourhood,
St Stephens, London)

I tell you, our race has thinned out. There are not many Belgians around here anymore and we can't communicate with the Moroccans. The social relationships have definitely changed for the worse. People don't have a chat anymore in the local shops because there are too many foreigners.

(75-year-old white woman, 55 years in the neighbourhood,
Old-Molenbeek, Brussels)

However, representation of areas as previously 'close-knit' itself require critical examination. Blokland (2003: 274), in her study of inner-city Amsterdam, suggests that older people's nostalgia for a 'lost community' reflects a tendency to play down social divisions that may have been apparent in the past. Indeed, our interviews suggest that people rarely referred to conflicts that may have been present in earlier days. One explanation may be that older people's views of their community reflect *current* needs and issues affecting their daily lives. Among those who refer to their locality as being more 'sociable' in earlier times, many commented on the limited number of close friends available within their area. Others reported difficulties in gaining access to social services and meeting places.

However, despite adverse comments, people would often express strong emotional feelings about their neighbourhood. Even those who held the strongest views about the 'loss of community' still reported a close identification with their neighbourhood. Many residents expressed their attachment though a reluctance to move to a new location: 'I don't fancy another area for living' (67-year-old black Caribbean man, 40 years in the neighbourhood, Moss Side, Manchester), and 'Listen, I was born here. I like this neighbourhood and I want to stay here until I die' (84-year-old white woman, 84 years in the neighbourhood, Old-Molenbeek, Brussels).

Next to this temporal dimension, older people's affiliation with place was also expressed in terms of a sense of belonging arising from integration within the neighbourhood. For example, older people who felt they could count on neighbours to receive help or support were more likely to express a sense of belonging to their community. They also tended to mention 'friendly neighbours' and proximate friends as reasons why they would not leave the area (see, also, Scharf *et al.* 2003; Verté *et al.* 2009). Among the Turkish population in Brussels, this was evident in the saying 'find your neighbour, choose your house', suggesting that trustful and supportive neighbours are the most important criterion for determining the choice of a home. Many older migrants, as well as long-term residents across all study areas, shared the conviction that 'looking out for one another' brought about 'a feeling of home' and a sense of 'safety' in the neighbourhood.

Managing urban space

Respondents also discussed various strategies for managing urban space, an issue especially significant for female respondents. In this context, Tonkiss (2005: 95) has argued that urban space can be seen as 'sexed' and 'gendered' in that women's spatial practices are constrained by what she terms 'geographies of violence and fear'. This was illustrated in our study through the spatial strategies that many older women had developed on the basis of their perceptions of safety and danger. Such strategies were further complicated by factors related to physical and psychological vulnerability, which accounted for a variety of ways in which women negotiated space. Older women's fears in public space were based primarily on feelings of vulnerability to unknown men, 'strangers' or groups of youth who were perceived as 'intimidating' (see also Holland *et al.* 2007 for similar findings): 'We have youth gangs here, and they wait for you on the corner or in the hallways. You never know who's coming behind you if you are alone as a woman' (70-year-old white woman, 30 years in the neighbourhood, Marollen, Brussels).

Against this, feelings of insecurity expressed by some respondents, both men and women, were not necessarily directly related to fear of becoming a victim of crime. Instead, they reflected more general uncertainties about their ability to cope with a changing environment. This was expressed for example by elders who experienced difficulties with walking in areas with heavy traffic or who feared they might fall without having someone to help them. In these examples, a discrepancy between the demand character of the environment and the capabilities of the person contributed to feelings of insecurity (Kahana 1982).

Analyses of the interviews with different migrant groups in the three English cities and Brussels also highlighted differences between men and women in terms of the use of public space (see, also, Buffel and Phillipson 2011). For example, older migrant men tended to have more informal gatherings with friends outdoors than migrant women. The mosques, cafes or teahouses, where the men met each other, were described as 'male spaces' in the interviews – by male as well as female participants. In this respect, our study points at the role of Islamic prescriptions, stipulating that men and women keep sufficient physical distance in public. This may sustain such 'male territoriality', since women are not allowed to enter a space where men are already present (Peleman 2003). A number of older migrant women in our studies, for example, avoided particular places in public space because they were afraid that the men would spread gossip about them and hence damage their reputations.

Against this, our findings also suggested that older migrant women succeeded in appropriating specific places, especially in private spaces such as their home or in the homes of family of friends where they had informal meetings amongst women. This was also the case in some semi-public spaces. For example, some women met to attend language courses or literacy programmes in community centres and did voluntary work such as cooking for students. Many Moroccan women also referred to a particular park, where they met other women and regularly took their grandchildren. They attached great importance to this park because it was seen as a 'female space', free of male control.

Spatial justice and age-friendly cities

This chapter has highlighted the complex way in which older people negotiate urban space, especially in the context of diverse age, generational and social groups. Laws (1997) highlighted what she termed the 'spatiality of ageing', exploring the degree to which spaces and places were age-graded, emphasizing the ageism of space where 'youth is everywhere'. Similarly, Holland *et al.* (2007: 39) in an observational study of an English urban town, concluded that: 'A striking finding is the extent to which older people involved in this study as interviewees or through observation, either perceived themselves as excluded or actively excluded themselves from public space for large stretches of the time'. Studies such as these as well as others reported in this chapter suggest that elderly men and women may find difficulties 'creating' space within cities. Global cities, it might be argued, raise tensions between a 'hyper-mobile' minority and those ageing in place. Conversely, de-industrializing cities (with shrinking populations) create problems arising from the collapse of an economic base which can maintain sustainable networks for different social groups. The challenge here then is creating an urban environment that supports the autonomy of the ageing body and the equal rights of older people with others to a 'share' of urban space. This issue will be especially important to implement at a local level, with a particular focus on improving the quality of urban design and promoting safety and inclusion as key features of urban living (Gehl 2010).

A major step in achieving the above aim will be linking the debate about achieving age-friendly environments with that of ensuring what Soja (2010) terms 'spatial justice' or what Fainstein (2010) refers to as *The Just City*. This argument builds upon the work of Henri Lefebvre (1971) and his research on citizenship and rights in an urban context. Lefebvre stressed:

> [t]he use-value of the city over its exchange value, emphasizing that citizens have a right to make use of the city, and that it is not just a collection of resources to enable economic activity. The uses of the city by citizens should be seen as valid ends in themselves, not merely as a means to produce economic growth.... The right to the city is the right to live a fully urban life, with all the liberating benefits it brings. [Lefebvre] believed the majority of city residents are denied this right because their lives are subordinated to economic pressures – despite being *in* the city, they are not fully *of* the city
>
> (Painter 2005: 9)

This last point applies especially well to older people, who may find that despite having contributed to an urban world in which they have spent most of their lives, it may present major obstacles to achieving a fulfilling existence in old age. On the one hand, ensuring the success of cities is viewed as essential for achieving economic security for nation states (Barber 2013). On the other hand, the reconstruction of cities is often to the detriment of those outside the labour market, especially those on low incomes. Achieving recognition of the needs of different generations within cities, and exploiting the potential of the city for groups of whatever age, will be central to implementing an age-friendly approach.

Conclusion: planning cities for all ages

Developing cities which meet the interests of all generations remains an important goal for economic and social policy. The future of communities across the world will in large part be determined by the response made to achieving a higher quality of life for their older citizens. A crucial part of this response must lie in creating supportive environments providing access to a range of facilities and services. However, the research and policy agenda will need to change in significant ways if this is to be realized: first, the issues raised by developing age-friendly communities within complex urban environments will require a more coherent link between research and policy than has thus far been achieved. Research on environmental aspects of ageing has an impressive literature to its name, yet it remains detached from analysing the impact of powerful global and economic forces transforming the physical and social context of cities. Remedying this will require, as has been argued in this chapter, closer integration with developments in disciplines such as urban sociology, urban economics and human geography. Understanding optimum environments for ageing must be seen as an interdisciplinary enterprise requiring understanding of the impact of developments such as the changing dynamics of urban poverty on older people; the consequences of urban renewal and regeneration; the influence of transnational networks; and changing relations between different class, gender, ethnic and age-based groups.

Second, in keeping with the approach taken in this chapter, given the rapid changes affecting many urban areas, new approaches to understanding older people's relationship to urban change – and city development in particular – are urgently required. In particular, there is a strong case for more research in 'urban ethnography' to capture the disparate experiences of those living in cities that are now experiencing intense global change and that are strongly influenced by complex patterns of migration on the one side and population ageing on the other. Sassen (2000: 146) pointed to the need for detailed fieldwork as a 'necessary step in capturing many aspects of the urban condition'. Urban sociology was founded (through the work of the Chicago School from the 1920s) upon intensive studies of experiences of urban life, particularly of disadvantaged and insecure people from different migrant populations. Ethnographies will bring to the surface the attitudes, motivations and experiences of older people who are 'ageing in place' and deepen our understanding about the way in which cities are changing, and about the positive and negative contributions that the changes have on the quality of daily life in old age (see for example Woldoff 2011).

Finally, incorporating issues about ageing in urban environments with the wider debate concerning spatial justice is also essential. Here, we would underline the relevance of Soja's (2010: 19) argument that the: 'geographies in which we live can have both positive and negative effects on our lives'. He writes: 'They are not just dead background or a neutral physical stage for the human drama but are filled with material and imagined forces that can hurt us or help us in nearly everything we do, individually and collectively'. He concludes:

> This is a vitally important part of the new spatial consciousness, making us aware that the geographies in which we live can intensify and sustain our exploitation as workers, support oppressive forms of cultural and political domination based on race, gender, and nationality, and aggravate all forms of discrimination and injustice

Ensuring spatial justice for older people is now a crucial part of this debate, with the development of an integrated approach to demographic and urban change representing a key task for research and public policy.

Acknowledgements

This paper has greatly benefited from discussions in the Economic and Social Research Council-funded network *Population and Ageing and Urbanisation: Developing Age-Friendly Citie*s (Grant No. R6751 B247). The authors would also like to thank colleagues in Manchester City Council's Valuing Older People team, in particular Paul McGarry, Rebecca Bromley, Sally Chandler and Sophie Handler. The authors also wish to acknowledge the support of the Fund for Scientific Research Flanders in Belgium and the Economic and Social Research Council in England (Grant No. L480254022). We would like to thank other investigators on this project, including Paul Kingston, Allison Smith, Tom Scharf, Aycha Bautmans, Anne-Sophie Smetcoren, Sema Sönmez, Noor Talpe, Naboue Vaassen and Dominique Verté, the community organizations who were involved in the study, the respective City Councils who were partners in this project and all the residents who participated in the research.

Note

1 The former industrial neighbourhoods located in Molenbeek gave rise to the nickname 'little Manchester' or 'le petit Manchester' in the nineteenth century. When this nickname is used today, it refers to the de-industrialization and the collapse of the area's former local economic structure.

References

Atlanta Regional Commission (2009) *Lifelong Communities: A Framework for Planning*, online, available at: www.atlantaregional.com/aging-resources/lifelong-communities (accessed 1 March 2014).
Barber, B. (2013) *If Mayors Ruled the World*, New Haven, CT: Yale University Press.

Blokland, T. (2003) *Urban Bonds*, Cambridge: Polity Press.

Buffel, T. and Phillipson, C. (2011) 'Experiences of place among older migrants living in inner-city neighbourhoods in Belgium and England', *Diversité Urbaine*, 11(1): 13–38.

Buffel, T., Phillipson, C. and Scharf, T. (2013) 'Experiences of neighbourhood exclusion and inclusion among older people living in deprived inner-city areas in Belgium and England', *Ageing and Society*, 33(Special Issue 01): 89–109.

Burdett, R. and Sudjic, D. (eds) (2011) *Living in the Endless City*, London: Phaidon.

DCLG (Department for Communities and Local Government) (2008) *Lifetime Homes, Lifetime Neighbourhoods: A National Strategy for Housing in an Ageing Society*, London: DCLG.

DETR (Department of the Environment, Transport and the Regions) (1998) *Updating and Revising the Index of Local Deprivation*, London: DETR.

Fainstein, S. (2010) *The Just City*, Cornell, NJ: Cornell University Press.

Fitzgerald, K.G. and Caro (2014) 'An overview of age-friendly cities and communities around the world', *Journal of Aging & Social Policy*, 26(1/2): 1–18.

Gans, H. (1972) *People and Plans: Essays on Urban Problems and Solutions*, London: Routledge.

Gardner, P. (2011) 'Natural neighborhood networks: Important social networks in the lives of older people', *Journal of Aging Studies*, 25(3): 263–271.

Gehl, J. (2010) *Cities for People*, London: Island Press.

Holland, C., Clark, A., Katz, J. and Peace, S. (2007) *Social Interaction in Urban Public Places*, York: Joseph Rowntree Foundation.

Izuhara, M. (ed.) (2010) *Ageing and Intergenerational Relations*, Bristol: Policy Press.

Jacobs, D. and Swyngedouw, M. (2000) 'Een nieuwe blik op achtergestelde buurten in het Brussels Hoofdstedelijk Gewest', *Tijdschrift voor sociologie*, 21(3): 197–228.

Kahana, E. (1982) 'A congruence model of person-environment interaction', in M.P Lawton, P.G. Windley and T.O Byerts (eds) *Aging and the Environment: Theoretical Approaches*. New York, NY: Springer, pp. 97–121.

Katz, B., Altman, A. and Wagner, J. (2008) 'An agenda for the Urban Age', in R. Burdett and D. Sudjic (eds) *The Endless City*, London: Phaidon, pp. 474–481.

Kelly-Moore, J. and Thorpe, R. (2012) 'Age in place and place in age: Advancing the inquiry on neighbourhoods and minority older adults', in K. Whitfield and T. Baker (eds) *Handbook of Minority Aging*, New York, NY: Springer, pp. 497–506.

Kesteloot, C. and Meys, S. (2008) *Atlas van Achtergestelde Buurten in Vlaanderen en Brussel [Atlas of Deprived Neighbourhoods in Flanders and Brussels]*, Leuven: KULeuven & Vlaamse interprovinciale werkgroep Sociale Planning.

Kesteloot, C., Vandenbroecke, H., Van der Haegen, H., Vanneste, D. and Van Hecke, E. (1996) *Atlas van Achtergestelde Buurten in Vlaanderen en Brussel [Atlas of Deprived Neighbourhoods in Flanders and Brussels]*, Brussels: Ministerie van de Vlaamse Gemeenschap.

Laws, G. (1997) 'Spatiality and age relations', in A. Jamieson, S. Harper and C. Victor (eds) *Critical Approaches to Ageing and Later Life*, Buckingham: Open University Press, pp. 90–100.

Lefebvre, H. (1971) *The Production of Space*, Oxford: Blackwell.

Mumford, L. (1956) 'For older people: Not segregation but integration', *Architectural Record*, 119: 109–116.

Newman, K. (2003) *A Different Shade of Gray*, New York, NY: New Press.

ODPM (Office of the Deputy Prime Minister) (2006) *State of the English Cities*, Urban Research Summary 21, London: ODPM.

Painter, J. (2005) 'Urban citizenship and rights to the city', project report, Durham: International Centre for Regional Regeneration and Development Studies, online, available at: http://dro.dur.ac.uk/5435/1/5435.pdf (accessed 26 June 2014).

Peace, S. (2013) 'Social interactions in public spaces and places: A conceptual overview', in G. Rowles and M. Bernard (eds) *Environmental Gerontology: Making Meaningful Places in Old Age*, New York, NY: Springer, pp. 25–53.

Peleman, K. (2003) 'Power and territoriality: A study of Moroccan women in Antwerp', *Tijdschrift voor Economische en Sociale Geografie*, 94(2): 151–163.

Phillipson, C., Bernard, M., Phillips, J. and Ogg, J. (2000) *The Family and Community Life of Older People*, London: Routledge.

Rodwin, V., Gusmano, M. and Butler, R.N. (2006) 'Growing older in world cities: Implications for health and long-term care policy', in V. Rodwin and M. Gusmano (eds) *Growing Older in World Cities*, Nashville, TN: Vanderbilt University Press, pp. 1–25.

Rogers, R. and Power, A. (2000) *Cities for a Small Country*, London: Faber and Faber.

Rowles, G. (1978) *Prisoners of Space? Exploring the Geographical Experience of Older People*, Boulder, CO: Westview.

Sassen, S. (2000) *Cities in a World Economy*, London: Pine Forge.

Satterthwaite, D. (ed.) (1999) *The Earthscan Reader on Sustainable Cities*, London: Earthscan.

Savage, M., Bagnall, G. and Longhurst, B. (2005) *Globalization and Belonging*, London: Sage.

Savage, M., Warde, A. and Ward, K. (2003) *Urban Sociology, Capitalism and Modernity*, second edition, London: Palgrave Macmillan.

Scharf, T., Phillipson, C. and Smith, A. (2003) 'Older people's perceptions of the neighbourhood: Evidence from socially deprived urban areas', *Sociological Research Online*, 8(4), online, available at: www.socresonline.org.uk/8/4/ scharf.html (accessed 17 June 2014).

Scharlach, A.E. and Lehning, A.J. (2013) 'Ageing-friendly communities and social inclusion in the United States of America', *Ageing and Society*, 33(Special Issue 01): 110–136.

Smith, A. (2009) *Ageing in Urban Neighbourhoods*, Bristol: Policy Press.

Social Exclusion Unit (1998) *Bringing Britain Together: A National Strategy for Neighbourhood Renewal*, London: HMSO.

Soja, E. (2010) *Seeking Spatial Justice*, Minneapolis, MN: University of Minnesota Press.

Tonkiss, F. (2005) *Space, the City and Social Theory*, Cambridge: Polity Press.

Townsend, P. (1957) *The Family Life of Old People*, London: Routledge.

UN-HABITAT (2010) *State of the World's Cities 2010/2011*, London: Earthscan.

Verté, D., De Witte, N. and De Donder, L. (2007) *Schaakmat of aan Zet? Monitor voor Lokaal Ouderenbeleid in Vlaanderen*, Brugge: Vanden Broele.

WHO (World Health Organization) (2007) *Global Age-Friendly Cities: A Guide*, Geneva: WHO Press.

Wiles, J.L., Leibing, A., Guberman, N., Reeve, J. and Allen, R.E. (2012) 'The meaning of "ageing in place" to older people', *The Gerontologist*, 52(3): 357–367

Woldoff, R. (2011) *White Flight/Black Flight: The Dynamics of Racial Change in an American Neighborhood*, Ithaca, NY: Cornell University Press.

Zukin, S. (2010) *The Naked City: The Death and Life of Authentic Urban Places*, New York, NY: Oxford University Press.

Index

Page numbers in *italics* denote tables, those in **bold** denote figures.